SOUTHEAST ASIA

Other Titles in
ABC-CLIO's
NATURE AND HUMAN SOCIETIES SERIES

NATURE AND HUMAN SOCIETIES

SOUTHEAST ASIA
An Environmental History

Peter Boomgaard

A B C ● C L I O

Santa Barbara, California • Denver, Colorado • Oxford, England

Library of Congress Cataloging-in-Publication Data
Boomgaard, P., 1946–
    Southeast Asia : an environmental history / Peter Boomgaard.
        p. cm. — (Nature and human societies series)
    Includes bibliographical references and index.
    ISBN-10: 1-85109-419-9 (hardcover : alk. paper)
    ISBN-10: 1-85109-424-5 (ebook)
    ISBN-13: 978-1-85109-419-6 (hardcover : alk. paper)
    ISBN-13: 978-1-85109-424-0 (ebook)
    1. Human ecology--Southeast Asia. 2. Nature—Effect of human beings on—Southeast Asia. 3. Human beings—Effect of environment on—Southeast Asia. 4. Southeast Asia—Environmental conditions. I. Title.

    GF668.B66 2007
    304.20959--dc22

                                    2006030280

10   09   08   07   10   9   8   7   6   5   4   3   2   1

Submissions Editor: Steven Danver
Production Editor: Martha Ripley Gray
Editorial Assistant: Alisha L. Martinez
File Management Coordinator: Paula Gerard
Production Manager: Don Schmidt
Media Image Coordinator: Ellen Brenna Dougherty
Media Editor: John R. Withers
Media Manager: Caroline Price

This book is also available on the World Wide Web as an e-book.
Visit http://www.abc-clio.com for details.
ABC-CLIO, Inc.
130 Cremona Drive, P.O. Box 1911
Santa Barbara, California 93116–1911

This book is printed on acid-free paper. ∞
Manufactured in the United States of America

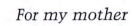

For my mother

CONTENTS

SERIES FOREWORD

Long ago, only time and the elements shaped the face of the earth, the black abysses of the oceans, and the winds and blue welkin of heaven. As continents floated on the mantle, they collided and threw up mountains or drifted apart and made seas. Volcanoes built mountains out of fiery material from deep within the earth. Mountains and rivers of ice ground and gorged. Winds and waters sculpted and razed. Erosion buffered and salted the seas. The concert of living things created and balanced the gases of the air and moderated the earth's temperature.

The world is very different now. From the moment our ancestors emerged from the southern forests and grasslands to follow the melting glaciers or to cross the seas, all has changed. Today the universal force transforming the earth, the seas, and the air is for the first time a single form of life: we humans. We shape the world, sometimes for our purposes and often by accident. Where forests once towered, fertile fields or barren deserts or crowded cities now lie. Where the sun once warmed the heather, forests now shade the land. We exterminate one creature only to bring another from across the globe to take its place. We pull down mountains and excavate craters and caverns, drain swamps and make lakes, divert, straighten, and stop rivers. From the highest winds to the deepest currents, the world teems with chemical concoctions that only we can brew. Even the very climate warms from our activity.

And as we work our will upon the land, as we grasp the things around us to fashion them into instruments of our survival, our social relations, and our creativity, we find in turn our lives and even our individual and collective destinies shaped and given direction by natural forces, some controlled, some uncontrolled, and some unleashed. What is more, uniquely among the creatures, we come to know and love the places where we live. For us, the world has always abounded with unseen life and manifest meaning. Invisible beings have hidden in springs, in mountains, in groves, in the quiet sky and the thunder of the clouds, in the deep waters. Places of beauty from magnificent mountains to small, winding brooks have captured our imaginations and our affection. We have perceived a mind like our own, but greater, designing, creating, and guiding the universe around us.

The authors of the books in this series endeavor to tell the remarkable epic of the intertwined fates of humanity and the natural world. It is a story only now coming to be fully known. Although traditional historians have told the drama of men and women of the past, for more than three decades now, many historians have added the natural world as a third actor. Environmental history by that name emerged in the 1970s in the United States. Historians quickly took an interest and created a professional society, the American Society for Environmental History, and a professional journal, now called Environmental History. U.S. environmental history flourished and attracted foreign scholars. By 1990 the international dimensions were clear; European scholars joined together to create the European Society for Environmental History in 2001, with its journal, Environment and History. A Latin American and Caribbean Society for Environmental History should not be far behind. With an abundant and growing literature of world environmental history now available, a true world environmental history can appear.

This series is organized geographically into regions determined as much as possible by environmental and ecological factors, and secondarily by historical and historiographical boundaries. Befitting the vast environmental historical literature on the United States, four volumes tell the stories of the North, the South, the Plains and Mountain West, and the Pacific Coast. Other volumes trace the environmental histories of Canada and Alaska, Latin America and the Caribbean, Northern Europe, the Mediterranean region, sub-Saharan Africa, South Asia, Southeast Asia, East Asia, and Australia and Oceania. Authors from around the globe, experts in the various regions, have written these volumes, almost all of which are the first to convey the complete environmental history of their subjects. Each author has, as much as possible, written the twin stories of the human influence on the land and of the land's manifold influences on its human occupants. Every volume contains a narrative analysis of a region along with a body of reference material. This series constitutes the most complete environmental history of the globe ever assembled, chronicling the astonishing tragedies and triumphs of the human transformation of the earth.

The process of creating the series, recruiting the authors from around the world, and editing their manuscripts has been an immensely rewarding experience for me. I cannot thank the authors enough for all of their effort in realizing these volumes. I owe a great debt, too, to my editors at ABC-CLIO: Kevin Downing (now with Greenwood Publishing Group), who first approached me about the series; and Steven Danver, who has shepherded the volumes through delays and crises all the way to publication. Their unfaltering support for and belief in the series were essential to its successful completion.

Mark Stoll
Department of History, Texas Tech University
Lubbock, Texas

PREFACE

What is the right time to write a reference book? The answer to that question is somewhat complicated. If the book is written too early, there might not be sufficient data available, so that many passages of the textbook will have to be speculative. The upside is that the potential writer is not buried under an avalanche of articles and monographs. If the book is written later, more data will have become available, which enables the author to write confidently about a well-researched topic. However, chances are that the volume of publications may have increased so much that the task of writing a textbook has become much more time consuming. Finally—and this may sound somewhat cynical—it may become "too late" for a textbook, because the topic is no longer as popular as it had been, and people might have moved on to a new paradigm. However, at the moment of writing, that latter possibility seems rather remote: environmental history is there to stay.

These ideas crossed my mind when the publisher asked me to write a textbook on the environmental history of Southeast Asia. It is a rather new field, and the number of specialist publications is not very large. However, there are many publications of a less explicitly environmental nature that contain stray references to environmental causes and effects. That is the case, for instance, with books and articles dealing with trade or agriculture. I knew beforehand, therefore, that I would have to read a lot, in return for rather slim pickings in terms of environmental historical information. I also knew that the many new publications to be expected after the manuscript is finished will unavoidably and undoubtedly prove me wrong on various questions, because I have had to generalize on the basis of insufficient information. And one does have to generalize when writing a textbook.

On the other hand, the production of a textbook at an early stage in the development of a new discipline or paradigm has its advantages. Perhaps the most important is that it might inspire students and scholars to embark upon the study of the new topic. Furthermore, it provides scholars with a point of reference and may focus their research.

When writing about a relatively new topic such as the environmental history of Southeast Asia, the best course of action is to seek the advice of other specialists with an interest in the history and natural environment of Southeast Asia—and that is precisely what I did. Through a combination of extraordinary good luck and some careful planning, I was able to team up with four other Southeast Asia specialists, forming a so-called research nucleus at the Netherlands Institute for Advanced Study in the Humanities and Social Sciences (NIAS), in Wassenaar, The Netherlands, during the academic year 2003–2004. The scholars were Robert Aiken, Greg Bankoff, John Kleinen, and Baas Terwiel. We all spent our sabbatical leave at NIAS, and while my four colleagues were working on their own related projects, they were also kind enough to suggest books and articles, read various draft chapters, and let me have their unsparing (in both senses of the word) comments. I owe them a debt of gratitude.

I am also very much beholden to the staff of NIAS for inviting our research nucleus to spend a year in the magnificent natural environment that is the setting of the institute, and for facilitating my research—particularly by providing an excellent library service, but certainly also by their unstinting attempts to create the right atmosphere for such an undertaking.

The research for this book can be seen as the logical sequel to a project that has taken up most of my research time over the last fifteen years or so. That is the EDEN project—EDEN being the acronym for Ecology, Demography and Economy in Nusantara, Nusantara being another word for Indonesia—undertaken by my main employer during that period, the Royal Netherlands Institute of Southeast Asian and Caribbean Studies (KITLV) in Leiden, The Netherlands. The project deals with the environmental history of Indonesia, for which most of the relevant printed material is to be found in the KITLV library. They also have excellent holdings on the (environmental) history of other Southeast Asian countries, and thanks for continually supplying me with fresh books and articles on which the present textbook is largely based are due to them as well. I am also grateful to the KITLV for allowing me to spend so much time on both the Indonesia project and its Southeast Asian sequel.

Among my colleagues at the KITLV, I wish to mention particularly Rosemary Robson-McKillop and David Henley; thanks go to them for their suggestions and comments, and for the long discussions we often had on topics dealt with in the present volume—although we often had to agree to disagree.

On specific chapters I solicited and received comments from various people. Leendert Louwe Kooijmans, Leiden University, was prevailed upon to comment upon Chapters 1 and 2, while Jan Wisseman Christie, formerly from the University of Hull, graciously consented to reading and commenting upon Chapter 3. Comments on Chapter 4 were provided by Geert Jan van Oldenborgh and Nanne

Weber, both from the Royal Netherlands Meteorological Institute (KNMI). David Henley commented upon Chapters 5 through 7, and Robert Elson, Griffith University, Brisbane, on Chapters 8 through 11. Wouter van der Weijden, Center for Agriculture and Environment (CLM), Culemborg, read and commented upon the introduction, Chapter 11, and the Conclusion. I am most grateful for all their comments and suggestions. It goes without saying that the responsibility for the end result is solely my own.

Finally, I wish to express my gratitude to the series editor, Mark Stoll, and to the various editors working at ABC-CLIO who, over the years, have gently but firmly kept me on track while writing this book.

—PETER BOOMGAARD
Wassenaar/Leiden

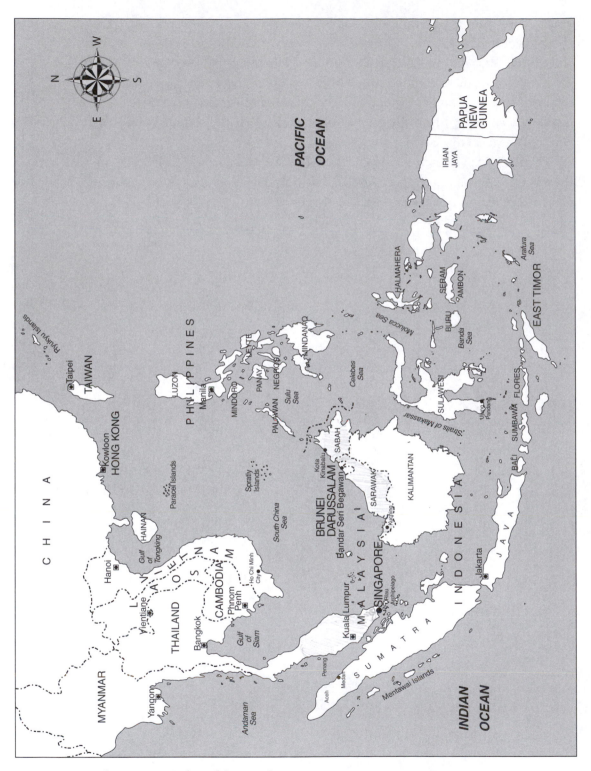

Southeast Asia, Political (present)

INTRODUCTION

Environmental history, a new subdiscipline of history originating in the 1970s, has come of age. There are two well-established international scholarly journals dedicated entirely to the field, confusingly called *Environment and History* and *Environmental History*. An overwhelming number of books on the topic are published each year, and national and international environmental history conferences are organized regularly. Also, a growing number of universities in Europe and the United States offer courses in environmental history, even if there are, to my knowledge, no established chairs in the field.

And yet, even at the time of writing (2006), many regions of the globe have barely been discovered by environmental historians—or perhaps one should say that the historians writing on those areas have yet to discover environmental history. Such was the state of affairs for Southeast Asia when in 1990 economic historian Malcolm Falkus wrote an essay on how little work had been done, and how much there could be done, regarding Southeast Asian environmental history. And that still appears to be the situation for many Southeast Asian countries, at least if one looks at the English-language publications. However, some countries have done relatively well, and I think it is fair to say that Indonesia is far ahead of the pack, followed at some distance by Malaysia and the Philippines. It may not be a coincidence that these are precisely the three countries in which large-scale logging has been such a conspicuous feature from the 1960s or 1970s. It is also remarkable that most of the environmental history of Southeast Asia is not being written by people from the region itself, but by Europeans, Americans, and Australians.

The fact that environmental history in many Southeast Asian countries has not really taken off forms a contrast with the situation in the neighboring countries of India and China. There the field is quite popular, and certainly in the case of India most environmental history research on the area is in the hands of nationals. It is possible—and this book presents some pointers to that effect—that these differences are linked to the differences in environmental problems, and environmental history, between the various regions.

There are many ways to write a textbook on the environmental history of Southeast Asia. The main problem is one of selection, as it is for every historian—even if some of them, biographers in particular, do not appear to be aware of that fact. In this case the publisher had set a rough limit to the total number of words or pages, so that decision was out of my hands—a good thing, given the many choices still to be made. Even in a field as young as that of the environmental history of Southeast Asia, one can never include all available information. Nor should one attempt to do so—a textbook is not an encyclopedia.

Selection, therefore, was not restricted to the question of how much factual information could or should be crammed into one book. Among the most important choices to be made was that of the approach to take to the field of environmental history. That in turn would influence the kind of topics, broadly defined, to be addressed, and would also certainly *be* influenced by the kind of topics that *could* be addressed, given the state of the literature and my expertise.

So what possible approaches are there? J. R. McNeill (2003) distinguished three main varieties of environmental history—one that is material in focus, one that is cultural and intellectual, and one that is political. Rather than summarizing what he wrote about the three varieties, I quote him in full:

> Material environmental history concerns itself with changes in biological and physical environments, and how those changes affect human societies. It stresses the economic and technological sides of human affairs. The cultural/intellectual wing, in contrast, emphasizes representations and images of nature in arts and letters, how these have changed, and what they reveal about the people and societies that produced them. Political environmental history considers law and state policy as it relates to the natural world. (ibid., 6)

The present volume largely represents the "material" variety or approach. Only in Chapters 8 through 11, on the period from 1870 to 2005, have the "cultural/intellectual" and the "political" aspects been dealt with in greater detail. That is partly a matter of the available literature, but also of personal familiarity and preference.

Going somewhat back in time, we find an alternative way of classifying the possible approaches to environmental history in an article written by Donald Worster (1990), who distinguished three "levels." The first is the study of the structure and distribution of natural environments of the past; the second one focuses on the technology of production as it interacts with the environment; the third is the level of ideas and ideologies. The present book emphasizes the first two levels (nature and technology) and dedicates less space to the third (ideas).

But let me elaborate on the structure of this study. The point of departure of this volume is that population growth is one of the main driving forces in human history, and certainly in environmental history. As long as people were few, their influence on their natural environment probably did not surpass that of other animals. That may have changed when *Homo erectus* (predecessor of modern humans, *Homo sapiens*) discovered how to make fire, perhaps, as we now assume, as long as 400,000 years ago. It certainly changed with the arrival of agriculture (the so-called Neolithic Revolution), a theme discussed further in Chapters 1 and 2. Ever since, unstoppable population growth—apart from temporary setbacks—has been typical for humanity, even if early growth rates were very low according to modern standards. Humans may not be the only larger animals to manipulate their natural environment, but they certainly are the only ones to have done so in a cumulative fashion, and to such overwhelming effect.

Not only was the human population increasing during most of the past, but population growth rates were slowly increasing as well. It is one of the main arguments of this book that the continuing increase of these rates has led to a progressively larger impact on the natural environment over time. In fact, it could be argued that the recent "demographic transition," dealt with in detail in Chapter 10, is the first (incipient) structural slowdown in population growth rates for a very long time.

As populations were growing, so did, from the Neolithic Revolution onward, the amount of land cleared for agriculture. Crops took the place of the original vegetative cover, in Southeast Asia more often than not forests, and agro-ecosystems replaced the natural ecosystems.

Next to population growth, economic growth induced by trade—and in particular "foreign" trade—is in this study regarded as a major motor of environmental change. There was certainly some trade between the various areas within the Southeast Asian region at an early date. But the factor that influenced the natural environment more than local trade was the increasing, and larger scale, demand for commodities that came from China, India, and Europe, particularly from the fifteenth century onward. Much later, demand came also from the United States, and recently from Japan and the newly industrialized countries as well.

As will be shown in this study, many "alien" crops were introduced in the region in the wake of the foreign traders, particularly after 1500. They were so successful, both as subsistence and export crops, that they transformed many Southeast Asian agro-ecosystems and landscapes.

Finally, other important influences on the environment were foreign migrants, foreign capital, and foreign technology, all of them originally linked to foreign trade and all of them instrumental in economic growth. Their importance increased from the eighteenth century and has gone on increasing—at least as regards capital and technology—up to the present time. They have been instrumental in the relatively recent establishment of large-scale industry, logging, mining, and fishing, all of them activities with an enormous impact on the Southeast Asian environment.

Summing up, then, it can be said that both population growth and economic growth are emphasized in this study. In earlier periods economic growth was brought about mainly by foreign trade, migration, agricultural developments, and, particularly, new crops, while in more recent times other factors have been more important, particularly industrialization, logging, mining, and fishing.

Of course, this study also tries to establish how and on what scale all of these activities have influenced local ecosystems and landscapes, but I have not attempted to look at those questions in great detail. From time to time local examples are given, but national or regional studies are the appropriate place for such detailed treatment.

In the four chapters dealing with the years between about 1870 and the present, there is more information on environmental policy and environmental thinking. That is partly the case because explicitly environmental thinking and policies are relatively recent phenomena. It should have been possible—and it would have been desirable—to write more about indigenous ideas about nature, but in fact there are not many recent studies on that topic. In addition, it is questionable whether we will ever be in a position, given the available historical sources, to write good empirical studies about indigenous conceptions of nature predating, say, 1900. In all likelihood, we will never be able to write a Southeast Asian counterpart to Keith Thomas's *Man and the Natural World* or to Harriet Ritvo's *The Animal Estate*.

I hasten to add that, in addition to examining the main factors that have influenced the natural environment in Southeast Asia, this study will also look at the ways in which the environment, in turn, has influenced population growth, agriculture, and the like. After all, I think it fair to say that all students of environmental history would agree on at least one thing: that humans and the environment influence each other mutually, and that changes in the one may be expected to lead to changes in the other. Change begets change. A beautiful illustration of this theme, which will be discussed extensively, is climate change; it is a difficult topic on which there is as yet no consensus.

I return to the problem of selection. Although it was understood from the start that the book would cover the entire time span of the human presence in the region—and more—another important choice, or rather complex of choices, still had to be made: regarding the role of chronology and periodization, and the relative weight to be given to the various epochs dealt with. In principle the book is chronologically ordered, but the narrative is interrupted once for a thematic chapter, on climate change.

I have given quite some thought to the matter of periodization. A truly "environmental" periodization would have been ideal, but because there are still so many uncertainties regarding, for instance, the influence of climate change, it would be very difficult to arrive at a solution that would please a considerable number of colleagues. In the end, therefore, I have opted for more traditional peri-od markers, related to things such as state formation and the growth of trade flows. The logic for the selected periodization is presented within the chapters themselves, to which the reader is here referred.

A few lines are also in order on the relative weight I have given to the histor-ical periods. As is so often the case in books covering a long time span, the aver-age number of pages per year increases as we approach the present. There are two reasons for this state of affairs: there is so much more information on recent times that one is naturally inclined to allot more space to them, and the fact that most students are more interested in what is usually called "modern" history. For envi-ronmental history in particular, there is a third reason—environmental change recently has occurred on a much larger scale and much faster than ever before.

However, in order to emphasize that Southeast Asia was not a sort of pristine Garden of Eden prior to 1870, there are three chapters on what is usually called the Early-Modern Period, here roughly the time span between 1400 and 1870. That period can be regarded, without too much exaggeration, as an earlier phase of globalization. During this phase, patterns in demography, agriculture, and trade were established that would lock future developments into particular path-ways. By giving rise to this "path dependence," they would have a lasting impact on the environment.

Just one chapter, albeit a rather voluminous one, is dedicated to the Pre-Modern Period (ca. 500 to 1400). That was the period of the early states in Southeast Asia; with the arrival of the state, new mechanisms that had an envi-ronmental impact came to the fore, although society then bore more lightly on the land than during the period that would follow. A single chapter, but a much shorter one, is also devoted to the period of the Neolithic Revolution and the Metal Ages, when incipient agriculture was already changing the landscape.

This book will begin with a short chapter dealing with the way in which the current geophysical structure of Southeast Asia came into being, and the arrival of the first hominids and the first (modern) humans. But before the reader embarks upon it, I should mention a number of the problems, pitfalls, and other peculiarities of the field that should be kept in mind when reading this volume.

Now that environmental history has come of age, it has, almost by defini-tion, also lost its innocence. Many notions and concepts often loosely used by environmental historians have been subjected to long and acrimonious debate over the last two decades or so, particularly in the United States, the result being "two decades of tortured language and undecipherable jargon" (Weiner 2005, 408). These discussions have not played much of a role in Southeast Asian envi-ronmental history, and I see no point in presenting a detailed summary. One notion, however, merits some scrutiny, and that is the term "nature."

If, in this book, I use the term "nature" or the expression "natural environment," I do not mean to imply that this nature is "pristine," unspoiled, or untouched by humans. During the period for which we have an abundance of written sources—say, from the seventeenth century—even supposedly impenetrable wilderness was seldom untouched by human hands. It might even be argued that, if they were impenetrable, they must in many cases have been secondary forests; old-growth forests often have soils with little vegetation under imposingly tall trees, the latter casting too much of a shadow to permit thick undergrowth.

Nowadays, one of the most worrying features of Southeast Asian forest exploitation is that the original vegetation is rapidly dwindling. In the past people were struck and even alarmed by just the opposite phenomenon—the surprisingly high rate of regrowth of tropical vegetation. Many foreign visitors who attempted to cut a trail through what they thought was pristine jungle were amazed to encounter ruins from remote epochs. The remains of ancient civilizations such as Sriwijaya and Majapahit became entirely overgrown by thick vegetation, having been all but forgotten by the time the European trading companies arrived.

Similar observations apply on a smaller scale. Slash-and-burn agriculture by definition manipulates the natural environment on a large scale, particularly when swidden agriculturalists, for a variety of reasons to be discussed later, preferred to cut old-growth forests. If a region was sparsely populated this preference could be indulged freely, which implies that even in the heart of Borneo "nature" was often less "natural" than most people may have assumed.

Another problem that writers and readers of environmental history must be aware of is that of environmental determinism. And if the setting of a story is tropical and colonial, as is the case with many chapters in this book, environmental determinism might soon be supposed to carry overtones of "orientalism" (the inclination, attributed to Western scholars, to essentialize certain features deemed typical of Asian societies) or even racism.

The basic element of the problem is that there are many older theories in which the differences between tropical societies and those located in the temperate zone are attributed wholly or largely to differences in climate and related features of the physical environment. In those theories, individual people and the societies they established were seen as determined largely by their natural environment. Historian David Arnold, writing about Indian history, formulates the problem as follows:

> The articulation of such extreme determinist views creates problems for the latter-day historian. Should these opinions be dismissed in their entirety as the product of an outmoded, self-justifying imperialism, or should it be recognized,

in the context of a new social history of India, that such major environmental events and calamities as the onset or absence of the monsoon did indeed profoundly affect the lives and livelihoods of the great majority of the Indian population? It ought, in theory, to be possible to separate the imperial rhetoric from the analysis of material reality. In practice, however, the distinction is often hard to maintain for the very sources themselves are imbued with a particular mind-set, a colonial way of understanding nature and representing its human consequences. (Arnold 1996, 173)

It goes without saying that in this book it has been attempted to do just that—separate the rhetoric from the factual information. It is not always an easy task, but one that has to be undertaken when writing an environmental history of this region. I think that an important step has been taken if we are aware that hazards are natural but disasters are not: they are often very much the product of political forces, a fact that modern disaster theory has taught us.

The problem of distortions brought about by a "colonial" point of view touches upon more issues than possible deterministic perceptions. The overwhelming majority of the historical sources dated between 1500 and 1950 were produced by Westerners, often administrators, clerics or missionaries, naturalists, the military, or anthropologists with strong links to the big merchant companies and later the colonial bureaucracies. And although these people did not necessarily represent the points of view of such institutions, they shared a Western worldview that had been influenced by the fact that Western people conquered and ruled over Eastern regions. Fortunately, such views are expressed mainly in theories or opinions, clearly recognizable as such, concerning the "nature" of "Orientals," and in most instances it is not all that difficult to separate facts from fiction and prejudice. What is less easily remedied is the fact that indigenous voices are very much under-represented throughout the Early-Modern Period and the earlier phase of modern history.

In contrast to the very rich historical record from the Early-Modern Period, the Pre-Modern Period is less well documented. Moreover, these sources, usually epigraphic ones, are often not easily consulted, and if they are, few people have the language skills required to read them. And no one, to my knowledge, has the language skills required to read all of the Southeast Asian sources dating from those years: there are just too many languages involved. We can only eagerly await the arrival of articles and monographs (in English) based on such sources, written by specialists who do read some of the languages. If it turns out that this book stimulates the production of such publications, it will have fulfilled one of its aims.

We are also much better informed about some topics than on others. There is, for instance, an abundance of literature about forest management and forest exploitation, particularly from the nineteenth century onward. We know much less about mining and fishing during the same period. It is, of course, to be hoped that this book will stimulate researchers to fill those gaps.

Another limitation imposed upon us by the historical sources is the fact that statistics are rare prior to about 1870, and for some areas even prior to the 1920s or 1930s. We would love to be much better informed about population growth and its components (mortality, fertility, and migration) prior to, say, 1870, but for most Southeast Asian countries, that will probably never become a possibility. The same applies, for instance, to rice harvests, amount of agricultural land per capita, percentages of surface area under forest cover, and the number of elephant or rhinoceros in the various countries.

Furthermore, quantitative and qualitative information is unequally distributed over the region. In principle we know much about densely settled areas, the centers of production of historical records, and much less about areas in which people were more thinly settled. Thus Laos and Cambodia are not well represented in this book, owing to a lack of data, while much is said about (parts of) Thailand, Vietnam, Indonesia, and the Philippines.

Talking about geography, it should be pointed out that the notion of Southeast Asia itself is not unproblematic. This study deals with the northern parts of Thailand and Vietnam, but not with Yunnan (an occasional reference apart) just across the Chinese border. We will be talking about Burma, but not about Bengal and Assam. This is evidently artificial and anachronistic, as the concept of Southeast Asia is a fairly recent one, separating areas that historically, culturally, linguistically, and, indeed, environmentally might better have been dealt with as undivided regions. We are the victims here of statehood, of modern borders, and of conventional ways of carving up the world.

Nevertheless, Southeast Asia is not just a mapmaker's invention. The area has many historical, cultural, and other features in common, not the least of which are environmental, as this book illustrates. There are certainly also, however, notable differences between the areas within the region.

The latter fact leads me to the following remarks. In this study I have attempted to be as comparative as possible. I am convinced that information is less expressive when presented in isolation. Information needs to be contextualized, and that can be done by contrasting it with similar information on other regions or periods.

Comparisons are made on various levels. In the first place, in a historical text comparisons over time are of course called for. Secondly, there were (and are) differences between peoples, landscapes, flora and fauna, crops, minerals, "modes of production," and character and degree of pollution within countries during the same period. Thirdly, such differences also exist between countries—for instance, in GDP per capita or proportion of land under forest cover. Finally, I compare Southeast Asia as a whole with other, larger regions, paying particular attention to China, India, and Europe.

It is through comparisons that such spatial and temporal differences show up most clearly. They are mentioned not only because they constitute factual information the reader might wish to have but also because such differences are at the basis of the dynamics of Southeast Asian environmental history.

When all is said and done, this is a book dealing with a neglected aspect of the history of a fascinating region. Many of us will be familiar with a number of picturesque features ("icons") of Southeast Asia—water buffaloes in wet-rice fields, irrigated mountain slopes, bamboo dwellings, coconut palms bordering sandy beaches, banana trees, ancient temples, tigers, elephants, birds-of-paradise, the rain forest—and with Southeast Asian commodities that are sensory feasts— cloves, nutmeg, teak, sandalwood. We also know of the darker side of Southeast Asian nature—El Niño droughts, tsunamis, volcanoes—and we are getting increasingly familiar with the less pleasing recent environmental aspects of the region, such as the smog blanket that now covers the big cities more or less continuously. This book links those features, putting them in historical perspective and thus filling a conspicuous gap in the historiography not only of Southeast Asia but also of the world at large.

BIBLIOGRAPHICAL ESSAY

Recent articles of a theoretical nature on environmental history in general are to be found in a special issue of *History and Theory* (42, no. 4, 2003); it opens with a useful historiographical overview: McNeill, J. R., "Observations on the Nature and Culture of Environmental History," pp. 5–43. Another (and shorter) attempt to summarize a number of recent debates is Weiner, Douglas R., 2005, "A Death-Defying Attempt to Articulate a Coherent Definition of Environmental History," *Environmental History* 10, no. 3, pp. 404–421.

Somewhat older theoretical/introductory articles are Worster, Donald, 1988, "The Vulnerable Earth: Towards a Planetary History." Pp. 3–22 in *The Ends of the Earth: Perspectives on Modern Environmental History*. Cambridge: Cambridge University Press; Worster, Donald, 1990, "Transformations of the Earth: Toward an Agroecological Perspective in History," *Journal of American History* 67, no. 4, pp. 1087–1106.

The 1990 article on environmental history regarding Southeast Asia mentioned in the text is Falkus, Malcolm, 1990, "Ecology and the Economic History of Asia (1)," *Asian Studies Review* 14, no. 1, pp. 65–87.

On China, see Marks, Robert B., 1998, *Tigers, Rice, Silk, and Silt:*

Environment and Economy in Late Imperial South China. Cambridge: Cambridge University Press; Elvin, Mark, and Liu Ts'ui-jung, eds., 1998, *Sediments of Time: Environment and Society in Chinese History*. Cambridge: Cambridge University Press; Elvin, Mark, 2004, *The Retreat of the Elephants: An Environmental History of China*. New Haven: Yale University Press.

On India, see Grove, Richard H., Vinita Damodaran, and Satpal Sangwan, eds., 1998, *Nature and the Orient: The Environmental History of South and Southeast Asia*. Delhi: Oxford University Press.

On Indonesia, see Boomgaard, Peter, Freek Colombijn, and David Henley, eds., 1997, *Paper Landscapes: Explorations in the Environmental History of Indonesia*. Leiden: KITLV Press; Boomgaard, Peter, David Henley, and Manon Osseweijer, eds., 2005, *Muddied Waters: Historical and Contemporary Perspectives on Management of Forests and Fisheries in Island Southeast Asia*. Leiden: KITLV Press [this volume also includes some articles on Malaysia/Singapore and the Philippines].

On attitudes toward nature in pre-1900 Europe, see Thomas, Keith, 1984, *Man and the Natural World: Changing Attitudes in England 1500–1800*. Harmondsworth: Penguin; Ritvo, Harriet, 1990, *The Animal Estate: The English and Other Creatures in the Victorian Age*. London: Penguin.

On determinism, natural hazards, and disaster theory, see Arnold, David, 1996, *The Problem of Nature: Environment, Culture and European Expansion*. Oxford: Blackwell; Bankoff, Greg, Georg Frerks, and Dorothea Hilhorst, 2004, *Mapping Vulnerability: Disasters, Development and People*. London: Earthscan.

For those readers who wish to know more about environmental history in general, the following recent and not-so-recent introductions could be of some value: Simmons, I. G., 1989, *Changing the Face of the Earth: Culture, Environment, History*. Oxford: Basil Blackwell; McNeill, J. R., 2000, *Something New under the Sun: An Environmental History of the Twentieth-Century World*. New York: Norton; Radkau, Joachim, 2000, *Natur und Macht: Eine Weltgeschichte der Umwelt*. München: Beck; Hughes, J. Donald, 2001, *An Environmental History of the World: Humankind's Changing Role in the Community of Life*. London: Routledge; Richards, John F., 2003, *The Unending Frontier: An Environmental History of the Early Modern World*. Berkeley: University of California Press.

Recent publications containing (implicit or explicit) global history pretensions with a strong environmental flavor are Schama, Simon, 1996, *Landscape and Memory*. London: Fontana; Diamond, Jared, 1998, *Guns, Germs, and Steel: A Short History of Everybody for the Last 13,000 Years*. London: Vintage; Marks, Robert R., 2002, *The Origins of the Modern World: A Global and Ecological Narrative*. Lanham, MD: Rowman and Littlefield; McNeill, J. R., and William H.

McNeill, 2003, *The Human Web: A Bird's-Eye View of World History*. New York: Norton; Diamond, Jared, 2005, *Collapse: How Societies Choose to Fail or Survive*. London: Allen Lane.

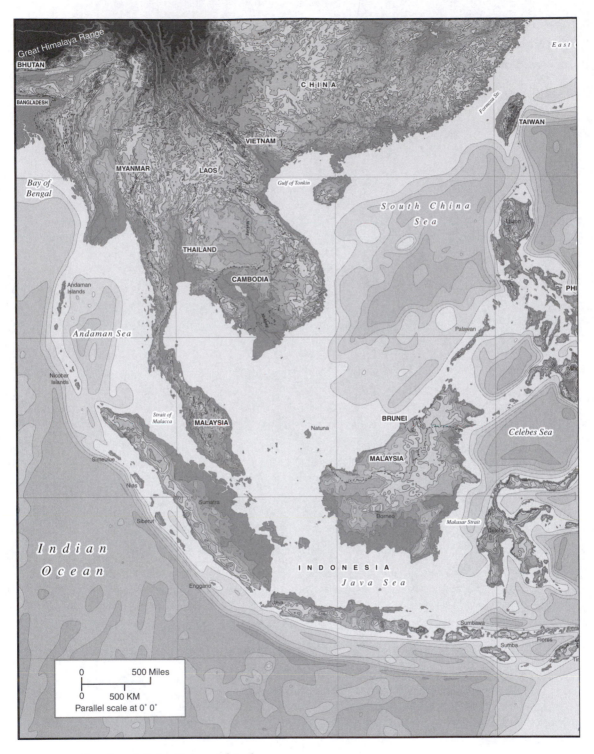

Southeast Asia, Topographical

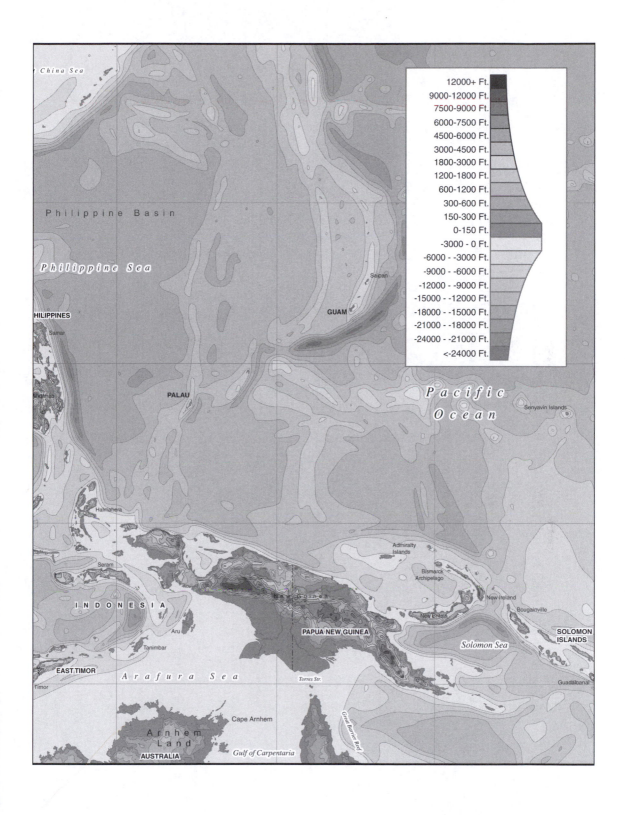

China Sea

Philippine Basin

Philippine Sea

HILIPPINES

Samar

ndanao

PALAU

Halmahera

Seram

I N D O N E S I A

Aru

Tanimbar

EAST TIMOR

Timor

A r a f u r a   S e a

Torres Str.

Cape Arnham

A r n h e m
Land

AUSTRALIA

Gulf of Carpentaria

Saipan

GUAM

*Pacific
Ocean*

Senyavin Islands

Admiralty
Islands

Bismarck
Archipelago

New Ireland

New Guinea

New Britain

Bougainville

PAPUA NEW GUINEA

Solomon Sea

SOLOMON
ISLANDS

Guadalcanal

Great Barrier Reef

	12000+ Ft.
	9000-12000 Ft.
	7500-9000 Ft.
	6000-7500 Ft.
	4500-6000 Ft.
	3000-4500 Ft.
	1800-3000 Ft.
	1200-1800 Ft.
	600-1200 Ft.
	300-600 Ft.
	150-300 Ft.
	0-150 Ft.
	-3000 - 0 Ft.
	-6000 - -3000 Ft.
	-9000 - -6000 Ft.
	-12000 - -9000 Ft.
	-15000 - -12000 Ft.
	-18000 - -15000 Ft.
	-21000 - -18000 Ft.
	-24000 - -21000 Ft.
	<-24000 Ft.

PART ONE

ORIGINS

1

BEFORE AND AFTER
THE ARRIVAL OF HUMANS

L et us begin by setting the stage for the arrival of humans. Of course, the region under discussion has a very long past prior to humans, and in this chapter a few glimpses at that past are offered. But as the theme of this book is the interaction between humans and their natural environment, that early past is not dealt with extensively. First, however, a few lines about the region itself, with some of its present-day features, are in order.

INTRODUCING THE REGION

Squeezed in between its enormous neighbors India and China, Southeast Asia in the past may not always have received the attention due to it. Even the term *Southeast Asia*, which seems to be such a logical name for the region, is of relatively recent (post-1940) origin. Nevertheless, there are good reasons for a separate volume in the Environmental History series, as the region shows many unique environmental features that make it stand out within Asian history. As a tropical region with a rich flora and fauna and with various ecological "hot spots," its basic environmental features are in many respects more akin to those of central Africa and Amazonia than to the natural environment of its Asian neighbors. Culturally and historically Southeast Asia's development differed greatly, of course, from that of central Africa and the Amazon, but that might make it all the more interesting to someone studying comparative (environmental) history.

Southeast Asia today consists of a number of independent states, ranging in size from very small to very large. Usually, a distinction is made between Mainland Southeast Asia and Island Southeast Asia, a convention that will be followed here. Burma (Myanmar), Thailand, Cambodia, Laos, and Vietnam constitute Mainland Southeast Asia.

Peninsular Malaysia and Singapore, though formally part of the mainland, are often regarded as belonging to Island Southeast Asia for environmental, cultural, and linguistic reasons. The other parts of Island Southeast Asia are the eastern states of Malaysia (Sarawak and Sabah), Brunei (all three being located on the island of Borneo, of which Kalimantan, part of Indonesia, occupies the remainder), Indonesia, East Timor (Timor Leste or Timor Loro Sa'e), and the Philippines. Of the island of New Guinea, the western half, which used to be called Irian Jaya and now goes by the name of Papua, is part of the Republic of Indonesia, while the other half (Papua New Guinea) is an independent state. The latter is not formally part of Southeast Asia, but in this book I will often refer to the island of New Guinea as a whole.

Given the cultural and linguistic similarities of its constituent parts, Island Southeast Asia is also loosely called "the Malay world." Biologists, particularly those with an interest in plant geography and zoogeography, often use the term "Malesia" for roughly the same area (Malaysia, Indonesia, the Philippines, and New Guinea). The reason is that, biogeographically speaking, the region has various features in common, although, as will be shown presently, there is also a great divide within the area.

THE NATURAL ENVIRONMENT

The whole of Southeast Asia is located within the tropics, lying between 25 degrees northern and 10 degrees southern latitude. This implies that temperatures are constantly high year round, and that the differences between daily maximum and minimum temperatures are small. However, temperatures drop at higher elevations, and frost occurs on the highest mountain peaks—for instance, in New Guinea. At higher latitudes seasonal variations in mean temperature as great as 13 degrees centigrade or more can occur.

Within Southeast Asia two climatic zones may be distinguished. The zone around the equator, the equatorial zone, covers roughly the region of Island Southeast Asia, with the exception of central and eastern Java, the Lesser Sundas (Nusa Tenggara), southern Sulawesi, and the western Philippines. The second zone, called the intermediate tropical zone, covers Mainland Southeast Asia and those parts of Island Southeast Asia not within the equatorial zone.

The equatorial zone is characterized by the absence of prolonged dry periods and hence by rainfall throughout the year, a pattern that is broken only in years with weather anomalies such as the ENSO phenomenon, short for El Niño–Southern Oscillation, discussed later in this volume. The type of vegetation corresponding to this rainfall pattern is that of the ever-wet and evergreen tropical forest, also called tropical rain forest. Soils are as a rule infertile clays, while nutrients are stocked in the biomass (trees) and, to only a much lesser degree, in the soil. Cutting the forest, therefore, leaves infertile soils that will leach rapidly. Such areas appear to have been less appealing to early humans

Sunset over the Irrawaddy River in Myanmar. Much of mainland Southeast Asia is defined by its large rivers like the Irrawaddy. (Corel)

(when they arrived) than the intermediate tropical zone, probably because of a combination of the lack of edible roots and tubers and the absence of large terrestrial mammals suitable for hunting; slash-and-burn agriculture in the ever-wet tropical rain forests is also difficult.

In the intermediate tropical zone, however, the rainfall pattern is governed by the monsoons, which means that a rainy season and a dry season alternate. In continental Southeast Asia the rainy season occurs during the summer; on the islands to the south of the equator, rainy season is during the winter. Where average annual rainfall is low, a long dry season is characteristic. The existence of a real dry season corresponds with the presence of monsoon forests that include a number of deciduous (leaf-shedding) trees. Such forests are more open than the rain forest, and roots, tubers, and terrestrial mammals are available in larger numbers. Thus hunting and slash-and-burn techniques could be applied more easily, while the creation of grasslands in order to attract game and to feed livestock was more feasible. The ground here may be quite fertile, particularly in areas with alkaline volcanic soils.

The role of such soils is beautifully illustrated in Indonesia, a country characterized by a large number of volcanoes. They form an arc, starting in Acheh in northern Sumatra and ending to the north of Menado in northeastern Sulawesi.

They cover western Sumatra, Java, Bali, Lombok, and the other Lesser Sundas, an arc running through the Moluccas and finally to Manado and points north. As a rule, areas without (many) volcanoes (eastern Sumatra, Kalimantan, most of Sulawesi, and New Guinea) used to be, and often still are, sparsely populated, while many of the volcanic areas (Java and Bali, for example) were and remain densely populated.

In Mainland Southeast Asia, which is located in its entirety in the intermediate tropical zone, the river valleys are (and were) much more densely populated than the mountainous areas. This area is largely defined by a number of large rivers, most of which originate in the Himalayas—the Irrawaddy, Salween, Chao Phraya, Mekong, and Red (Hong) River—separated by mountain chains. The mountains have their own vegetation zones (lower montane, montane, subalpine), which are cooler and often wetter than the lowland zones. This implies that mountains in areas where the lowland woods are (deciduous) monsoon forests can be covered with ever-wet evergreen forests. However, above the 1,000-meter line, no prehistoric settlements are in evidence.

In addition to the tropical rain forests and the monsoon forests, mention should be made of two other vegetation formations that were once important: mangrove forests covering many coastal zones, and swamps (peat swamps, freshwater swamp forests, and swamps in the river deltas). The many differences in precipitation, temperature, topography, elevation, and soil types have encouraged biotic diversity in the region.

NATURAL CHANGES IN THE ENVIRONMENT

Lest it be thought that only humans can change the environment, a phenomenon that we are all too familiar with, a few sentences on natural change may not go amiss. This is all the more important as these changes explain certain otherwise incomprehensible features of the origin and distribution of fauna and flora in the region, and even the early presence of humans.

The first major natural change that should be dealt with here is that of the so-called continental drift. Most of Southeast Asia is positioned on a continental shelf called Sundaland. The only areas that are not part of this shelf are the Philippines, eastern Indonesia (Sulawesi, the Lesser Sundas minus Bali, Moluccas), and New Guinea. The Philippines and eastern Indonesia constitute an area that has been called Wallacea (after the famous naturalist Alfred Russel Wallace), while New Guinea (and Australia) are part of another shelf, called Sahulland. Sahulland and large sections of Wallacea were originally part of a large continental mass (Gondwanaland), far to the south of their present position; they drifted northward over millions of years. By the time that the first hominids (nowadays also called hominin) entered the region, more than a million years ago the process had come to resemble the current situation. Proof of this development, however, can still be found in the enormous differences between the faunas and floras of Sundaland and Sahulland, with Wallacea as a transitional zone.

Sundaland has a rich, highly diverse Asiatic flora and fauna, characterized, among other species, by large placental mammals (elephant, rhinoceros, banteng and seladang, Asiatic buffalo, deer, pig, monkey, gibbon, orangutan) including large predators (tiger, leopard). Sahulland, in contrast, has an almost entirely different, relatively poor Australian flora and fauna, of which the marsupials (kangaroo, wombat, cuscus [*Phalanger*], Tasmanian devil) are typical representatives. The only two types of placental mammals in Sahulland in prehistoric times—rodents and bats—must have somehow crossed the sea without the help of humans. The flora and fauna of Wallacea, often called "impoverished" by naturalists like Wallace, contain elements of both Sunda and Sahulland.

As Sahulland was still separated from the rest of Southeast Asia by large stretches of water—though it was closer than ever before to Sundaland when the first humans arrived in Java a million or more years ago—the first people never reached New Guinea or Australia. Humans did not come to those areas until between 40,000 and 60,000 years ago. Thus, because of the special geological history of Sahulland, both its original inhabitants and its flora and fauna are different from those of the rest of Southeast Asia. To a somewhat lesser extent, that applies to Wallacea as well.

The second major natural change dealt with here is that of the glacial and interglacial periods. Generally speaking, during glacial periods overall temperatures were much lower: the polar ice caps grew considerably; and sea levels and water circulation, therefore, were much lower; and the equatorial zone contracted. During interglacial periods the earth heated up; the ice caps reverted to their former size; and the water level of the oceans rose again. The drop in temperature in the lowlands of Southeast Asia during glacial periods was probably rather small, but in the highlands it may have been as much as 8 degrees Celsius, which would have had consequences for human habitation if those areas had been inhabited.

Equally or even more important was the fact that sea levels dropped so much that the entire part of the Sunda shelf that is now underwater would have been exposed during some periods. This implies that the Malay Peninsula, Sumatra, Java, and Borneo were no longer separated, having all become joined to the mainland. In Wallacea most of the Philippine islands would have been joined together as well, while in Sahulland, New Guinea and Australia had become one continent. However, Sundaland remained separated by sea from Wallacea, and both were separated from Sahulland.

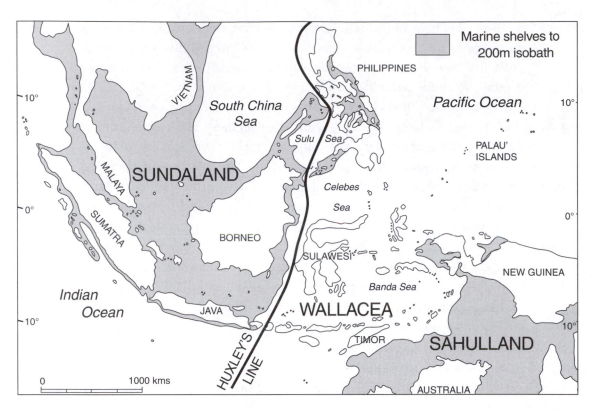

Southeast Asia, Biogeographical

The last time that the Sunda and Sahul shelves were exposed was a period of a few millennia around 21,000 years ago. Sea level has been estimated to have been 100 meters or more below present levels. Prior to that, a drop in sea level seems to have taken place every 100,000 years or so, going back perhaps as far as 2 million years. Interglacials took place at similar intervals, with the last sea level maximum occurring possibly between 6000 and 3000 BP, when it may have been almost 6 meters higher than its present level. This means that the coast of most areas would have been farther inland than today, while all islands must have been considerably smaller. However, the raised marine deposits in Asia upon which the notion of this maximum is based might reflect tectonic movements rather than an actual rise in sea level. The glacial troughs in temperature and sea level were quite short, perhaps some 10,000 years in duration, while the rise in temperature after such a minimum was rapid. It seems plausible that there was a connection between the first stages of agriculture and the last episode of climate amelioration after the temperature minimum some 18,000 years BP.

The upshot of all this is that periods of insular isolation and connection with the mainland have alternated during the last 2 million years or so in Island Southeast Asia. This has implications for the distribution of flora and fauna.

During periods in which the islands of western Indonesia were connected to Mainland Southeast Asia, they experienced an influx of fauna and flora from the mainland. During periods of high sea levels, when the islands were isolated from the mainland, extinctions seem to have been more frequent, and some large animals appear to have undergone a process of dwarfing.

It is also plausible that the area covered with tropical rain forests around, say, 1500 CE was interspersed during the glacial episodes with large corridors of monsoon forest or even savanna vegetation, which are generally held to have been more congenial to early humans.

Java, one of the best-researched areas, may serve as an example. There is no evidence of mammals in Java 2.4 million years ago, when the first clearly defined sea level lowering linked to a glacial period occurred. The island was probably entirely underwater around that time, starting to emerge only in the wake of tectonic and volcanic processes during the Late Pliocene (roughly 2 million years ago). This corresponds to the Satir Fauna (2.0–1.5 million BP)—called, as usual, after the place of the dig—which contains an elephantoid *Sinomastodon bumiajuensis*. There is evidence of dwarfed elephantoids like *"Elephas" indonesicus* dating from around the same period. There are also a giant tortoise and a number of herbivores, such as hippopotamus and deer, often encountered on islands.

The Trinil Fauna (ca. 900,000 BP) contains the first solid evidence for the presence of humans (*Homo erectus*), who were apparently in the company of a tiger subspecies, a rhino, a buffalo (*Bubalus*), and various endemics (animals not to be found elsewhere), suggesting rather isolated conditions. One of these was another elephantlike creature, *Stegodon trigonocephalus*. During this early period of mammalian colonization (2.4–0.8 million BP), sea levels appear to have fluctuated around 70 meters below present levels.

The Kedung Brubus Fauna (800,000–700,000 BP) appears to represent a major faunal immigration event, resulting in a maximum number of mammals. This episode coincides with the onset of a new type of sea level fluctuation linked to glacial events, with lower minima and therefore broader "corridors," and with open woodland conditions. During this stage the "real" elephant (*Elephas*) first entered Java. There was a hyena (since extinct), another rhino (also extinct), a tapir (which became extinct only in more "recent" times), a deer, and a wild boar.

A major change in environmental conditions in Java, represented by the Punung Fauna (125,000–60,000 BP), must have taken place toward the end of the Pleistocene. Several animals from the Kedung Brubus Fauna had disappeared, while other species were now found for the first time. The Punung Fauna represents—also for the first time—humid forest conditions in Java, witnessed by the presence of large numbers of primates, such as gibbons and the orangutan, of which the latter has since died out in Java. The replacements must have taken

Orangutans have seen their habitat change widely in the past due to climatic variation. The orangutan is still to be found on the islands of Sumatra and Borneo. (Corel)

place between the onset of the glacial maximum (that is, with temperatures at their lowest level) of 135,000 BP, when conditions in western Java were very dry, and the period around 70,000 BP. It can even in all probability be narrowed down to the period between 80,000 and 110,000 years ago, when the climate was warm and humid, with sea levels around 50 meters below the present level, sufficiently low to leave an overland corridor in the west that would have permitted migration. The Punung Fauna possibly also includes *Homo sapiens*, modern man.

During or following the latest glaciation—a number of millennia centered around 21,000 BP—conditions in Java became much drier and seasonality increased. The result was that the orangutan disappeared from the Holocene fauna (from 10,000 BP), along with a large number of other species, while an increase in grass pollen and a drop in fern spores reflect accompanying floral changes.

THREE "WAVES" OF HUMANS

When the physician Eugene Dubois, a scholar with an interest in paleoanthropology, found a part of a skull near a place called Trinil in Java, he thought that he had discovered the "missing link" between apes and humans. Therefore, he called this being *Pithecanthropus erectus*. For a short period the thought could be entertained that Java was the cradle of humanity, and not the Middle East, as many people then believed. Nowadays we no longer use the name *Pithecanthropus erectus*, and we regard this early Asian humanlike creature as a variety of *Homo erectus*, a hominid originating in Africa some 1.7 million years ago. The hypothesis that the early evolution of humanity took place in Africa, from where *H. erectus* spread over Eurasia, now supported by most scholars, is usually called the Out of Africa One hypothesis.

As it is difficult to provide exact dates for these early hominids, estimates for the age of the earliest remains of *H. erectus* in Java are quite divergent. The same is true for claims regarding early hominids in China. In both cases, estimates of 1.8 million years or more have been presented. This implies an appearance of *H. erectus* in Java and China prior to the hypothetical origin in Africa, which is highly improbable. Most scholars agree that a date between 0.5 and 1.3 million years ago for the earliest Java man is the most plausible.

The earliest traces of *Homo sapiens* in the Indonesian Archipelago can be dated to perhaps around 60,000–50,000 BP. One of the hitherto unanswered questions is whether these modern people were the descendants of the local *H. erectus*, or whether the latter had died out and *H. sapiens* was—again—introduced from outside (and again, with an African origin). Those who favor the latter solution (replacement) are the adherents of the so-called Out of Africa Two hypothesis, while those who oppose this view may be thought to belong to the regional continuity school. There are now also scholars who favor the notion of a combination of recent migrants (*H. sapiens*) who may have genetically assimilated elements of the population that was already present (*H. erectus*).

It has been generally assumed that *H. erectus* never reached Australia and New Guinea. That part of the world was not populated until 60,000–40,000 BP, when *H. sapiens* crossed a considerable expanse of sea, probably during a glacial trough. According to Peter Bellwood, one of the main experts in this field, this was the first time in history that a group of humans had crossed such a distance over open sea. However, there is now evidence that *H. erectus* did at least reach Wallacea, perhaps some 700,000 to 800,000 years ago, as stone artifacts, possibly dating from around that time, have been discovered on the island of Flores (Lesser Sundas).

An even more spectacular discovery occurred between 2003 and 2005, when on the same island bones of altogether nine individual hominids of an entirely new type were found. This *Homo floresiensis*—dated between 95,000 and 12,000 years ago (and therefore partly contemporary with *H. sapiens*)—was much smaller than *H. erectus*, with a brain of barely 400 cubic centimeters, while *H. erectus* had between 600 and 1,000 cc of brain tissue. It is assumed that *H. floresiensis* was a dwarfed form of *H. erectus*, another example of the process of dwarfing on an island mentioned earlier. These recent discoveries put *H. erectus* closer to the Sahul shelf than had before been thought possible.

The *Homo sapiens* populations of Australia–New Guinea (Sahul area), one continent when it first became populated, and the Sunda area started to drift apart biologically. The Sahul people, developing in isolation, have not undergone great changes up to the present, although there has been more change among the Papuans than among the Aboriginals. The Sunda people underwent cranial and facial gracilization, a term referring to increasing slenderness of build. This was no doubt largely attributable to the Mongoloid gene flow, to be dealt with presently, but perhaps also to local selection. Gracilization of the human skull since the Late Pleistocene is a worldwide phenomenon, usually associated with climatic changes and the development of agriculture.

After the first wave—we have, of course, no idea of the numbers involved, and the term "wave" might be a misnomer—of hominids (*H. erectus*) around 1 million BP and the second one (*H. sapiens*) 60,000–50,000 BP, a third wave started, possibly around 6000 BP. *H. sapiens* was by then the only hominid, so this was a case of one subspecies of *H. sapiens* occupying—or rather slowly invading by long drawn-out stages—the territory of another subspecies. The biological grouping that undertook this expansion has been called Southern Mongoloid, as their area of origin is the southern part of what is today China. They took the place of or interbred with a grouping that had developed locally from the second wave, and that may be called Australo-Melanesian.

The Southern Mongoloids have dominated Southeast Asia ever since, but remnants of the Australo-Melanesians are still to be found in New Guinea (Papuans), and in even smaller numbers in the central and northern Philippines, in parts of Thailand and Peninsular Malaysia, and on the Andaman Islands

(Negritos). There are noticeable differences between the Southern Mongoloids and the Australo-Melanesians in terms of clearly visible physical characteristics, such as the much darker skin coloration of the Australo-Melanesians, shorter stature, and other inheritable features. However, part of Eastern Indonesia (the Moluccas, the eastern Lesser Sundas) shows a cline between these two groupings, representing, therefore, a very gradual transition between the two, something that is also supported by linguistic evidence.

The fact that most of the equatorial zone in Southeast Asia is inhabited by relatively light skinned people, while the inhabitants of the equatorial zone in Africa are dark skinned, is testimony to the relatively recent arrival of the Southern Mongoloids in Island Southeast Asia. In contrast, the dark-skinned Negritos nowadays do not live close to the equator, which suggests that these ever-wet rain forest areas were hardly inhabited at all (except for their coastal fringes) prior to the arrival of the Southern Mongoloids. The Negritos may always have been inhabitants of the intermediate tropical zone, while their relatively short stature suggests adaptation to the warmer and more humid conditions of the Holocene, when the forests became denser and animal proteins harder to acquire.

In addition there are linguistic differences, although they do not always coincide with biological differences. Broadly speaking, however, the Southern Mongoloids were and are speaking Austroasiatic and Austronesian languages, while the Australo-Melanesians were not. The northern branch of the Southern Mongoloids gradually invaded Mainland Southeast Asia, where now Austroasiatic languages predominate. The southern branch of the Southern Mongoloids occupied Island Southeast Asia (and large parts of Oceania) over the millennia, and in Malaysia, Indonesia, and the Philippines Austronesian languages are now spoken by an overwhelming majority of the people. Most Papuans do not speak Austronesian languages, while among the Negritos only those of the Andaman Islands are still speaking their original language.

The clinal zone in Wallacea aside, the Australo-Melanesians seem to have disappeared from the face of the earth in the areas where the Southern Mongoloids established themselves. As was the case with the "replacement" of *Homo erectus* by *H. sapiens*, it is not clear what happened to the Australo-Melanesians when they were "overrun" by Southern Mongoloids. The most plausible explanation is that, at least in the latter case, the numbers of the invading Southern Mongoloids, who were agriculturists, were much greater than those of the foraging Australo-Melanesians inhabiting the region originally. Biologically speaking, therefore, the genes of the original inhabitants were a drop in the ocean constituted by the gene pool of the newcomers. In the only region in Southeast Asia where agriculture developed almost certainly independently,

New Guinea, the newcomers were never able to replace the original population, no doubt because there they were not as numerous.

In Mainland Southeast Asia, the original population of foragers also disappeared virtually without leaving a trace, after the Southern Mongoloid agriculturalists had established themselves in the region. But that is a topic dealt with in more detail in the next chapter.

PRE-AGRICULTURAL ACTIVITIES AND ARTIFACTS

In the last section it was pointed out that the people living in Southeast Asia prior to the third "wave" of invaders were foragers—that is, hunters and gatherers. Like pre-agriculturists the world over, these were prehistoric people, people who had not yet mastered the art of writing. Therefore our knowledge of their societies must be based on archaeological evidence. Basically, this observation applies to the entire period prior to the beginning of the Common Era in Southeast Asia. However, we are relatively well informed for the period since 40,000 BP, when radiocarbon dating can be utilized. Artifacts are overwhelmingly restricted to *Homo sapiens*. Evidence for artifacts that might be associated with *H. erectus* is rare—and moreover weak, as it is beset by dating problems, partly related to volcanic upheaval. This lack of artifacts is a bit of a puzzle, as human-made objects usually accompanied *H. erectus* in other areas. However, many Southeast Asian places where artifacts could have been found are now covered by the sea, and it would be rash to say that they are entirely absent; it is too early to offer firm conclusions.

All available evidence, however, suggests that we are dealing with a very sparsely populated area in this period. It is, therefore, rather unlikely that human hunting prior to the Holocene (ca. 10,000 BP) was solely responsible for the various local (megafauna) extinctions that have been recorded. It is plausible to assume that climatic variation during the Late Pleistocene also had something to do with it.

Examples from Indonesia and Vietnam are the tiger and the orangutan. The tiger is an animal not to be found in Borneo during the historical period, but of which fossilized remains appear to have been discovered. When during the Late Pleistocene sea levels were rising, Borneo was cut off from the mainland, while the climate became hotter and wetter. As the ever-wet rain forest in its mature-phase form is not a habitat favorable to tigers (they prefer ecotones between forests and grasslands, or arable lands, because of the presence of wild boar and deer), these climatic changes seem to have led to their local extinction. Remains of the orangutan have been found in cave deposits in northern Vietnam dating to approximately 18,000 years ago, but the animal has become locally extinct since;

it is now to be encountered only in Sumatra and Borneo. It would appear that the drier and cooler conditions and the loss of rain forest around that time led to the disappearance of the orangutan, a typical rain forest dweller, from Mainland Southeast Asia.

Many tools of the foraging population of Southeast Asia were made of stone and bone (and, of course, wood, which has long since vanished). Generally speaking, much of the material mentioned here has come from caves and rock shelters, where human burial sites have also been encountered. There are also coastal shell middens, which are often to be found in inland areas, as sea levels have dropped since people lived there.

A basic flake industry was ubiquitous from 40,000 BP until well into the Bronze Age (from ca. 2000 BCE). Pebble tools have been found in northern Vietnam, indicated by the term "Son Vi," to be dated between 23,000 and 18,000 BP. The best known stone tool industry in Mainland Southeast Asia is the one called Hoabinhian (type site Hoa Binh, Vietnam), producing pebble tools, dated from some 13,000 years ago. Evidence from the Malay Peninsula suggests that the Hoabinhians were almost certainly the ancestors of the Negritos. There are also indications that during the Holocene (from 10,000 BP onward), the Hoabinhians started the colonization of the rain forests of the Malay Peninsula and northern Sumatra. At roughly the same time, there are signs of increasing population density in the monsoon areas more to the north, particularly in western Thailand and northern Vietnam, reflecting successful adaptation to the post-glacial climate changes.

In the Niah Caves in Sarawak, Borneo, flake and pebble tools accompanied by bone implements have been found dating from the period between 40,000 and 30,000 BP. The inhabitants of the caves were hunters or scavengers, and the faunal remains (bones) found there indicate that the natural environment during the period was rain forest, quite similar to that of today. Here, therefore, we find human occupation at the coastal edge of the rain forest; evidence for long-term human presence in the interior of the rain forest is lacking in the pre-agricultural period.

Plant remains dating from 13,000 BP or earlier have been found in eastern Timor, including Job's tears, the ingredients for betel chewing (betel vine and areca nut), and candlenuts. It is assumed that these plants were collected, not cultivated.

The Hoabinhian pebble tool culture is not to be found in the Philippines or in Indonesia beyond northern Sumatra. In those areas the basic flake industries of the Late Pleistocene type were continued into the Holocene without technological change until the arrival of Neolithic tools (polished adzes). However, as a variation on this basic theme, in parts of the Philippines,

Sulawesi, and Java we encounter the production of small blades and bladelike flakes from around 8000 BP. The most important of these industries was the Toalian of southern Sulawesi. The microliths produced in these industries suggest bands of hunters with arrows or spears hunting in the monsoon forest areas of Island Southeast Asia. Signs of this industry have not been found in the equatorial parts of the region, but, of course, finding such artifacts in tropical rain forest is not easy.

What else do we know about the societies of these prehistoric cave-dwelling and tool-making hunters and gatherers? It seems safe to assume that they were egalitarian, living in small groups of highly mobile, slowly reproducing nuclear families, and that not only the size but also the number of those groups was quite small. All that would change, in the long run, with the so-called Neolithic Revolution, the transition to agriculture, to be dealt with in the following chapter.

With such small numbers, foraging hominids and humans might not have had more of an impact on their natural environment than other mammals of comparable size, had it not been for their ability to make fire. It is now generally accepted that *Homo erectus* was already able to do so. When, in addition, *Homo sapiens* developed speech and fairly sophisticated hunting weapons, the scene was set for a much greater human impact on the environment. In hindsight we are tempted to see the origins of agriculture as the beginnings of humans' role as the great environmental manipulators, but, potentially, many elements of human expansion were already visible at an earlier stage.

What makes humans as a species stand out is also that there does not appear to be a mechanism that keeps population numbers within limits set by the natural environment, as is the case with other mammals. Among most mammals, population numbers can grow as long as there is enough food, but as the amount of food is finite (although subject to variation over time), numbers will eventually stagnate and even fall as a result of food scarcity. The term "equilibrium" used to be employed for this phenomenon, but biologists are now reluctant to apply that notion, as the processes involved appear to be more complicated than was once thought. Nevertheless, something approaching equilibrium appears to have been involved.

Variation in numbers according to the availability of food appears to apply to humans as well, but only in the short run. If we look at longer term developments, we see that there has always been a long-term trend of gradual population growth. There are variations around this trend, temporary ups and downs, but there is no ceiling, as is the case with other mammals. This is the corollary of humans' unique ability to manipulate their environment to such an extent that

Lake Toba in northern Sumatra, occupies part of the Toba caldera, which was created 74,000 years ago when Mount Toba erupted. Toba erupted with such force that the skies were darkened for six years. (Charles & Josette Lenars/Corbis)

it is significantly altered and adapted to their needs—and this on such a scale that the term "environmental change" must be used, a notion that would be difficult to apply to the actions of nonhuman species, even though some of them also manipulate their natural environments.

If speech, fire, and weapons had already shifted the balance between humans and other large—and occasionally much larger and stronger—mammals in favor of *Homo sapiens*, the limitless growth of the numbers of humans rendered the competition between their species and the others even more uneven. Recently, the suggestion has been put forward that *Homo sapiens* almost perished before really getting started, and the reason for this is to be found in the region's past. Around the year 74,000 BP, Mount Toba, a volcano in northern Sumatra—now just a lake—erupted with such force that the skies remained dark for six years. *Homo sapiens* may have been reduced to just a few thousand individuals. It was a narrow escape, and perhaps the last almost successful attempt by "nature" to rid itself of this peculiar brand of primate.

BIBLIOGRAPHICAL ESSAY

The literature on the topics dealt with in this chapter (and the next) is dominated by two scholars: Peter Bellwood for Island Southeast Asia and Charles Higham for the mainland. For this chapter I have made use of Higham, Charles, 1989, *The Archaeology of Mainland Southeast Asia: From 10,000 B.C. to the Fall of Angkor*. Cambridge: Cambridge University Press; Bellwood, Peter, 1997, *Prehistory of the Indo-Malaysian Archipelago*. Honolulu: University of Hawai'i Press; Bellwood, Peter, 1999, "Southeast Asia before History." Pp. 55–136 in *The Cambridge History of Southeast Asia*. Vol. 1: *From Early Times to Circa 1500*. Edited by Nicholas Tarling. Cambridge: Cambridge University Press. A short, recent introduction is Bellwood, Peter, and Ian Glover, 2004, "Southeast Asia: Foundations for an Archaeological History." Pp. 4–20 in *Southeast Asia: From Prehistory to History*. Edited by Glover and Bellwood. London: RoutledgeCurzon.

A recent, more general text, in which what happened in Southeast Asia can be seen in a global context, is Scarre, Chris, ed., 2005, *The Human Past: World Prehistory and the Development of Human Societies*. London: Thames and Hudson.

For a very short section on geography and environment, see Richards, John F., and Elizabeth P. Flint, 1994, "A Century of Land-Use Change in South and Southeast Asia." Pp. 18–19 in *Effects of Land-Use Change on Atmospheric CO_2 Concentrations: South and Southeast Asia as a Case Study*. Edited by Virginia H. Dale. New York: Springer.

An interesting study, mentioned here for its focus on New Guinea, is Diamond, Jared, 1998, *Guns, Germs and Steel: A Short History of Everybody for the Last 13,000 Years*. London: Vintage.

For a detailed overview of Java's flora and fauna during this period, see Bergh, Gert D. van den, John de Vos, and Paul Y. Sondaar, 2001, "The Late Quatrenary Palaeogeography of Mammal Evolution in the Indonesian Archipelago," *Palaeogeography, Palaeoclimatology, Palaeoecology* 171, pp. 385–408.

On the recently discovered Flores man, see Brown, Peter, et al., 2004, "A New Small-bodied Hominin from the Late Pleistocene of Flores, Indonesia," *Nature* 431, pp. 1055–1061; and Morwood, Michael J., et al., 2004, "Archaeology and Age of a New Hominin from Flores in Eastern Indonesia," *Nature* 431, pp. 1087–1091.

On the development of skulls, see Storm, Paul, 1995, *The Evolutionary Significance of the Wajak Skulls*. Leiden: Nationaal Natuurhistorisch Museum.

The story about the eruption of Mount Toba is taken from "The Proper Study of Mankind," *Economist*, December 2005/January 2006, pp. 3–12.

2

EARLY AGRICULTURE AND
THE METAL AGES

THE NEOLITHIC REVOLUTION

Depending on the area, agriculture appeared in Southeast Asia between 4000 and 1000 BCE (6000 and 3000 BP). For this first phase of agriculture, prior to the use of bronze and later iron, the term "Neolithic" (Late Stone Age) or, more dramatically, "Neolithic Revolution," is conventionally employed. It goes without saying, however, that there was no sudden break between the period in which all humans were foragers (hunter-gatherers) and the time of the early agriculturalists—a transition, moreover, that took place at various moments in time in different places.

Typical of early agriculture is the presence of domesticated plants (crops) and animals (livestock and pets). However, these domestications were a long time in coming, and there was a long, drawn-out phase during which foragers started manipulating and protecting certain plants and animals long before they could be called fully domesticated. It should be pointed out that taming and domesticating are not the same thing. Taming occurs when an individual animal is captured by humans and then trained to live with or work for people, usually in some kind of confinement. Domestication takes place when animals are subjected to selective breeding, thereby stimulating some hereditary features (weight, color, aggressiveness, and the like) and discouraging others. It is a process that takes a great many generations, as it did with the transformation of wolves into dogs. Similar processes took place when plants were domesticated, although the terminology employed is different.

Other typical features of early agricultural societies usually include the presence of pottery and of villages. However, they do not always appear in the same order or in the same combination, at least initially. Living in fixed places (villages) is called sedentism, and sedentism preceded agriculture (at least the crop cultivation part of it)—hence the term "sedentary revolution." The presence of pottery implies sedentism, but not necessarily agriculture. There are, for instance, sedentary foragers, particularly under circumstances of high biodiversity and abundant aquatic resources. There are also mobile (nonsedentary) agricultural-ists, whom we usually call pastoral nomads. The latter did not play much of a role in the history of Southeast Asia, though they did in some of the mainland border areas.

There are strong indications that there have been two regions in the world where an autonomous Neolithic phase of sedentary agriculturists-cum-stock-breeders ("agro-pastoral system") was preceded by one in which broad-spectrum foragers had developed pottery, stored surplus foodstuffs, kept domesticated dogs, buried their dead in cemeteries, and had become semisedentary. Those areas are southwestern Asia (the Fertile Crescent) and central and southern China.

Nevertheless, when all is said and done, it can be argued that there is a broad tendency for agriculturalists to live together in semipermanent settlements and to produce pottery for the storage of harvested crops. At a later stage, shifting cul-tivators are dealt with; they lived in settlements of a certain permanence, but often not more than for a few years.

Not long ago it was generally accepted that agriculture started independent-ly in four centers (hearth areas): southwestern Asia, central and southern China, Mexico, and the Andes, all cradles of great civilizations. Today there are more candidates, although most of them are contested. However, a strong case has been made for New Guinea, a region that does not fit the concept of a "cradle of a great civilization." It is now accepted that the New Guinea highlands were a primary center of plant cultivation, where taro, yam, sugarcane, *Pandanus*, and a banana species were domesticated, at least by 4000 BCE but perhaps much ear-lier; there are traces of drainage ditches dating from around 7000 BCE.

However, possibly owing to its isolated position, New Guinea does not appear to have had much influence on the rest of Southeast Asia (with the excep-tion of parts of the southern Moluccas and the Lesser Sundas)—unless one assumes that cultivated yam and taro were introduced in the Indonesian Archipelago and the Philippines from New Guinea (see also below). Both in Mainland and Island Southeast Asia agriculture was introduced from China south of the Yangtze River, another primary center of plant cultivation, of which the coastal zone of northern Vietnam may have been a part. The lower Yangtze region also appears to have been the area where rice was first cultivated, which took place under conditions somewhat warmer than those of today. By around 5000 BCE wooden villages based on the cultivation of rice had sprung up there. Their technology included pottery and carpentry, including the use of stone adzes, wooden and bone agricultural tools, boats, paddles, spindle whorls for weaving, matting, and rope. In addition there is evidence for domesticated pigs, dogs, chickens, and perhaps cattle and water buffalo. Millet was domesticated in central China but seems to have reached southern China at an early stage.

The explanation of the transition from hunting and gathering to agriculture has always been subject to much speculation, as the archeological record has not so far yielded unequivocal answers (and probably never will). Nowadays there are those who link the processes of domestication to climate fluctuations. Climate amelioration during the early Holocene would have favored large-grained cereals, which then would have been collected in increasing quantities by foragers. This productive environment was then a stimulus for sedentism on strategically located sites where storage (in pottery) of the collected grains developed. That led to population increase, eventually to environmental pressure and subsistence stress, and finally to agriculture. During cooler stages, which might have resulted in lower cereal (rice) yields, the hunter-gatherers would have had to start protecting and cultivating these plants—which may be regarded as the beginning of agriculture.

Another popular theory is that of "affluent foragers," whose subsistence was guaranteed, enabling them to experiment with crop and animal manipulation. This idea may be combined with the notion of "competitive feasting"—that is, the idea, based on the recent ethnographic record, that wealth in many societies has been used as a means to compete with one's neighbors for status and power. We then arrive at a model of affluent foragers who "invented" agriculture as a means to enhance their status by means of conspicuous consumption.

The spread of agriculture has sometimes been linked to population pressure, but for Southeast Asia there is no archaeological evidence to support such a hypothesis. In all likelihood it was the other way around: once agriculture had become successful, the number of people began to grow.

One of the most important crops in Southeast Asia is rice, and that has been the case for at least a few thousand years. Rice grew naturally in the Yangtze River area, and it is now supposed to have "radiated" from its region of origin to (most of) Southeast Asia. We are not entirely sure that rice was *not* present in many Southeast Asian areas, but in much of the recent literature that is assumed to have been the case.

What may have played a role in the spread of (wet) rice is that in all probability early wet-rice cultivation depended on the presence of naturally wet habitats—swamps and natural flood plains. As the extent of such habitats in the primary cultivation areas was supposedly limited, people soon started to look for other such environments in the intermediate tropical zone—the area where rice cultivation originated and where rice yields were highest—rather than create artificial environments for rice. The same naturally wet habitats were also suitable for taro, one of the principal domesticated roots and tubers in early Southeast Asia.

However, when the rice cultivators reached the tropical zone, they were in trouble. Rice in its homeland is accustomed to a dry season, while it was originally also sensitive to changes in day length (photoperiodicity)—two conditions that could not be met in the tropical zone. Rice, therefore, seems to have dropped to the status of a minor crop when the Southern Mongoloids reached Borneo and other tropical areas, and it took some time before varieties suited to the local climate had been developed. In contrast, when islands like Bali and Java, with a typical monsoon climate, had been reached, wet-rice came into its own again. But we are getting ahead of the chronological narrative.

NEOLITHIC MAINLAND SOUTHEAST ASIA

The Neolithic record is much better for Mainland Southeast Asia than it is for Island Southeast Asia. That fact might be linked to a real difference in occupation during this phase, perhaps because a tropical rain forest environment was less conducive to Neolithic settlements than areas with a monsoon climate—or possibly because there was no need to go there. However, there is also the possibility that it is much more difficult to find Neolithic remains in the tropical rain forest zone.

Starting then with Mainland Southeast Asia, we find the earliest signs of agricultural societies in northern Vietnam. These finds, dating from the fifth millennium BCE onward, include, in addition to burial sites, pottery, stone adzes, and the bones of deer, cattle, pig, and dog, although it is not clear to what extent these animals were domesticated (the dog aside, of course). More abundant material came from the Phung Nguyen culture of the area usually called Bac Bo, the lower Red River valley and delta, a culture dating from around 2000 BCE, where evidence has been found for the cultivation of rice in addition to implements used for spinning. This points to the production of textiles, another feature of many agricultural societies.

Despite claims for Neolithic remains of a very early date in Thailand, hard evidence for such remains predating 3500 BCE is not available. The archeological record improves considerably, however, around 3000 BCE, when several sites on the Khorat Plateau in northeastern Thailand appear to have been occupied. The best known sites are Non Nok Tha and particularly Ban Chiang. The rich finds have included adzes, shell beads, pottery, rice chaff temper in pottery shards, bones of domesticated cattle (probably *Bos javanicus* or *B. gaurus* rather than *B. indicus*), pig, dog, and chicken, as well as the remains of hunted or collected animals such as shellfish, turtles, crocodiles, and various mammals including rhinoceros. The Ban Chiang people seem to have been growing rice in seasonally flooded soils. An analysis of the skeletal remains found in burial sites in Ban Chiang suggests a life expectancy of 31 years. Compared with today's figures this is, of course, very low, but it does not compare badly with Neolithic Europe, or

Ceramic pots in the Ban Chiang Museum in 1974 showing red-on-buff geometric designs. Occupying northeastern Thailand roughly 3,000 years ago, the Ban Chiang people showed evidence of already growing rice in flooded soils. (Courtesy of Michael Pietrusewsky)

with figures for the nineteenth century. Not much is known about the major causes of death, but it has been suggested that malaria may have played a role in the high mortality rate.

It is plausible to assume that the people of the Khorat Plateau during the third millennium BCE were recent migrants into a hitherto very sparsely inhabited region, who had brought their Neolithic technology with them from northern Vietnam or southern China. An alternative hypothesis is that they came from the Gulf of Thailand, where the impressive site of Khok Phanom Di, with archaeological deposits dating from the period after 2000 BCE, has been excavated. Today an inland location, it used to be located near the coast, where rice could be cultivated in swamps, while products from the sea were also part of the diet. The analysis of cores taken from sites nearby suggests the presence of rice around 2500 BCE, but that might have been wild rice.

The site of Khok Phanom Di also yielded burials provided with rich gifts, dated after 2000 BCE, which suggests the existence of rich individuals or perhaps even wealthy kin groups in the society. One female burial suggests a link between fine pottery and wealth. Another striking feature of these burials was the high perinatal and infant death rate, which has been attributed to a high incidence of malaria, a disease linked to the swampy coastal environment. When the seas slowly retreated again (between 2000 and 1500 BCE), infant and fetal mortality rates dropped, but then the proportion of older children dying increased; perhaps because the number of malaria mosquitoes dropped and children survived longer, only to die from anemia.

Finally, mention should be made of various sites from southern Thailand and Malaysia dating from 2000 to 500 BCE, indicated by the term "Ban Kao culture." Pottery similar to that of Khok Phanom Di has been found there.

THE NEOLITHIC IN ISLAND SOUTHEAST ASIA

Turning now to Island Southeast Asia, we start again in southern China. It is now accepted by many scholars that people from southern China introduced their Neolithic culture to the island of Taiwan, perhaps as early as 4300 BCE, but almost certainly by 3500 BCE. There we find, among other artifacts, pottery, stone adzes, stone hoes, stone net sinkers, stone bark-cloth beaters, and spindle whorls of clay. The net sinkers suggest fisheries; the bark-cloth beaters indicate that textile (from hemp fibers?) had to compete with bark-cloth. Large numbers of burials (with high proportions of infant and fetal deaths) were uncovered, in addition to house foundations.

It is not clear whether cereals (millet, rice) were already part of the culture at this stage. Evidence for the early presence of millet is entirely linguistic, while rice may not have been present prior to 2500 BCE, when stone reaping knives became common. Another interesting feature is that a burial site yielded stained human teeth, indicating the use of the betel quid, a typical feature of "Malay" culture.

According to linguistic reconstruction, a large hypothetical language superfamily, Proto-Austro-Tai, to be found in southern China many millennia ago, developed into Proto-Austronesian on the island of Taiwan. (According to the linguistic reconstruction of this language it included words for cultivated rice, millet, and sugarcane, in addition to domesticated pigs and dogs.) This language, in turn, developed into Proto-Malayo-Polynesian, which is the root of all Austronesian languages spoken today in the Philippines, Indonesia, Malaysia, and Oceania. Proto-Malayo Polynesian, which may have been spoken in the central or southern Philippines, included, according to its reconstruction, words for taro, yam, banana, sago, betel-chewing, breadfruit, and chickens, in addition to the words already encountered in the Proto-Austronesian language. All of this does not mean that words absent from the reconstruction were not part of the language; it means only that there is no proof that they were.

In the archaeological record we find traces of Neolithic cultures (assumedly introduced by Neolithic groups of Austronesian-speaking Southern Mongoloids from Taiwan) in several places in the Philippines, ranging from northern Luzon and Palawan to the islands of Negros and Masbate. These cultures may be dated from 2500 BCE in northern Luzon and onward, perhaps even from 3000 BCE in Palawan. In the more southern areas of the Philippines the Neolithic assemblages appear to date from around 2000 BCE and later. Most of the remains are pottery related; the finds include house foundations, bark-cloth beaters, many artifacts made from shells, including shell tools (perhaps a local tradition), and, again, stained teeth. Clay spindle whorls are not found south of Luzon, and stone reaping knives have not been found at all. It is, of course, possible that other types of knives, such as bamboo knives, were used to cut cereals. However, it is also possible that, as has been mentioned above, cereals had been dropped from the agricultural repertoire of the immigrants as they came closer to the equator, with roots and tubers (taro, yam) and tree crops yielding the carbohydrates that the cereals produced in Taiwan and China.

Turning now to Indonesia and northern Borneo, we find possibly the oldest Neolithic remains (perhaps as early as 2500 BCE) on the Talaud Islands (to the north of northeastern Sulawesi), but that is a rather insecure date. Much better documented are the early Neolithic societies of Sabah, northern Borneo, to be dated from 2000 BCE. Shell items, including tools, have been found there, as is the case in the Philippines; such conspicuous objects as agate blades and awls and small chips of obsidian have also been discovered. The latter are the most remarkable item of this complex, as the obsidian can be traced to the place where it was mined, a location in northern New Britain, Melanesia, called Talasea, at a distance of roughly 6,500 kilometers (over sea). It is tempting to suppose that we are dealing here with people of the sea, perhaps both sailors and traders, a reminder that many of these Southern Mongoloid migrants may have turned to agriculture but at the same time preserved a tradition of seafaring. After all, the first groups sailed from China to Taiwan, while their descendants crossed the sea between Taiwan and the Philippines and then started island hopping between the various Philippine islands and from the Philippines to Talaud and Borneo.

Pottery found in west-central Sulawesi resembles that from a later Neolithic phase in Sabah, but it cannot be dated; southwestern Sulawesi has produced remarkable Neolithic pottery that is probably not much older than 1500 BCE. Similar pottery was found in the northern Moluccas, on a small island near Halmahera. A cave from the same area has yielded bones of domesticated pigs and dogs, datable to the period between 1200 and 300 BCE.

Finally, similar pottery was dug up in caves in eastern Timor (Lesser Sundas), dated between 2500 to 2000 BCE. In addition to the bones of domesticated pigs—

introduced from Java or Sulawesi—remains have been located of cuscuses, civets, and macaques, all of them introduced but not in a domesticated state. This is an interesting example of early manipulation of the fauna by introducing wild or tame animals (which are not the same as domesticated animals). Remains found there of dogs, cattle, and goats are dated after 1000 BCE and are perhaps even more recent. Among the plant remains there is a tantalizing grain of foxtail millet, also dated after 1000 BCE. In an older layer, dating from before 3000 BCE, remains were found of the cereal Job's tears, the betel vine, and the areca nut (the latter two being ingredients of the betel quid).

All the locations mentioned thus far are early Neolithic sites that can be roughly dated. For Sumatra and Java, however, such reliably dated material is largely lacking. For Sumatra the archaeological record is very poor, and for Java the many well-made adzes, including those made of semiprecious stone, cannot be dated reliably. This also applies to the extensive working floors for adzes and stone bracelets in central and western Java. An estimate based on linguistic reconstruction is that Neolithic Austronesian speakers of Southern Mongoloid biological affinity, coming perhaps from Borneo, may have reached Java as late as 1000 BCE.

The pollen record for Sumatra and Java suggests human disturbance perhaps as early as 4500 BCE, but permanent clearance may have occurred much later, perhaps after 1500 or 1000 BCE, which might imply that Austronesian speakers with a Neolithic tool kit reached Sumatra around the same time as their relatives supposedly reached Java. However, it is possible that Sumatra (and western Borneo) were also settled at an early stage by Neolithic groups from southern Thailand or Malaysia.

Finally (and disregarding the Austronesian migration to Madagascar in the far west, and Oceania in the far east), the Austronesians reached the Malay Peninsula, where, as we have seen, the Neolithic Ban Kao culture (of Austroasiatic speakers) had arrived earlier. The latter, often collectively called Orang Asli in Malaysia, can be found, to this very day, in the interior of the Malay Peninsula; the coastal Malays are the descendants of the Austronesian immigrants of a later date.

THE METAL AGES IN MAINLAND SOUTHEAST ASIA

The term "Metal Ages" covers the Bronze Age and the Iron Age. In Mainland Southeast Asia those two phases can be clearly distinguished. Bronze was introduced, in all likelihood from China, probably after 2000 or 1500 BCE. Iron, probably also from China, arrived around 500 BCE. From then on the two metals appear together in the area. In Island Southeast Asia the two phases cannot be distinguished, and copper and iron arrived on the scene around the same time, ca. 500 BCE.

Conventionally, the Metal Ages are supposed to end with the arrival of written sources, when history proper begins (as opposed to prehistory, for which there are no written sources available). However, given the fact that developments in Southeast Asia were so uneven, with a prehistoric phase locally extending far into the Modern Era, there is not much sense in looking for a cutoff point. Suffice it to say that the influence on Southeast Asia of the Indic and Chinese civilizations, together with their writing traditions, should be dated around the beginning of the Common Era.

The presence of artifacts made of bronze—an alloy of copper and tin—and iron represents in the first place technological improvement. People developed the technology of turning metal ores into copper, tin, and iron, and then into metal objects. Technological improvement resulted in more sophisticated instruments of agriculture and war, and also more durable vessels. Thus the productivity of land and labor increased, as did the ability of humans to manipulate and increase their impact on the natural environment.

At the same time, bronze and iron are usually associated with a concentration of wealth and status among the few, and therefore increasing social, economic, and political differentiation between lineages and between individual families and people. Although Neolithic groups may not have been entirely egalitarian, it is generally assumed that inequality increased with the arrival of bronze and iron production, as some people succeeded in controlling the metallurgy and the distributive systems. Bronze and iron implements led to higher productivity of land and labor, while bronze ornaments enabled the rich and powerful to show off their wealth to a much larger degree than before, thus acquiring more prestige. And what is the point of wealth if it cannot be displayed? Ornate vessels have been taken to represent feasts and rituals, which themselves may have become more lavish when income per capita and conspicuous consumption—to use two modern terms—of the rich were on the increase.

There are also clear indications that supralocal trade flows became more important with the development of bronze and iron artifacts on the one hand, and the demand for luxury items on the other. It seems to be a universal law throughout most of human history that "exotic" traded items are attributed a higher prestige (we often speak of "prestige goods") than locally produced ones; looking at burial gifts, this principle may be seen in operation during the period under discussion. This would have reinforced the importance of trade (often called "exchange" at this stage), probably conducted largely over seas and along rivers.

Returning to Mainland Southeast Asia, the archaeological record is restricted largely to northern Thailand and northern Vietnam. In northern Thailand the site of Ban Na Di, near the Neolithic site of Ban Chiang, is one of the most important Bronze Age locations so far unearthed. The inhabitants were engaged in copper casting, and items found include socketed axes, spearheads, fishhooks, bracelets, beads, bells, and bowls. The site includes many burials, where in addition to bronze items small shell beads have been found, as well as large numbers of unbaked clay figurines of cattle, pigs, dogs, other animals, and humans. The same site also yielded silk impressions in the corroded surfaces of bronze items, but it is not clear whether this silk was produced locally or imported from China.

Iron appeared in this area for the first time around 300 BCE. Iron objects found here include spearheads, knives, neck rings, and bracelets. While in these riverine lowlands agriculture and livestock keeping may not have differed much from Neolithic practices, there have been suggestions that during the Iron Age rice fields were used more permanently and intensively, employing water buffalo for trampling or even plowing the fields (although plowshares have been found for this period only in northern Vietnam). Production must have grown at least apace with the expanding population. In the more mountainous areas, agricultural settlements have not been found prior to the introduction of iron, which seems to suggest that in Mainland Southeast Asia such areas were largely empty or used only by foragers prior to the Common Era.

Finally, and in keeping with what has been said above about increased inequality during the Metal Ages, it has been suggested for northern Thailand during this period that differences in the size of settlements started to develop. The presence of some settlements much larger than others, and perhaps encircled by moats (probably built for water control in the dry season and perhaps also for defensive purposes), might suggest some form of ranking and dependence between the larger and the smaller communities. It is also suggested that some form of political integration was to be found here, independent of influences from India or China.

DONG SON

In northern Vietnam, the early Bronze Age started around the same time as in northern Thailand, during a phase called Dong Dau, with an assemblage broadly similar to that of Ban Na Di, beginning around 1500 BCE. Among the bronze artifacts to found there, we can mention spearheads and arrowheads in addition to socketed axes, shaft-hole sickles, knives, and fishhooks.

However, the name that has made this area and period famous among art collectors and (art) historians alike is that of the Dong Son metallurgical style, named after the main excavation site. The Dong Son style represents the bronze- and iron-using phase that found its epitome in the large Dong Son bronze kettle-drums with their peculiar shape and intricate decorative patterns, possibly invented around 600 or 500 BCE. Its area of origin is the Red River valley and the

A bronze drum of the Dong Son civilization in Vietnam. Dong Son style is an indigenous Southeast Asian art form famous among art collectors. (Erich Lessing/Art Resource)

adjacent coastal regions of northern Vietnam (Bac Bo). The Dong Son phase represents the first appearance of iron in Southeast Asia.

The Dong Son bronze kettledrum is an indigenous Southeast Asian art form, not borrowed from elsewhere, and the roughly 200 drums that have been found so far have been discovered throughout Mainland and Island Southeast Asia (but up to now not in the Philippines). The largest drums are just under a meter high and have a weight of 100 kilograms. They are heavily decorated, with both geometric patterns and realistic representations of animals, warriors, boats, and houses. The drums have often been found in burials, and it has been hypothesized that they were considered as prestige goods of the highest value by local rulers (*regalia*). As there are sometimes differences between the drums discovered in Mainland Southeast Asia and those found in Indonesia, it is possible that the latter were made to order for some Indonesian rulers, but that is entirely speculative. The tradition of casting these magnificent drums seems to have continued until the third century CE, or well into the historical period of Southeast Asia, when Vietnam had become part of the Chinese Empire.

Apart from the artistic value of these kettledrums, and their testimony to the craftsmanship of the bronze casters who produced them, they are also valuable as historical sources in their own right. For the first time in Southeast Asian history (or rather protohistory, the period for which there are written sources in addition to archaeological ones), we are given glimpses of daily life in the form of pictures or illustrations. Although some of the seemingly realistic designs have been taken to represent aspects of the afterlife, such as the journey of the soul of the deceased in a boat, most illustrations appear to reflect daily life. So we find houses on stilts and with a concave roof, with peaked gables supported by poles, reminiscent of the houses of the Torajas, still to be found in central Sulawesi, and those of the Toba Bataks in Sumatra. There are cocks and peacocks perching on the roof, and under the floor of the houses there are domestic animals. In addition to cattle, chickens, yapping dogs, and perhaps a rodent and a pig, there are also deer, and in one case a young elephant. Although it would seem that elephant and deer were never domesticated, which implies selection of desirable features over a long period of time, these illustrations seem to show that they were tamed, something that has been reported from historical times as well.

We also encounter scenes of feathered warriors with elephants and horses, and various hunting scenes, with men hunting deer with a long stick on which there is a loop that was thrown over the deer's head, and with archers hunting tigers, which in turn are stalking deer. There is also a scene of a man and his dog fighting a tiger, armed with a sword or a knife and a stick. There are many boats (without sails) of a type that can still be observed in the southern Moluccas in Indonesia, sometimes with an archer. We see cooking pots, people presumably pounding rice in a mortar, and people shooing birds away (from the ripening grain, or because they represent death?). Besides the tiger and the domesticated or tamed animals, there are many more representations of nature, such as various birds, fish, crabs, lizards, and trees (palm trees and trees on stilt roots). Regarding people, we find dancers and musicians in addition to the warriors and hunters, and also people who may be Chinese.

Finally it should be mentioned that the surface of the drum—that is, the part that is beaten, which is called the tympan—is usually flat (apart from the patterns in low relief), with the exception of many drums found in eastern Indonesia, of which the rims of the tympan are decorated with (usually) four three-dimensional frogs or toads. It has been hypothesized that these may have had something to do with rain and fertility (some of the frogs are copulating).

The Plain of Jars, near Phonsavan, Laos, is one of the few remainders of the Metal Ages in Southeast Asia. (Chris Hellier/Corbis)

Among the other Dong Son artifacts mention should be made of a hoard of more than a hundred bronze plowshares, found inside one of these drums. This was found in Co Loa, the traditional capital of the Au Lac Kingdom from 257 BCE. Here traditional historiography and archaeology seem to meet, and it would appear that the Dong Son style was part of an early state complex that may have included canal-irrigated rice fields (with the possibility of double cropping), and a population of almost 1 million people according to a Chinese census of northern Vietnam taken in the year 2 CE. It is also clear that there were frequent trade contacts between southern China and northern Vietnam, as witness, for example, Chinese Han dynasty coins and a Chinese cast-iron hoe found in the latter region.

Before turning to Island Southeast Asia, the Sa Huynh culture of southern Vietnam ought to be mentioned. It has been linked to an Austronesian-speaking people, the Chams, of Indo-Malaysian origin, who may have arrived there from Peninsular Malaysia or Borneo. Their culture would appear to have been firmly established by 600 BCE. Their jar burials resemble those of the Indonesian Archipelago, dealt with presently. They relied more on iron and less on bronze than did the Dong Son people, while the bronze items found at the Sa Huynh sites were mainly of a decorative nature. Other items found are gold beads, silver wire, glass, agate, and carnelian beads that may have come from India. It is possible that this culture played a role in the transmission of metal-working techniques (particularly iron) to Island Southeast Asia.

Items imported from India, in addition to Roman and Chinese ones, have been found in Oc Eo in southern Vietnam, dating from the second century CE onward. Among the Indian items is a Buddha head. Long-distance trade in this area was evidently picking up.

As was said earlier, the Metal Ages in the remainder of Mainland Southeast Asia are poorly documented. It may suffice to mention one remarkable location, the Plain of Jars in Laos, infamous because of the terrible fighting that went on there during the Vietnam War. The name is based on the large (on average 1.5-meter-high) stone jars that contained cremated human remains associated with bronze and iron grave goods.

THE METAL AGES IN ISLAND SOUTHEAST ASIA

Whereas in Mainland Southeast Asia the Bronze Age started about 1,000 to 1,500 years earlier than the Iron Age, in Island Southeast Asia the two metals were introduced more or less simultaneously, possibly between 500 and 200 BCE. These estimates are based on dated materials from Peninsular Malaysia, Java, Sabah (northern Borneo), Timor, and the Philippines. Bronze and iron were introduced from the Dong Son culture in northern Vietnam and possibly from the Sa Huynh culture, in southern Vietnam, as well.

As we have seen above, Dong Son kettledrums have been found in many areas of Island Southeast Asia. According to a list, dated 1988, of the Dong Son kettledrums of the so-called Heger I type, six of these drums were found in Peninsular Malaysia, fifty-five in Indonesia, and none in the Philippines. Within the Indonesian Archipelago, twenty-two drums were encountered in Java and seven in Sumatra. Hardly any drums were found on the large islands of Borneo and Sulawesi. An astonishing eight kettledrums were unearthed on the tiny island of Gunung Api, near Sumbawa, one of the Lesser Sunda Islands, while eleven specimens were found in the Moluccas, mainly in the southern Moluccas (none have been found in the northern Moluccas), and three in New Guinea. If one looks at these figures, one thing is clear: the number of kettledrums is not in proportion to the surface areas of the islands where they were found. The much higher number found in Java than, for instance, in Sumatra suggests a link with population density, which would throw an interesting light on the Lesser Sundas (particularly Sumbawa) and the southern Moluccas during this early period. It would appear that, as a rule, the drums found in Sumatra and Java represent "early" drums, while those found in Eastern Indonesia can be called "late" on the basis of stylistic characteristics. Of course, we have no idea whether the "early" drums came early to the area and the "late" ones late, but it is certainly possible. Whatever else this distribution may suggest, it clearly bears witness to extensive early trade contacts and gift exchange between Mainland and Island Southeast Asia, a network that may have been focused on the export of spices from the Moluccas.

Bronze and iron technologies, therefore, came from outside Island Southeast Asia, but fairly soon the people from the area had started metalworking centers of their own, as witness the stone and clay molds found in Java, Bali, Sabah, and the Talaud Islands, to the northeast of northern Sulawesi. This technology may have been in place from the early first millennium CE.

All of the kettledrums mentioned above were likely produced in northern Vietnam and exported to Island Southeast Asia. They do not, therefore, reflect local Malaysian or Indonesian craftsmanship. However, there are two types of bronze kettledrum that are indigenous to Indonesia, although their design was evidently based, directly or indirectly, on that of the Dong Son drums—the Pejeng kettledrums from Bali and the so-called *mokos* from the small island of Alor, one of the Lesser Sundas. Although these drums have not been dated as yet, a Bronze Age origin is not impossible.

Various other bronze objects, many of them exquisitely crafted and apparently with a ceremonial function, appear to have been made in the Indonesian Archipelago. Among them are a number of ceremonial axes from Savu and Roti (Lesser Sundas), highly ornamented flasks, and clapperless bells, found in Makassar, southern Sumatra, Madura (off eastern Java), and in Selangor and Johor (Malay Peninsula), in addition to one of each in Cambodia—all this in addition to the usual implements of agriculture and war, such as socketed axes and spearheads, and jewelry (bracelets). Sometimes bronze figurines of uncertain origin have been found, such as the water buffaloes in Priangan, western Java, and warriors-cum-horses in eastern Java. As the warriors (some armed with spear and ax, others archers) appear to be Tartars, they may have been made in China. If we could be sure that these figurines had been made locally, it might help us in dating the appearance of water buffalo and horse in Java (although the former may have been indigenous).

Remarkable iron tools of local manufacture have been found in Peninsular Malaysia, including axes, knives, and sickles, all with shaft holes. The tools appear to be of low-carbon steel, deliberately reheated on charcoal after the first smelting process. A link with the Dong Son or Sa Huynh culture is likely.

Given the data presented here, it is probably fair to say that the arrival of the Bronze Age was a great advance for humans in a technological, social, cultural, and artistic sense, representing a rather pronounced break with the past. Exploitation of various natural resources went into higher gear, while the impact on the natural environment must have increased sharply.

MEGALITHS AND JAR BURIALS

Bronze and iron are not the only typical features of the Metal Age (also called the Bronze-Iron Age) in Island Southeast Asia. Indonesia in particular was characterized during this period by a remarkable megalithic tradition. These remains have been found mainly in southern Sumatra, Java, Bali, and other Lesser Sundas. It has been suggested that the megalithic tradition has an Indian origin, a notion partly based on the many finds of Indian objects in and around the megaliths.

Megaliths should be seen as representing territorial claims, usually in connection with the institutionalization of burial rituals. It is tempting to interpret this tradition as an indication of increased "territorialization" of the people involved, perhaps in connection with a growing tendency toward rice cultivation on permanent fields. If that is true, the megalith tradition would be an important marker in the environmental history of the region, symbolizing a type of land use that would lead to large-scale permanent changes in the landscape, dividing the natural environment into a "civilized" world of agriculture and a "wild" world of mature and secondary forest growth.

The most conspicuous megalithic remains were (and are) encountered in southern Sumatra, particularly on the Besemah Plateau, but less remarkable stones have been found in adjacent Lampung. The two most famous megaliths are the Batugajah, or elephant stone, and the Batu tatahan, or sculptured stone. The Batugajah shows two men in military dress who are attempting to tame or ride an elephant. Each carries a typical kettledrum on his back, a feature that strongly suggests that we are dealing here with a Metal Age monument. The Batu tatahan shows two men with a kettledrum and two buffaloes, while the head of a dog and perhaps a crocodile are visible as well. The men have unusual ears, somewhat pointed, and one of them has an almost feline look, which might suggest that they are "weretigers" (tigers that can assume human shape, or the other way around). That would link this megalith with another well-known stone from the Besemah area, which represents two copulating tigers, with a human head between the paws of the tigress. Some reliefs show people fighting tigers or snakes. Also encountered in the area were slab graves and megalithic chamber graves with polychrome wall paintings of humans and animals, among which are water buffaloes and a fighting cock.

In Java early Metal Age finds have been discovered in connection with slab graves, and in eastern Java, Bali, Sumbawa, and Sumba with remarkable carved sarcophagi. However, like the slab graves of Peninsular Malaysia, some of these burials date from historical times, as witness the datable Chinese ceramics that have been found there. One of the stone sarcophagi from Bali has a lid shaped like a buffalo. In western Java there are a number of terraces with stones that may represent humans, and were often seen by the local populations as petrified pre-Islamic princes and their warriors. These sacred places seem to be pre-Indic complexes, but they have not been dated as yet.

One of the sites with Metal Age assemblages in Bali, coastal Sembiran, has yielded ample evidence for early trade contacts with India. The same site produced evidence for rice growing. Indian contacts from roughly the same period can also be inferred from artifacts found in Buni, an archeological site on the western coast of northern Java.

Finally, it should be mentioned that it seems likely that an indigenous jar burial tradition in Island Southeast Asia formed part of the early Metal Age complex. Secondary burials in large jars or pottery bone boxes, perhaps to be dated from 200 BCE onward, have been found in Sabah (northern Borneo), Talaud (Sulawesi), the northern Moluccas, the central and southern Philippines, and the Lesser Sundas (Sumba). In many of these areas such burials continued into ethnographic times (that is, the period in which anthropologists started to record such things, or from the late nineteenth century). Beads, bracelets, and metal artifacts have often been found as grave goods in or around these jars. In most areas where Indian, Islamic, or Chinese influences made themselves felt later on, these burials were replaced by inhumation.

Jar burials on the Talaud Islands showed some evidence of betel staining and for tooth evulsion of females. Agate and carnelian beads found here are probably of Indian origin. There are also indications of local metal casting. Imported Indian beads and evidence of local casting were also found in Sabah. On Negros Island, in the Philippines, bones of pigs and chickens have been found with the jar burials. In some of these burial sites high-necked flasks with geometric and anthropomorphic designs have been encountered, but these are also found in some areas in association with inhumations (Bali, Lombok).

CONCLUSION

Surveying the data on crops and animals, it seems clear that the cereals rice and (foxtail) millet were domesticated in China, and therefore outside Southeast Asia, although the area of primary domestication of rice may have included the adjacent northern tip of what today is Vietnam. Yam and taro, the two most important roots and tubers of the region, may have been first domesticated in New Guinea, as seems to have been the case with sugarcane and a specific banana species. It is not clear whether taro, yam, and sugarcane were independently domesticated in the parts of Island Southeast Asia that were occupied by Austronesian speakers, or whether the domesticated forms of those plants had arrived earlier from New Guinea.

So far not much has been said about fruits, as staple crops (cereals, roots, and tubers) have been studied much more intensively and are therefore much better documented than other foodstuffs. Nevertheless, it should be mentioned that there are good reasons to assume that various fruit trees were domesticated in Southeast Asia. However, in some cases it is not clear whether the area of origin was China alone, or whether it included the adjacent areas of Southeast Asia. Candidates for domestication in Southeast Asia are bilimbing, citrus (various species), durian, langsat, mangosteen, rambutan, rose apple, snakefruit, and starfruit.

As regards livestock and commensal animals, there is evidence that pigs, dogs, (humped) cattle, water buffalo, chicken, and certainly horses arrived from China in their domesticated state, but a case has also been made for Southeast Asian origins for domesticated chickens and water buffaloes. Moreover, it is not impossible that, for instance, pigs were domesticated independently in two locations, say, China and Indonesia. We can be certain, however, that the so-called Bali cattle, the domesticated form of the banteng (*Bos Javanicus*), was originally domesticated in Java.

Why were the Southern Mongoloids and their domesticates more successful than the people of New Guinea in their expansion? Biologist Jared Diamond lists three factors: no cereals were domesticated in New Guinea, because there were no wild ancestors available; no domesticable large mammals were present; and the roots and tubers that were locally domesticated (yams and taro) were unsuitable for higher elevations, thus keeping population numbers down.

A somewhat similar picture emerges when one looks at the leading bronze- and iron-producing centers in Southeast Asia. In the first place, compared with India and China, the presence of high-grade iron ores in Southeast Asia was rather limited. In the second place, within Southeast Asia, Indonesia had almost no iron, in comparison with Malaysia and Mainland Southeast Asia; the northern Philippines, however, were rather rich in iron. Thirdly, the mainland was much better endowed with copper than was Indonesia, but that does not apply to the Philippines, where high-grade copper ore was abundantly available. Tin was not available in the Philippines and Indonesia, with the exception of Bangka and Belitung, islands off the coast of Sumatra, whose deposits were not discovered until the eighteenth century. The late arrival of the Bronze and Iron Ages in Indonesia, therefore, seems to have had a strong environmental component, but that does not appear to be the case in the Philippines. The early arrival of bronze and iron in Thailand and Vietnam was doubtlessly partly influenced by the proximity of southern China. Long-distance exchange networks took care of the spread of people, artifacts, and technology to areas farther removed from the centers where bronze- and iron-producing technologies were first developed. Areas that were marginal to these networks appear to have remained relatively unaffected, even if they possessed the necessary ores.

It would appear that there was more local variation in iron tools than in bronze ones in Southeast Asia, which suggests that ironworking could be more easily carried out by small, local workshops than could the more complex bronze casting technology, and perhaps also that iron was more abundantly available (although not everywhere) and that the demand for (less expensive) iron tools was greater.

Finally, it should be pointed out that most sites to be found in Southeast Asia during the period from 1000 BCE to 1000 CE seem to have been located in areas of strong seasonal rainfall (monsoon climate). This implies that some form of irrigation appears to have been necessary in most cases.

It is also likely that most sites were located in a fairly narrow altitudinal band between the lowlands and the uplands. In the case of Java, this was between 100 and 400 meters. Lower areas were probably too wet and higher areas too steep. For these mid-altitude settlements, a lowland-upland dichotomy, so often found in the literature, does not appear appropriate. One supposes, though, that coastal settlements, although almost certainly located on sand ridges, may have been located at levels lower than 100 meters.

Evidently, the Neolithic Revolution and the Metal Ages led to a growing distance between *Homo sapiens* and the other species. Humans domesticated a range of animals, adapting the gene patterns of the latter to their own needs, thus establishing human domination over nonhuman species and harnessing their energies to mankind's endeavors.

Humans also started to manipulate various plant species, turning them into agricultural crops and thus creating a new phenomenon—agro-ecosystems. This led to a gradual change of the landscape, which, as human numbers were growing more or less constantly—although with ups and downs—must have led to considerable environmental change. That was particularly the case in areas in which permanent "wet" rice cultivation—a topic dealt with in the next chapter—became customary. In those areas, forests would no longer grow back— unless, of course, wars, epidemics, or volcanic eruptions led to the end of the existing human settlements, something that was far from rare during the history of the region. However, forest regrowth was no longer a routine affair in the permanently settled areas. Thus, the rift between "nature" and "culture" came into being.

This development was reinforced with the arrival of the Metal Ages, when bronze and iron tools enabled people to exploit their natural environment more effectively, thus leading to more environmental change. Higher production and more efficient weapons were conducive to increasing social stratification, which in turn generated a growing demand for "exotic" trade goods. This stimulated the growth of exchange networks in which prestige goods circulated, leading in turn to a desire for increased production of exchangeable commodities. As many prestige goods ended up as burial gifts, the accumulation of such goods was probably rather restricted; that, on the one hand, kept the up-and-coming elite relatively poor and unable to acquire a large following (and therefore not yet able to develop into "monarchs"), while on the other hand the demand for trade goods, and hence levels of resource exploitation, must have remained relatively high.

At the tail end of the period we have been dealing with in this chapter—the first centuries CE—long-distance trade with Rome, India, and China was clearly in evidence, although the quantities involved appear to have been modest. Nevertheless, this was a development that, as the next chapters show, would play an important part in shaping the future of the region, including the way it would deal with its natural environment.

BIBLIOGRAPHICAL ESSAY

The literature on the topics dealt with in this chapter (and the previous one) is dominated by two scholars—Peter Bellwood for Island Southeast Asia and Charles Higham for the mainland. For this chapter I consulted Higham, Charles, 1996, *The Bronze Age of Southeast Asia*. Cambridge: Cambridge University Press; Bellwood, Peter, 1997, *Prehistory of the Indo-Malaysian Archipelago*. Honolulu: University of Hawai'i Press; Bellwood, Peter, 1999, "Southeast Asia before History." Pp. 55–136 in *The Cambridge History of Southeast Asia*. Vol. 1: *From Early Times to Circa 1500*. Edited by Nicholas Tarling. Cambridge: Cambridge University Press. A very short, recent introduction is Bellwood, Peter, and Ian Glover, 2004, "Southeast Asia: Foundations for an Archaeological History." Pp. 4–20 in *Southeast Asia: From Prehistory to History*. Edited by Glover and Bellwood. London: RoutledgeCurzon.

More general texts, in which what happened in Southeast Asia can be seen in a global context, are Vasey, Daniel, 1992, *An Ecological History of Agriculture, 10,000 B.C.–A.D. 10,000*. Ames: Iowa State University Press; Diamond, Jared, 1998, *Guns, Germs and Steel: A Short History of Everybody for the Last 13,000 Years*. London: Vintage; Morrison, Kathleen D., and Laura L. Junker, eds., 2002, *Forager-Traders in South and Southeast Asia*. Cambridge: Cambridge University Press; Yasuda, Yoshinori, ed., 2002, *The Origins of Pottery and Agriculture*. New

Delhi: Roli Books; Scarre, Chris, ed., 2005, *The Human Past: World Prehistory and the Development of Human Societies*. London: Thames and Hudson.

On the origins of crop agriculture and animal husbandry, see Maloney, Bernard, 1993, "Climate, Man, and Thirty Thousand Years of Vegetation Change in North Sumatra," *Indonesian Environmental History Newsletter* 2, pp. 3–4; Caras, Roger A., 1996, *A Perfect Harmony: The Intertwining Lives of Animals and Humans throughout History*. New York: Simon and Schuster; Harris, David R., ed., 1996, *The Origins and Spread of Agriculture and Pastoralism in Eurasia*. London: UCL Press; Louwe Kooijmans, Leendert P., 1998, *Between Geleen and Banpo: The Agricultural Transformation of Prehistoric Society, 9000–4000 BC*. Amsterdam: Stichting Nederlands Museum voor Anthropologie en Prehistorie. [20th Kroon lecture]; Hill, R. D., 2004, "Towards a Model of the History of 'Traditional' Agriculture in Southeast Asia." Pp. 19–46 in *Smallholders and Stockbreeders: Histories of Foodcrop and Livestock Farming in Southeast Asia*. Edited by Peter Boomgaard and David Henley. Leiden: KITLV Press.

On early agriculture, water, and rice, see Rigg, Jonathan, ed., 1992, *The Gift of Water: Water Management, Cosmology and the State in Southeast Asia*. London: School of Oriental and African Studies; Hayao, Fukui, 1999, "A Comparative Study of the Dry Areas of Southeast Asia," *Indonesian Environmental History Newsletter* 12, pp. 1–7; Hayao, Fukui, 2000, "Historical Cities and Agriculture in Tropical Asia: A Hydrological Examination," *Journal of Sophia Asian Studies* 18, pp. 27–38. Rice and irrigation are dealt with in more detail in the next chapter.

On metals and the Metal Ages, see the classics by Hoop, A. N. J. Th. à Th. van der, 1932, *Megalithic Remains in South Sumatra*. Zutphen: Thieme; Heekeren, H. R. van, 1958, *The Bronze-Iron Age of Indonesia* [Verhandelingen KITLV 22]. 's-Gravenhage: Nijhoff.

More recent studies on this topic are Bernet Kempers, A. J., 1988, *The Kettledrums of Southeast Asia: A Bronze Age World and Its Aftermath*. Rotterdam: Balkema; Bronson, Bennett, 1992, "Patterns in the Early Southeast Asian Metals Trade." Pp. 63–114 in *Early Metallurgy, Trade and Urban Centres in Thailand and Southeast Asia*. Edited by Ian Glover, Pornchai Suchitta, and John Villiers. Bangkok: White Lotus.

PART TWO

THE ERA OF STATE FORMATION

3

FIFTH TO
FIFTEENTH CENTURIES

The factor that most captures the imagination of historians of Southeast Asia who study the fifth to fifteenth centuries is the process of state formation. That process, in turn, was usually thought of as having been linked to two other factors: the movement of groups of people (Burmans, Tai, and Viet pushing Pyu, Mon, Khmer, and Cham aside) and the influence of Indian culture—including language, script, art, religion, and notions of statehood—as well as, to a much lesser extent, that of the other Asian empire, China. However, over the last few decades historians have started to emphasize two other factors related to state formation: the growth of international trade and the increasing importance of wet-rice agriculture.

Of the wealth of "states" to be found in the literature, this chapter deals mainly with the most important ones. Funan, in the area of present-day Cambodia and southern Vietnam, is usually regarded as the first early state, to be found in the sources from the third century CE and past its peak in the sixth century. After approximately the year 600, the big names are Angkor, also in Cambodia; Champa, in what is now southern Vietnam; Srivijaya in southeastern Sumatra; and Mataram in central Java. After around 1000, Angkor and Champa remained important, being joined by Dai Viet (northern Vietnam), Pagan (Burma), and various eastern Javanese states, with Singhasari as the most important. Finally, after ca. 1300, in addition to the surviving states of Champa and Dai Viet, we encounter Ayuthaya (Thailand) and Majapahit (Java). By 1500 only Ayuthaya had survived as a more or less unitary state. Majapahit came apart under Muslim attacks, and Champa was conquered by Dai Viet, a unified Vietnam soon going under as a result of perennial feuding between north and south.

This is also the first period to which the term "historical" can be applied, as from the second and third centuries CE onward there are written sources available from Europe and Asia (mainly Southeast Asia itself and China). Apart from the European ones, most sources dating from this early period are written in the Indian language Sanskrit or in Chinese, both languages that originated outside Southeast Asia; this reinforces the notion that there was a link between state formation processes in Southeast Asia and the two Asian empires of India and China (India, however, it should be noted, was not an empire throughout the period being discussed here). Inscriptions in vernacular languages date from the seventh century onward in Mainland Southeast Asia, from the seventh century in Sumatra, and from the ninth century in Java. Local scripts used for the vernacular languages were local adaptations of an Indic script (Brahmi), except in northern Vietnam, where adapted Chinese characters were used. Until the fourteenth century all Southeast Asian sources consist of inscriptions in stone or metal, usually called epigraphic material, while the sources from China are state chronicles and other products of the central Chinese bureaucracy, in addition to materia medica and geographical descriptions.

Although, therefore, there are more historical sources for the period under consideration than for the prehistoric period, the quality of the sources leaves much to be desired for those with an interest in environmental history. That applies not only to the sources themselves but also to the books and articles based upon them, as most scholars with an interest in pre-1500 Southeast Asia did not write with environmental historical questions in mind. In fact, sometimes we seem to know more about the environmental factors that played a role in shaping prehistoric societies (and the impact of humans on the environment) than we do about such factors between the years 500 and 1500 CE. This state of affairs might be related to the fact that the latter period has been studied by scholars with a background in languages, while archaeologists and prehistorians are often trained as scientists or at least as social scientists with a strong leaning toward the natural sciences.

CHIEFDOMS, STATES, AND EMPIRES

A book on environmental history is hardly the place for a detailed account of the rise, decline, and fall of the (early) states of Southeast Asia. There is no shortage of books on the topic, to which the reader may be referred. Here we deal only with states such as Funan, Champa, Srivijaya, Angkor, Pagan, and Majapahit in very broad outline, concentrating on their main features and emphasizing the ones that are relevant to the field of environmental history.

Before dealing with these polities, a short discussion about terminology is in order. The first question to be dealt with is, What do we mean when we talk about "states" in this context? There is a convenient tradition within political anthropology, usually indicated with the term "evolutionary theory," that has left us a series of terms for consecutive developmental stages. These terms are "band," "tribe," "chiefdom," and "state." Bands and tribes have been called acephalous societies—literally, societies without a head—while chiefdoms and states do have a structural position for a chief or king. All of these terms come with their own problems, of course, but they are handy as indications of scale and complexity.

The first chapter of this book dealt with societies of hunters and gatherers, people who were organized into bands and tribes. In the second chapter, dealing with the Neolithic and the Metal Ages, the first chiefdoms appeared. However, bands and tribes remained important, probably even the dominant forms of sociopolitical organization. A similar observation applies to the present chapter: although states have now made an appearance, chiefdoms and tribes have not disappeared. In some areas they even survive today, though in adapted forms. It is a convenient convention (and one easily remembered) to think of the Southeast Asian states coming on stage after year 1 CE, but mainly after the fifth century.

Some authors, unfortunately, use the term "empires" for polities like Srivijaya, Vietnam, Angkor, and Majapahit. The question is one mainly of scale. For South, Southeast, and East Asia, one is inclined to apply the term "empire" to (parts of) India and China only. What purpose is served by employing the term "empire" for both a real empire like China and a polity like Srivijaya, an early state at best and possibly at times not much more than a loose confederacy of ports of trade? Even polities like Majapahit and Angkor, probably the most centralized states in Southeast Asia between 1000 and 1500, were too small and too loosely structured to be called empires. In this chapter, therefore, the term "empire" will be reserved for China, India, and Rome.

We will also encounter the notion of "early states." They were more loosely structured than "mature" or "classical" states, and for that reason they have also been called "segmentary states" or "galactic polities." On the basis of archaeological, epigraphical, and other written sources, it is possible to recognize a number of regional centers in Southeast Asia during the first millennium CE. They were often fortified places, surrounded by a wall with or without a moat, usually near a river. Often these centers formed a cluster from which one place emerged as a supraregional center. Such clusters were seldom stable for a long time, undergoing processes of fission and fusion.

Early states were based on dyadic relations between the ruler of a supraregional center and a number of regional rulers. Such dyadic links were highly personal patron-client relations, based on the wealth, prowess, and charisma of the supraregional leader. From the latter, protection, honor, and gifts were expected, while the client regional ruler would reciprocate with material and spiritual support, including manpower in times of war. For the ruler residing in the supraregional center the term usually employed in the English-language literature is "king" or "monarch," and early states are often called kingdoms.

Three factors appear to have contributed most to the fissiparous nature of the (early) Southeast Asian states. In the first place a geophysical (or "environmental") factor, the problem of communications. Unless a coast or a river connected two regional centers, the cost of military intervention in the case of a breakaway center was considerable, and often prohibitive. If, therefore, a regional center decided to leave the kingdom, it could often be done with impunity. Another factor was the bilateral nature of Southeast Asian kinship. Although there were exceptions, most ethnic groups in Southeast Asia recognized (and still recognize) descent through both the male and the female line. This created a much larger group of claimants to the throne than would have been the case if descent groups had been formed following the male or the female line only. Although it was fairly rare, women also could become rulers in their own right. In addition to the bilateral descent rule, the fact that Southeast Asian rulers had large harems and therefore numerous progeny boosted the number of potential heirs to the throne. Wars of succession, therefore, or at least attempts to carve out new and independent polities, were fairly frequent.

The third factor is that it was often difficult to match income and expenditure of the royal houses of the (early) states of Southeast Asia. A strong king had to be a generous king, who would bestow liberal gifts on the regional leaders and the religious establishment. This gift-giving (so-called merit-making) doubtlessly had a strong religious aspect, but it represented at the same time a conspicuous display of wealth. In fact, it could be argued that the king had to show that he was richer than everybody else. It must have been often the case that the "regular income" of the rulers (including compulsory labor and produce in kind) in these early states was insufficient for such expenditure. The most obvious—and quickest—way in which rulers could supplement their income was to go to war, either in the form of raiding (over land or water) or territorial expansion. In addition to an economic rationale, wars and raids also served to demonstrate the king's prowess.

Apart from precious metals and other rare objects, raids probably yielded mainly war captives who could be kept as serfs or sold as slaves, and livestock. A phenomenon that may be considered intermediate between a raid and a territorial war was the forced migration of large numbers of people from the invaded territory to the heartland of the ruler, where the newcomers may have been settled as bonded labor.

If this kind of expansion occurred under more or less normal circumstances, one can imagine that loss of income owing to weather anomalies (droughts, floods, typhoons, tsunamis), harvest failures, volcanic eruptions, and epidemics also triggered expansionary responses. A drop in income and therefore of donations could mean that the king's charisma would be in doubt, which ultimately would lead to regional leaders breaking away in search of a new overlord. The most likely response to such a potential threat to the integrity of the realm was the launching of hostile expeditions into neighboring territories. Thus raiding parties and wars would add their burden to those imposed by the natural calamities. If the natural disasters had already weakened the realm too much for such hostile incursions to be organized, one might expect the kingdom to come apart.

However, by a certain stage in the development of these states, their central rulers had built up so much religious and secular prestige and power that the net flow of gifts started to go the other way: from the subordinate functionaries and the subjects to the ruler. A combination of supernatural fear of the "divine" ruler, a regular show of force, and large ceremonies kept people in awe of the monarch. But even then, rulers needed stable income levels for their everyday needs, including the pomp and circumstance that no king could do without.

Lower-level sociopolitical formations (so-called village federations) exhibited the same cycles of fission and fusion, which is well documented, although that is for a later date, for highland Burma and highland Sulawesi. There, wealthy individuals and lineages attempted to increase or maintain their standing by organizing large-scale feasts of merit that involved the exchange of prestige goods and slaughtered animals. However, the kin groups that sponsored these attempts were not sufficiently wealthy to practice conspicuous consumption and at the same time remain rich enough to continue impressing their followers (neighbors, dependents) with their liberality. Thus they had dissipated the very means they needed to establish or maintain their higher (hereditary) status. Their relative poverty was caused by the fact that their communities were usually too far from ports of trade to enable them to demand high prices. Thus a low level of wealth generated by involvement in trade would appear to have nipped aristocratic tendencies in the bud, and to have stimulated the persistence of more egalitarian communities. In this respect we are reminded of the potlatch ceremonies of the North American Kwakiutl Indians.

Although it is argued here that the premodern or precolonial state in Southeast Asia would remain structurally weak, a more detailed study of the regional polities between the years 500 and 1500 reveals signs of increased integration and centralization. Originally, the ruler's court was essentially a patriarchal household, whose rule was the affair of the chief's lineage. This "king" was not much more than a primus inter pares among local chiefs. Gradually we witness the appearance of some sort of "administrators" not recruited from the royal household, which seems to justify the use of Max Weber's term "patrimonial state" for this polity (that is, a bureaucracy staffed by the personal followers and dependents of the ruler, including members of his family). The next step was the expansion of royal authority through a stepwise integration of the areas adjacent to the original core area. In the wake of this process, various Southeast Asia states witnessed the development of quite elaborate hierarchies of patrimonial officers. Finally, the East Javanese state of Majapahit, at its height in the fourteenth century, may be taken to represent the culmination of the precolonial state formation processes in this area. This is reflected mainly in the following features: the replacement, in the expanded core area (most of East Java), of regional lords by members of the king's family and the extension of the dynasty's hegemony over most of Java, in addition to the establishment of tributary relations with many Indonesian areas outside of Java. Similar developments had led to the centralized state of Angkor (Cambodia) in the twelfth century, thirteenth-century Dai Viet, and fifteenth-century Ayuthaya. As far as centralization and integration are concerned, it may be concluded that some kind of convergence appears to have occurred between the premodern "patrimonial states" of Southeast Asia and the contemporary "bureaucratic states" of Western Europe.

It has been assumed—as was said earlier—that when states had reached such advanced levels of integration and centralization, their rulers usually no longer needed to spend so much of their income on gift giving. Instead of expensive material gifts they could confer titles and other "symbolic capital" on their subjects and allies, although one assumes that they would continue donating less virtual gifts to the temples. In theory, the shift to symbolic capital in the more centralized states should have limited the ruler's expenditure and therefore his thirst for extra income. Yet it does not appear to have reined in the inclination of the Southeast Asian states to wage constant war upon each other, or the inclination of regional centers to hive off.

A peculiar feature of all of these states was their dependence on religious institutions such as temples, stupas, and monasteries, a theme that is dealt with in more detail presently. It has been argued that a network of religious foundations, financed largely from donations made by the ruler and the regional aristocracy, made up for the lack of a large-scale bureaucracy in the modern (Weberian) sense of the term ("patrimonial state" versus "bureaucratic state"). It has also been argued, with Angkor as the classic example, that overly ambitious temple building programs were the undoing of many of these states.

Within the group of Southeast Asian polities dating from the period under consideration, many scholars have made a distinction between inland and coastal states. Almost all inland states were to be found in Mainland Southeast Asia (with Mataram in Java as the main exception), while the coastal states were to be found mainly in Island Southeast Asia (with Champa, in Vietnam, as the exception). Majapahit, in Java, is regarded as a mixture of the two types of state, being both a coastal and an inland state.

The inland states have also been named "hydraulic states," as the presence of irrigated rice cultivation is one of their main features; the term "hydraulic," however, has been a source of much misunderstanding, as will be explained

Borobudur, a Buddhist stupa, Central Java, is one of the large religious monuments built by early Southeast Asian states. (Corel)

shortly. These are also the "monumental" states with the "sacred cities," such as Borobudur and the Prambanan complex of Central Java, and Angkor Wat and Angkor Thom in Cambodia. Most of the Classical States (Pagan, Ayuthaya, Angkor, Mataram) were inland kingdoms, although some of them had coastal state features as well.

The coastal states had a dendritic physical structure: one "city" (town would be a better word) dominated a river with its various branches and the river's hinterland. As contact between two rivers over land was difficult or downright impossible, trade between two adjacent river systems had to be conducted along the coast between the two establishments that dominated these systems. Srivijaya is an example where one river system (the Musi, with Palembang at the base of its estuary) dominated various other systems for a time, but where at a certain moment the hegemony shifted to another system (the Batang Hari, with Jambi-Malayu). It has been argued that coastal states had more difficulties regarding the formation of larger, centralized polities than did inland states, because these riverine polities formed a series of enclaves that could not easily be dominated by one of them. In addition, adjacent river systems—or so the theory

Ruins of a temple at Angkor Wat, built during height of the Khmer Empire during the 12th century CE. It's been argued that overambitious temple-building was the undoing of the Khmer polity. (Corel)

goes—often produced more or less the same trade items, which thwarted attempts at monopolization, and gave traders the option to go to the harbor with the lowest prices, which kept the income of the local ruler at a modest level. Good examples of coastal states are Srivijaya (which did rather well, considering), Samudra-Pasai (also Sumatra), Malacca (Malay Peninsula), and Champa.

INDIANIZATION AND TRADE

In many areas of Southeast Asia, the arrival of early states was heralded by inscriptions in Sanskrit written in Indic script. Kings, states, and cities were known by their Sanskrit names; Hinduism and Buddhism were imported from India (and China) and became the religions of the states and many of their subjects; and temples, stupas, and monasteries were built in India-inspired styles. This has led to an earlier scholarly tradition linking the beginning of state formation in the area to a process of "Indianization." Two caveats should be introduced here, however, the first being that the northern part of Vietnam was not Indianized but Sinicized, and the second that large areas escaped either largely or entirely the cultural influence from India (and, for that matter, China). The Philippines, Sulawesi (Celebes), and New Guinea are areas that spring to mind immediately, but many upland areas in Mainland Southeast Asia were also hardly influenced.

Today, we no longer accept the idea that the spread of culture and religion from India to Southeast Asia was the cause of state formation in the latter area, although evidently Indian culture heavily influenced the early states of the region. Indian statecraft was adopted by the rulers of the early states, and in the course of time notions from Hinduism and Buddhism boosted their status to that of "divine" monarch—if not in this life, then certainly after death—with the king as link or mediator between human society and the cosmic order. However, such notions seem to have been enhanced, not created, by Hinduism or Buddhism.

What, then, is the current explanation for the rise of the early states in Southeast Asia? Many scholars now argue that there is a strong link between developments in international trade flows and the rise and fall of states. As was to be expected, Southeast Asia, wedged in as it is between India and China, was involved in the trade between and with those two regions. It was also influenced by mercantile patterns that originated in the Roman Empire, particularly the eastern half, roughly coinciding with what we now call the Middle East, including Persia (Iran). It seems that, as early as the beginning of the first century CE, more than a hundred ships from the Roman Empire traded with India. Originally, much of the trade between the Roman Empire and China followed the overland route (silk route), but after around 150 CE disturbances in Central Asia led to an increased importance of the sea route. The rise of Funan, in the Mekong Delta, is supposed to have been linked to these developments. To be sure, that was not the first instance of international trade flows in the area. As we have seen in the last chapter, bronze kettledrums produced in what is now Vietnam were traded with the Indonesian Archipelago somewhere between 200 BCE and 300 CE, and local trade networks must have been much older still. This goes to show that international trade did not so much "create" early kingdoms as it did give already developed chiefdoms the kiss of life, so to speak. The first condition, therefore, for the rise of an early state, is, not surprisingly, the presence of an advanced (some would say complex) chiefdom.

Between the rise of Funan at the start of the period discussed here and that of Malacca at the end, a coming and going of (early) states is witnessed, something that was caused by several factors, among which shifting mercantile flows were very important. They, in turn, resulted from technological changes in boat building, the arrival of new groups of traders (such as Arabs and Tamils), and, of course, shifts in demand and supply of commodities. Political factors, however, should not be forgotten. The expansion of the Mongols led to the destruction and creation of states, and wars—both on land and water—were fought over the domination of harbors, which also led to the demise of some states and the strengthening of others. If one port of trade thus disappeared, another would profit from the displacement of commodity flows, as happened when Javanese polities took the place of Srivijaya.

Dependence on long-distance trade varied considerably between states. It ran the gamut from total dependence, as in the cases of Srivijaya and Malacca, to marginal importance, as in the case of Angkor. It seems likely—indeed almost tautological—that coastal states needed maritime commodity flows, whereas inland kingdoms did not. However, it can and will be argued that even inland states depended partly on trade flows for their expansion and consolidation. The origin of states in (for instance) eastern Java is perhaps not to be explained by the presence of international trade flows; the flowering of the most important one among them, however—Majapahit—probably is.

AGRICULTURE AND HYDRAULICS

The "classic" precolonial states of Southeast Asia conjure up images of impressive ancient monuments against a backdrop of rice fields. This applies to Pagan, Ayuthaya, Angkor, and Vietnam as well as Mataram and Majapahit. It does not apply to Srivijaya or Malacca.

Particularly in the case of Srivijaya, this is somewhat of a riddle. An early state, which dominated trade between Europe, India, and China, as well as between the Indonesian Archipelago and other areas for so long, left virtually no traces in the Sumatran landscape. When European historians around 1900 discovered that such a state had existed in Sumatra in the remote past, it came as a huge surprise. We should probably assume that Srivijaya, despite its more recent fame, was in all likelihood never much more than a federation of ports of trade, of which one town, located in the area of the present-day city of Palembang in southeastern Sumatra, was the primus inter pares (at least most of the time). There is not much evidence of a rice-producing immediate hinterland, but the upland valleys of western Sumatra and central Java may have functioned as such, transporting their rice to feed the capital of Srivijaya by river and by sea transport, respectively. If that is true, people have been looking for Srivijaya's "ecological footprint" in the wrong direction. The permanent population of the city of "Palembang" in that period (seventh to eleventh centuries) was probably not large. Most of the large states dealt with in this chapter were polities based on densely populated valleys, where labor-intensive wet-rice agriculture was being practiced.

Although the historiography on early food crop cultivation in (Southeast) Asia may not be voluminous, it is certainly complex, and scholarly opinion has changed considerably over time. Some thirty to forty years ago many scholars believed that rice was originally a "dry" crop, coming from India, having spread to Southeast Asia at a relatively slow pace and thereby taking the place of more "primitive" cereals such as (foxtail) millet and, above all, roots and tubers such as taro and yam. According to various scholars, basing themselves loosely on Karl Marx and Karl Wittfogel, there was a causal link between the need for large-scale irrigation for wet-rice cultivation and the origins of the "despotic" Southeast Asian state. They assumed that the construction of large-scale irrigation works could not be undertaken by individual villages. The triumph of irrigated ("wet") over "dry" rice was regarded by various scholars as a relatively late development, spanning the last 500—or as some would even argue, 200 or 150—years.

The new orthodoxy is that rice was originally a "wet" crop, a plant growing naturally in marshy environments, which seems to have been first domesticated in southern China or northern Vietnam, from where it spread to the (other) Southeast Asian countries. The earliest form of rice cultivation was doubtlessly wet as well—receding floodwater agriculture around lakes and in river valleys, and (floating) swamp rice cultivation in coastal areas. Those types of wet-rice agriculture are supposed to have played an early role in Funan and in Champa, respectively, and some of it has survived up to the present, for instance, in Cambodia (Lake Tonle Sap) and Borneo. It is also clear that during the first millennium CE bunded fields for irrigated rice cultivation and the buffaloes and ploughshares that belong to the same wet-rice complex were to be found in many areas of Mainland and Island Southeast Asia. Broadcast sowing seems to have been more common in the mainland areas, whereas the more labor intensive method of transplanting was to be found in Java and Bali. By the ninth century, irrigated rice cultivation in Java provided the largest single source of tax income. The role of the state in the development of wet-rice cultivation, according to current opinion, was much more modest.

One of the debates still raging among scholars interested in Southeast Asian agriculture relates to the question of why and how the shift from "dry" to "wet" rice occurred. Generally speaking, in terms of productivity of labor (production per hour worked), rice grown as a dry-land crop by shifting cultivators produces a much larger crop per unit of time worked than does irrigated rice on permanent, bunded fields. In a comparison between slash-and-burn agriculturalists in Kalimantan (Borneo) and Javanese *sawah* (wet-rice field) cultivators made some twenty years ago, returns to labor for the first system were between 88 and 276

Terraced rice fields in Southeast Asia were a response to population increase. Growing demands for food meant that rice had to be grown on lands not naturally flooded by lakes or rivers. (Corel)

percent larger than in the latter one. The obvious question is why people would shift from a system with a low labor demand to one with a much higher degree of labor intensity.

The question is, in fact, more complicated, however, because it must be assumed that cultivators first shifted from wet-rice produced by receding flood-water agriculture to dry rice under a regime of shifting cultivation (also called slash-and-burn agriculture and swidden agriculture). That question is usually not even posed, let alone answered. The answer should probably be that the natural environments available for receding floodwater agriculture (lakes, some lands adjacent to rivers) were gradually filling up with people. When population increase continued—though low by modern standards—rice had to be grown on lands not naturally flooded by lake or river, as a noninundated crop under shifting cultivation, like millet or yams. It is also possible that cultivators preferred somewhat higher and drier areas—the mid-altitude level that was mentioned in the last chapter—for different reasons, such as fewer problems with flood control or a lower incidence of malaria.

Returning, then, to the question of why cultivators at a later stage changed over from dry to wet-rice, we usually encounter two answers in the literature. The first is the older, linking this shift to the Hydraulic/Despotic State hypothesis, under the assumption that the state would have forced its population to lay out wet-rice fields and the concomitant irrigation works. Thus the state would have killed several birds with one stone: people living in permanent (often nucleated) settlements are much easier to control than slash-and-burn agriculturists, particularly as regards taxation and the performance of corvee labor (statute labor), while yields per acre would be higher and more secure, leading to higher tax returns. The second answer, found in the literature since the late 1960s, was a response to Ester Boserup's thesis that population growth was the driving force behind technological innovation in general and intensification in agriculture in particular. Scholars then started to assume that population growth might have been behind the shift from dry to wet-rice in Southeast Asia.

Neither of these two explanations is very convincing. Given the fact that low population densities prevailed even at the end of the period we are dealing with in this chapter (probably about five people per square kilometer on average around 1500), the population growth argument does not appear plausible. It would also be difficult for any premodern Southeast Asian state to force large numbers of its population to construct irrigation works and to lay out complexes of bunded fields, given such low population densities.

Swidden agriculturists in recent times did not as a rule appear to be particularly eager to start laying out sawahs, arguing that it was just too much work, despite the fact that wet-rice grown on permanent, bunded, irrigated fields produces more secure harvests in times of droughts and higher yields. It should be pointed out, however, that these swidden cultivators did not depend on their rice crop alone. In addition to being agriculturists, they were also horticulturists who practiced agro-forestry and foragers who acquired all kinds of nonrice foodstuffs (other crops, game, foraged fruits, nuts, roots, and tubers) and tradable commodities (so-called Non-Timber Forest Products, or NTFPs) in addition to their rice crop. This could be easily done in the time that their labor-extensive rice cultivation left them. So who would want to be a settled peasant with a heavy workload?

In a number of recent studies it has been argued that some swidden cultivation systems are not much less labor intensive than some irrigated wet-rice systems. However, it would appear that in those studies the labor input of livestock has not been included. What to do with livestock in these calculations is, indeed, a difficult methodological issue. Do we count livestock hours as human labor hours, or do we account for the amount of labor that went into raising the animals? More will be said about livestock presently. It has also been argued that nonirrigated wet-rice systems, such as swamp rice agriculture, have the same low labor intensity as swidden cultivation systems.

WET-RICE, THE STATE, RELIGION, SLAVERY, LIVESTOCK, AND THE MARKET

So why did wet-rice fields take the place of swiddens in so many areas, and often at so early a date that population pressure could hardly have been the cause? There are seven factors that may have played a role, alone or in combination. They will be presented in no particular order, as we have no idea regarding their relative importance.

The first factor is that people tended to flock together in uncertain times. People living in relatively large, concentrated, nucleated settlements would have had to walk long distances to their fields if they practiced slash-and-burn agriculture; cultivated areas were usually left fallow after only one or two years under cultivation, when soil fertility had dropped considerably and weeds had become so prolific that too much labor had to be spent on weeding. Under those circumstances the construction of irrigation works and the laying out of bunded fields would have been a logical step. As the societies we are discussing were certainly characterized by frequent conflicts at almost all levels, and violence of one kind or another was fairly ubiquitous, it stands to reason that people were living in concentrated settlements. This process was reversible: there is evidence (dating from a later period, but illustrating the same principle) from Burma under British rule that the Pax Imperica (law and order) imposed by the colonial rulers made for an increase in dispersed settlements in sparsely populated areas, and therefore for a drop in the proportion of permanently cultivated, irrigated fields.

A second factor is the role of the state. Although the idea of a so-called Hydraulic State, in the sense of a despotic state that undertook to initiate, execute, and coordinate the construction of hydraulic works and the sawahs that go with them, has by now been discarded by most scholars, this does not mean that the state, or rather the ruler, played no role at all. In Ayuthaya and Pagan monarchs were involved in the laying out of large-scale waterworks, perhaps because works on such a scale were beyond the capacity of local communities. It also would appear that in most kingdoms dating from the period under discussion, the monarch actively encouraged the laying out of irrigation works and bunded fields. This was done mainly by means of donations of land and people to religious institutions, as well as by arranging tax exemptions for temples and monasteries. Such tax exemptions, like the well-documented *sima* grants from Java dating from the ninth century onward, were granted to religious establishments that would take it upon themselves to convert wasteland to wet-rice–growing areas. Control over waterworks was usually in the hands of villages, occasionally in those of bodies that could be called water users associations (such as the *subak* of Bali or the *karuin* and *tuik* of Burma), and sometimes in those of temples (Bali). The state, therefore, did encourage rather than force its subjects to switch from swiddens to sawahs.

A third factor is the role of religious establishments (temples, stupas, monasteries, and water temples connected to holy waters). Hinduism and Buddhism reached Southeast Asia during the period under consideration, mainly from India and Sri Lanka but also via China. Both religions were props for the state formation process, facilitating the transition from chiefdoms to early states by strengthening the legitimation of the monarch. Local (ancestral) temples were usually connected to the state temples in a tributary relationship, thus creating a network covering much of the (core) territory of the state. Kings stimulated the liberality of local aristocrats to religious foundations, thus strengthening the temple network and at the same time weakening the economic strength of potential aristocratic rivals.

The literature appears to underestimate the economic role of the religious foundations, particularly as regards the spread of wet-rice cultivation. As we have seen, kings, nobles, and other rich people gave gifts to religious institutions. With a view to their status in this world and their well-being in the next, they donated land, people, livestock, and valuable commodities (such as gold and silver) to all kinds of temples, making the Hindu and Buddhist "clergy" as a collective the largest landowner in many of those states. Occasionally this led to conflicts between "church" and state, as was the case in Pagan, where the king reduced the amount of land owned by the Buddhist clergy periodically. However, under normal circumstances the religious establishments were instrumental in the expansion of agriculture into the more remote corners of the realm, thus making those areas safe for habitation. They were also instrumental in the construction of irrigation works and the laying out of wet-rice fields.

It should be mentioned here that the notion of the king as owner of all the land in the realm is becoming rather outdated, as also is the notion that all land was owned by clan or tribe. It becomes increasingly clear that land, at least wet-rice land, in the more densely settled areas was owned by individual families, who could sell, mortgage, lease, or otherwise alienate it at pleasure (it seems likely that wet-rice cultivation and individual land tenure reinforced each other). Wasteland and forests were often "owned" by a local community (village, clan, or lineage), but land under shifting cultivation often must have been subject to more individual (family) claims, albeit that the lineage may have had a residual claim. The king—or for that matter any other dignitary, therefore—could not assign land to temples unless he owned it, and it is often stated in the epigraphical sources that the lands given to such an institution had been bought first. In Java locally minted gold and silver coins were available for such transactions from the ninth century; copper coins—first from China, then minted locally— were available from around 1000. However, instead of giving land, kings and local functionaries could and did assign the rights to the tax returns of certain lands to a temple, without the land itself changing hands.

There are examples of temples paying laborers for the construction and upkeep of irrigation works (for example, in Pagan), but temples also gave advances in the form of livestock and seed to the peasantry on their lands (Angkor). However, what appears to have happened even more frequently is that they used bonded labor for such tasks, a topic discussed presently.

Many large-scale waterworks appear to have been constructed primarily with a religious purpose in mind. The most famous example in this respect is that of the large artificial lakes (*baray*) of Angkor. Up to about 1980 it was believed that these lakes constituted the center of a complex and widespread irrigation system, but since then it has been argued that evidence for connecting channels between the lakes and the arable lands is lacking, implying that the purpose of the lakes was mainly, or perhaps even entirely, religious. However, even more recently (2000), a more thorough investigation of the barays revealed that off-take structures did exist after all, which takes us back to square one. It is another matter entirely how important irrigation with baray water really was: the area of retreating floodwater agriculture embodied by the Tonle Sap Lake, located close to the barays, was so much larger than the lands that could be irrigated with water from the barays that the additional inundated acreage thus acquired might have been hardly relevant. Still, the principle of baray irrigation is now back in the arena.

Generally speaking, religious institutions may have been the most important single source of the "social production" of a surplus that was to be used for conspicuous consumption—that is, production to be used entirely for "merit making" and "competitive feasting" type ceremonies. It has been suggested that even ordinary peasants, of their own free will, were increasingly producing more than was needed for subsistence requirements, not because they were interested in a higher income as such but because they wanted to show off their wealth, or at least accumulate merit in view of their afterlife. It may have been one of the reasons why agriculturalists switched from swiddens to sawahs, even if they were not being paid for it by a temple or were not unfree temple laborers.

Thus we have arrived at the fourth factor to be mentioned in connection with the spread of wet-rice cultivation—the presence of bonded labor in considerable quantities. In much of the literature on Southeast Asia one encounters unfree people, often referred to as slaves or serfs. However, the meaning of this phenomenon for the Southeast Asian economies and societies in the period of indigenous state formation has seldom been analyzed. In fact, these societies can be described as consisting of three orders—the aristocracy, the commoners, and the bondsmen and -women (sometimes the clergy is mentioned as a separate, fourth order). There was a large variation in "degrees of freedom," ranging from the rather mild institution of debt bondage (essentially someone working off a debt) to chattel slavery of people who could be bought, sold, and even sacrificed, and whose status was hereditary.

There appears to have been a strong link between religious institutions and bondage. Temples almost invariably owned slaves, and donations made to temples by wealthy people often included bondsmen. Temples were given the land, as well as the livestock and the bonded labor needed to work it, which were used to clear new arable lands and convert swiddens to sawahs. Temples had granaries with which to take care of large numbers of people in difficult times, which enabled them, for instance, to accept recent war captives, who did not have the means to make a living on their own, as labor. It is also quite possible—and well attested for the Early-Modern Period—that the presence of granaries led destitute individuals to give themselves and their children voluntarily into bondage to a temple in times of famine.

Kings and noblemen owned slaves and serfs as well, and they may have been put to the same use as were the temple slaves, but that is much less well documented. We do know, however, that the king of Srivijaya, in an inscription dated 683, referred to the cultivators of the royal domain as "my bondsmen," the same people who also formed the nucleus of his army. Royal bondsmen are also supposed to have been involved in the construction of temples in central Java during the eighth and ninth centuries.

What might have been more important is that rulers occasionally seem to have resettled populations from other areas to their own core lands (Java, tenth century; Burma, eleventh century; Dai Viet/Champa, eleventh century). Here, one assumes, some initial force must have been used by the state to move these people, but they must also have been given advances in their new surroundings in order to tide them over, which put them on the same footing as peasants who were given livestock and seed by the temples provided that they laid out wet-rice fields. If the rulers had invested some capital in such compulsory migrants, it seems likely that their status would have been that of serfs. Taken together, the data on bonded labor suggest that it may have been a powerful tool in the hands of temples and rulers with an interest in creating large settlements with permanent, irrigated fields.

The presence of bonded labor as such was linked to the type of state to be found in Southeast Asia during this period. The ruler of a rather weak, (early) state, with its centrifugal tendencies, had to wage war and organize raids on a regular basis in order to show his prowess and to acquire slaves and commodities that he could bestow upon his followers and upon the religious establishments.

The fifth factor to be mentioned here is more speculative. It hinges on the supposition that the amount of labor invested in the acquisition and upkeep of livestock, mainly water buffalo and cattle, was a major issue in the choice between slash-and-burn and irrigated permanent-field agriculture. As we have already seen, the difference in rice yields per hour worked between these two systems may have been small or even negligible if hours spent by livestock are not counted.

The point is that, in Southeast Asia during the period we are dealing with, and sometimes even up to the present age, large numbers of livestock were kept mainly for sacrificial purposes. In many societies—for instance, in upland Burma, Sumatra, and Sulawesi—people practicing slash-and-burn agriculture possessed large numbers of buffalo that were of no practical use at all, apart from their function in rituals and feasts (and possibly for their manure). These herds constituted "living" capital: wealth on the hoof. We do know that large numbers of livestock were already being slaughtered in Hindu rituals in western Java and eastern Borneo in the fifth century, and such practices continue locally today. In other words, there were always animals available to trample the mud of the rice fields or to draw the plow. In itself, that was not enough to make people switch from swiddens to sawahs, but the presence of "idle" livestock may have contributed to the transformation.

Sixth, one supposes that a switch to wet-rice—and other labor-intensive sawah crops—might have been stimulated by a combination of (export) trade, markets, and (moderate) taxation. Most of the literature tends to equate rice with subsistence production, and that is no doubt true in a great many cases. However, rice cultivation was by no means restricted to production for the peasant's own consumption, with a small surplus ending up on the local market. In the case of Java, for instance, rice was one of the main export crops as early as the ninth century, and in a list of commodities stored in the warehouse of the port of Mananjung, eastern Java, dating from the mid-eleventh century, rice is the first item mentioned. In the Brantas delta region of eastern Java, double cropping of rice was common by the early fifteenth century, something that can be done only on sawahs. As that is also the area where the export and import trade of Java was concentrated around that time, the link between trade and double cropping seems plausible.

A dense network of local markets linked to the ports of trade and serviced by specialized transporters, traders, engrossers, and merchants orientated toward international trade, in combination with a system of moderate taxation and the availability of gold, silver, and copper currencies, were mutually supportive factors. The same network provided import goods that wet-rice–producing peasants could buy for their extra money, as well as industrial products that had been locally made from imported raw materials, such as iron, copper, and bronze. Dealing with these issues we should think in terms of a labor-consumer balance, which means that peasant-cultivators would increase their labor input only if doing so produced at least a proportionate and ensured increase in wealth or prestige. The combination of moderate taxes, which forced the agriculturalists to take their produce to the nearest market in order to convert it into money, and the presence in the same market of desired goods on which part of that money could be spent, may have tipped this balance in favor of wet-rice cultivation. This combination of factors is well attested for Java from the ninth through the fifteenth centuries, and elements of it can be traced in other wet-rice–producing kingdoms as well, though not always all at the same time.

Under the circumstances mentioned here, people would concentrate around centers of commerce, which would lead to relatively high population densities, which, in turn, would stimulate irrigated wet-rice agriculture, as was the case with people flocking together for safety reasons, an argument given above. In these cases, therefore, the growing population density argument would appear to be valid.

Finally, it has been suggested by Richard O'Connor that the expansion of irrigated wet-rice cultivation was stimulated by Burmans, Tais, and Vietnamese— skilled irrigation experts—migrating from upland valleys or a piedmont where they had harnessed fast-flowing perennial streams to wet-rice cultivation. They migrated to the lowlands, where, beside house gardens, rain-fed and retreating floodwater rice cultivation was, in this view, predominant. Obviously, this argument does not apply to the Indonesian and the Philippine archipelagoes.

Summing up, it could be said that the switch from swidden to sawah was stimulated by the state, which created incentives and the legal and administrative framework in which markets and traders could prosper; by religious establishments, the existence of which stimulated increased social production; by the growth of international trade, which was boosted by the policies of both state and "church"; by the presence of "captive" factors of production such as bonded labor and livestock; and by the introduction locally of new technologies. It is a much more complex model, therefore, than either the despotic/hydraulic state or the population growth hypothesis, even if both the state and population growth did play a role locally and at different moments in time.

Returning now to our original question of which factors stimulated the arrival and development of (early) states in the region, it can be said that states and wet-rice cultivation were mutually supportive. The state created incentives in many ways, among which mention should be made of tax incentives, laws governing trade and merchant communities, coinage, road construction (Angkor, Dai Viet, Majapahit), and protection. The state was, in turn, supported by taxes based on wet-rice cultivation. We could argue that without foreign trade, this positive feedback mechanism would not have existed, or at least would have been much weaker. Both the ruler and his aristocracy as well as the clergy and their establishments generated an enormous demand for rare objects (gold, silver, copper, and bronze items, ceramics, textiles, rare woods) and rare substances (perfumes, spices, drugs) that could be met only by supply from outside, for which a marketable (rice) surplus was a necessary condition.

Export-led growth, therefore, appears to have been one of the motors of "sawahization" and state formation. Another motor of both was the growth of the number of religious establishments with Hindu or Buddhist features.

ENVIRONMENTAL CAUSES

Is it possible to say anything about the environmental factors that influenced the formation of early states in Southeast Asia? Geographical position, often in combination with geophysical properties, including climate, appears to be a key to a number of developments. For instance, the fact that India and China produced empires, whereas Southeast Asia did not, could be at least partly explained by the lack of very large (navigable) rivers flowing through the large plains in the latter region. In addition, Southeast Asia is located almost entirely in the tropical and savanna zones, whereas those parts of India and China where empires developed are located largely in temperate maritime climatic zones. Tropical and savanna zones are hotter, perhaps more subject to droughts and floods, and are characterized by a higher parasite load than is the temperate zone.

In a similar vein, differences in state formation within Southeast Asia can be perceived to be linked to such factors. The ever-wet tropical areas are not prime areas for sedentary agriculture, almost a condition for the creation of states: in them, leaching and erosion are very strong once the original forest cover disappears; owing also to a carbohydrate-poor environment, foraging populations had been rather thin on the ground; and a prolonged dry season, necessary for sun-ripening in cereal production, is absent. At the same time, extensive flood plains with their slightly more elevated uplands (see below) were largely lacking there. The fact that Srivijaya, where several other conditions for the creation of a state were met, did not develop beyond the stage of an early state, might be linked to the absence of large areas that with the technology of that epoch could have been converted to wet-rice fields. These factors may have played a similar role in Malaysia, Borneo, Sulawesi, the Moluccas, the Philippines, and New Guinea, all of them areas where stable and large (early) states, despite some local attempts, were either entirely absent, of a fragmentary, fissiparous nature, or very late in coming during the period under consideration.

The presence of rivers in Southeast Asia—although not as large or navigable as those of India and China—has been a positive factor as regards state formation. Although, as we have seen, people seldom lived very close to the river, they did live in the somewhat more elevated areas nearby, which enabled them to use the river for irrigation, transportation, and for drinking, cleaning, and waste disposal. The large wet-rice cultivating states almost invariably developed around river systems. If, in addition to such a river, the state had one or two good natural harbors at its disposal, located at positions suited to the mercantile flows of the moment, that strengthened both agricultural productivity and the state.

However, the presence of a good harbor was a blessing only so long as there was a concomitant trade flow. When changes occurred in the global trade patterns (Europe, Middle East, India, China), harbors would be passed by, their people turning to piracy or hiving off from the inland state to which they had been linked, as happened in Burma and Java at the end of the period dealt with here. Clearly, then, the position of harbors was not a factor that in itself can explain the flowering of a state; it is, however, conducive given the right circumstances.

There are several environmental features that alone or in combination go a long way toward explaining the presence or absence of large-scale irrigated rice agriculture. One of these features, explaining a west-to-east gradient in the presence of rice verses roots and tubers from Sumatra to the Pacific, is rainfall variability, which increases from west to east. Another feature encountered in the region where rice cultivation does occur is that the largest concentrations of wet-rice fields are traditionally to be expected in areas with volcanic soils or regularly flooded alluvial soils that are neither too wet nor too dry, and not too steep nor indeed too high. The exploitation of very wet low-lying areas, for instance, had to wait until large-scale drainage works could be carried out, which started (as a rule) in the (late) nineteenth century with the colonial state.

Volcanoes, of course, are not only benevolent. Data on volcanic eruptions during the period under consideration are rare, but there is some evidence that the switch from the Javanese court after ca. 900 from central to eastern Java was partly motivated by a number of eruptions of Mount Merapi.

A final remark should be that the lack of a particular type of resource is not necessarily a bad thing in economic terms. For instance, a lack of tool-making stone and metal ores, as was the case in Java, can lead under the proper conditions to a lively trade in those products, locally stimulating production to be exchanged for those items. It has often been argued, by historians and nonhistorians

Mount Merapi sends out hot gas clouds, as seen from Borobudur in Indonesia. Merapi, active in recent years, is one of the most dangerous volcanoes in the "Pacific Ring of Fire." Some evidence points to the Javanese court relocating circa 900 CE due to eruptions by Mount Merapi. (Weda/EPA/Corbis)

alike, that a natural environment too rich in easily collected resources, and thus not posing a challenge to the entrepreneurial spirit of early humans, did not constitute a good starting point for social change and economic growth.

In most of the cases mentioned here it could be argued that environmental conditions created "favorable" circumstances that then could be made use of for agriculture, trade, and state building. Shifting migration streams (for example, the Tai) and the changing trade flows that were at least partly responsible for the timing and scope of the responses to the environmental situation, were strongly influenced by events outside the area (China, India, Europe). In that sense, globalization is nothing new; it is not a twentieth century phenomenon, nor one that came to the area with European expansion. Therefore, I consider the period under consideration to be the first (of four) "phases" of globalization that would be instrumental in co-shaping the environmental history of the region.

ENVIRONMENTAL EFFECTS

It is generally accepted that serious and more or less continuous population growth started with the Neolithic. Semisedentary agriculturalists, so the argument goes, have higher food security and higher fertility than foragers. Although after the beginnings of sedentism the palaeodemographic record often shows a slight dip in the life expectancy at birth (as reconstructed from burial sites), probably for the most part because of new illnesses taken over from livestock, with which people were now living closely, that effect seems to have worn off at a later stage, in all likelihood because of acquired immunity.

It is also assumed by scholars that wet-rice cultivators have higher population growth rates than shifting cultivators, but it is less clear why. Greater food security may well have played a role, but it also possible that, in addition, people married earlier and therefore had more children; also, they were possibly less inclined to practice abortion and infanticide. There is an intriguing story about northern Vietnam in the first century CE, then under Chinese occupation. The Chinese administration appears to have ordered all men between twenty and fifty years of age, and all women between fifteen and forty, to get married, apparently because considerable proportions of those age groups were still single. This may have been a fairly general pattern among those people in Southeast Asia who were not wet-rice cultivators. There are also indications that competition among lineages of swidden cultivators centered around bridewealth payments (see below), which, other things being equal, may have had a dampening effect on nuptiality. But all this is rather speculative, and chances are that it will always remain so. If this hypothesis is accepted, population growth rates would have risen slowly but surely, as the proportion of wet-rice cultivating agriculturalists was in all probability steadily increasing throughout the period.

Be that as it may, growth rates were still low by modern standards, perhaps in the neighborhood of 0.1 or 0.2 percent, the growth rate also obtained in most areas of Southeast Asia between 1600 and 1850 or even 1900. If these percentages are taken as our point of departure, and if we accept Anthony Reid's estimate for 1600 of 22.5 million people, we should not expect to find many more than 5 million people and not fewer than 1.5 million in the region around the year 1 CE. As northern Vietnam alone is supposed to have had 1 million inhabitants according to a Chinese census taken at around that time, the latter figure is almost certainly too low, which implies a growth rate closer to 0.1 percent. But even that estimate may be too high, given another estimate for annual population growth in Asia as a whole of just under 0.05 percent on average between the years 1 and 1600. A similar growth rate for Southeast Asia would imply the presence of 10 million people around the year 1, a figure that seems to err on the high side. Summing up, then, we can say that Southeast Asia at the beginning of the Common Era may have had between 5 and 10 million inhabitants, and an average annual population growth rate of between 0.05 and 0.1 percent between then and 1600.

Although, therefore, population growth rates were low according to twentieth-century standards, the increase in numbers of people and the concomitant amounts of land cleared for agriculture were considerable. Given the large numbers of livestock mentioned in the sources at an early stage, we may also assume that considerable areas were turned into pasturage. Most of the "wasteland" cleared for arable lands and pasture originally must have been under forest cover, which implies considerable deforestation. These twin phenomena—land clearing and deforestation—must be regarded as the main environmental effects during the period. If we add the construction of dams, reservoirs, feeder canals, and the laying out of bunded fields, often grouped in picturesque, amphitheater-type complexes on the lower slopes of hills and mountains, it will be clear that vegetative cover, soil composition, and landscape all changed beyond recognition. Agro-ecosystems had taken over in many areas from the original floral and faunal communities.

There are indications that some of the irrigation systems have been subject to siltation, something well known from other "hydraulic" civilizations as well (for example, Mesopotamia). One would assume that, in theory, such a process should have been reversible, but in practice it seems to have been much harder to clean up a system when it was clogged with silt than it was to construct one. Siltation of irrigation systems has been cited as one of the causes of decay of the classical states of Southeast Asia.

It can be said, then, that both population growth and increased sedentary cultivation—particularly irrigated rice cultivation—appear to have been instrumental in the arrival and development of the early state. The state, in turn, aided and abetted by the religious establishments, consciously sought to expand the population and the cultivated, particularly irrigated, area of the realm. Thus a positive feedback mechanism had come into being of upward-spiraling population numbers, areas under permanent cultivation, and flowering states. As we have seen, this mechanism would also be fueled by long-distance trade, part of the first globalization phase mentioned earlier.

Having paid so much attention to wet-rice cultivators, it should not be forgotten that outside the core areas of the states mentioned here, large groups of foragers and swidden cultivators were still to be found, probably for the most part in the highlands and in the marshy and insufficiently drained lowlands, perhaps having been pushed out of the mid-altitude areas by the more successful wet-rice cultivators. We sometimes forget that the former—and the swidden cultivators in particular—also had an impact on their natural environment. Slash-and-burn agriculturists, for instance, were not only involved in subsistence agriculture; they also produced crops (such as pepper) for the international market. It also appears that they were stockbreeders (horses, cattle, and buffalo), which they did on a commercial basis (that is, for the lowland market). The wish to acquire necessities such as salt aside, the motor for this production seems to have been the trade in status goods, which, in combination with political titles conferred by lowland rulers, were inducements for status competition, including the acquisition of prestige goods to be used as bridewealth. For all of these products they had to clear a piece of wasteland, often primary or secondary forest, every one or two years, when the fertility of the soil started to drop and the growth of weeds became problematic.

But according to recent scholarly opinion even foragers, reputed to have minimal requirements, manipulated their natural environment, not only for subsistence reasons (such as the manipulation of natural sago stands) but also in order to satisfy market demand for NTFPs such as valuable woods, rattan, nuts, gums and resins (damar, camphor, benzoin), and bird feathers and game (hunted meat), which they traded for carbohydrates and salt (and sometimes, one assumes, for prestige goods). In addition, they gathered plants that we now regard as crops but that in that period were still wild plants, such as long pepper, cubeb, and cloves. It is also quite likely that they used fire to create ecotones attractive to game. In fact, it could be said that the earliest Southeast Asia exports were all NTFPs, implying an important commercial role for foragers during the first centuries CE.

We should not forget the "foragers of the sea," the fishermen and pearl divers who probably constituted a not unimportant proportion of the population, groups that have been sadly neglected by the scholarly literature on this period. Freshwater fishers were important in terms of food, but not regarding durable trade items. Nor should we forget the miners, about whom we are equally ignorant, but who must have had an impact on their environment.

Basically, China, India, and Europe were exporting metals and industrial products, in exchange for forest products and some products of the sea (pearls, shells), only gradually being supplemented by cultivated crops, in addition to metals (gold, silver, copper, tin, and lead). We do know that many of these groups of foragers, swiddeners, fishermen, and miners survived the attempts of premodern states to convert them to wet-rice producers, and by 1500 they may have made up between one-third and one-half of the population of the entire region, covering a surface area of probably more than half the regional total. Their share of the total population was, of course, decreasing continuously, but as the total population was growing, their absolute numbers dropped at a lower rate.

What all of this meant in terms of land use has never been considered in detail, but it must be supposed that the shift toward permanent field agriculture long before population densities would have forced people into more labor intensive forms of agriculture led to a reduced impact of swidden cultivation on the natural environment, and therefore probably to larger areas of limited human impact. By concentrating populations in some areas, the states, the religious establishments, and the international market forces acting in tandem actually may have been instrumental in preserving more "nature" in a relatively pristine state than would have been the case if swidden agriculturalists had been left to multiply unhindered. It has to be admitted, however, that the overall average annual rate of population growth must have been slightly higher because of the same policies that led to a larger share for wet-rice cultivators, which would have canceled out part of the positive environmental effects.

One of the environmental effects of the expansion of agriculture is, of course, erosion. Erosion is a natural process—think of topographic names like Red River (Vietnam), Yellow River (China), and Black Water (Malaysia)—but it is speeded up by human intervention. Although one might be inclined to believe that more intensive forms of agriculture are characterized by higher rates of erosion than swidden agriculture, that is not necessarily true. In fact, the reverse may be true if we compare irrigated wet-rice cultivation and slash-and-burn agriculture, particularly when the latter is carried out on fairly steep slopes. There is an interesting suggestion in the recent literature that erosion owing to dry land agriculture in upland areas in the Thai-Malay Peninsula between 400 and 1500, feeding the city-states of Kedah and Satingpra, led to the accretion of rather flat coastal areas, which were then used for wet-rice agriculture. It is a notion that could be applicable to more areas within Southeast Asia.

Another possible effect that should be looked into is that the accretion of swampy mud flats in coastal areas owing to erosion as a result of the expansion of agriculture may have led to an increase in malaria in the coastal zone, but perhaps to a drop in malaria in the zones that used to be coastal but, because of the accretion, had become inland areas. As swampy and muddy coastal areas are also preferred habitats for mangroves, which, in turn, are breeding grounds for fish and other edible or at least tradable maritime life-forms, the effects on fisherfolk of such shifts may have been considerable.

Agricultural intensification was not the only change to be observed in agricultural practices. Because of a combination of increased demand from traders, taxation pressures emanating from the state, and perhaps conscious attempts at risk-spreading and a more evenly distributed utilization of labor over the year, Southeast Asian agriculture and livestock breeding were more diverse at the end of the period than at the beginning. Examples are black pepper and safflower dye, products originating in India, but after 1000 being grown in Java and exported in large quantities to China. In the fifteenth century Sumatra became known as an important producer of pepper for export. Cotton and Indian indigo came to Java probably as early as did pepper and safflower and from the same region, while coriander, fennel, tamarind, and sesame were introduced, via India, from the Mediterranean. Mung beans may have come from India, and garlic from China. Pepper, a perennial upland crop, might very well have had a profound influence on social relations among the swidden cultivating upland dwellers, leading to an increase in "social production," thus leaving its mark on the natural environment.

Similar observations apply to livestock. The water buffalo and banteng-type cattle are indigenous to the area (and domesticated there, at least as regards the latter), but humped cattle came from outside (in some areas probably even prior to the beginning of the Common Era), as did horses, goats, and sheep. It stands to reason that especially the larger types of livestock left a mark on the landscape as much as did pepper.

BUILDING

Finally, a short note on construction is in order, a sector of the economy strongly influenced by environmental factors, and one with important environmental consequences. One of the typical features of Southeast Asia during this period is the clear dichotomy between the construction of temples, stupas, and the like in stone or brick, and the building of houses (in Island Southeast Asia even palaces) in timber or bamboo.

The temples of Srivijaya, Ayuthaya, early Angkor, Champa, and many parts of eastern Java were erected mainly in brick; those of later Angkor, central Java (Mataram), and parts of eastern Java were built in stone (sandstone, laterite, andesite). We must assume that the choice of building materials was determined largely by their presence nearby, although one wonders why Angkor shifted from brick to stone, as depletion of the resources needed for the production of brick does not appear plausible. Nevertheless, it is possible that in densely populated areas of the Khmer kingdom the firewood needed to produce bricks had become difficult to get hold of. Large-scale brick production, therefore, could have had a serious environmental impact even at an early stage of development. In the case of stone, the builders must have depended largely upon the natural occurrence in the neighborhood of easily worked types of stone, such as sandstone and volcanic stone. That the exploitation of such materials might have altered the landscape considerably is suggested by a local legend near the Angkor temples, according to which an entire hill had been quarried out of existence. One is tempted to ask whether the proximity of such a source might have determined the place where temples were built.

The materials used throughout Southeast Asia for the building of houses and regionally also for palaces during the period under consideration (and later) were, as we have seen, timber and bamboo, in addition to various vegetative materials for thatching and the construction of "walls," ceilings, partitions, and the like. There are various questions regarding this state of affairs, one of them being why people, and particularly the more wealthy ones, did not build in more durable materials, which would also have been more capable of withstanding the fires that were a regular occurrence in many of the larger settlements. In some areas, such as the Philippines and some parts of the Indonesian Archipelago, the frequency of earthquakes would appear to be a likely explanation, but one that cannot be used for the other regions. In other cases, such as some of the estuaries of the rivers of the Malay Peninsula, Sumatra, and Borneo, the raw materials were not as easily available, and the soils may have been too marshy for stone or brick buildings, leading to a situation in which the houses of places like Brunei and Palembang were built entirely on stilts or even on rafts that moved with the tides. Whatever the reasons, given the high frequency of fires just mentioned, the habit of building in timber and bamboo must have been a serious source of deforestation around areas where relatively large numbers of people were concentrated—in other words, towns and cities.

A related point about "cities" is that they seem to have been largely absent in most of Island Southeast Asia prior to 1500. This may be an artifact of the archaeological sources—mainly the fact that most buildings would have been constructed from perishable materials. However, there are also various reasons to believe that this purported absence may have been real. Arguments for this state of affairs to be found in the literature include tax policies in Java that limited the growth of urban concentrations, favoring instead a proliferation of smaller, nucleated settlements, in addition to the presence of a dense network of rural markets. Nor was there the need for people to live in dense agglomerations, felt in Mainland Southeast Asia because of limited transport possibilities. Another feature is that the various functions that cities in, for instance, Europe or China used to have are not necessarily to be found in one place in Island Southeast Asia. The former often combined the functions of religious center, palace, marketplace, harbor, and military establishment, while in Island Southeast Asia such functions might be spread out over several locations. Finally, rulers often left their residence when some "unlucky" event had taken place in their capital, starting anew somewhere else. That feature was not restricted to Island Southeast Asia; in Thailand a new city was built after an epidemic had broken out (Ayuthaya, 1351) and another one when a new dynasty came to power (Bangkok, 1782). All of these elements militate against very large concentrations of people. To these factors we can add the just mentioned possibility that timber may have become scarce near a town or city because of the constant rebuilding resulting from the many fires.

Thus a strong motor of economic growth, as were the cities in China and Europe, may have been absent in (Insular) Southeast Asia. As the example of the supposed high rate of timber consumption shows, this was not necessarily a good thing in environmental terms.

BIBLIOGRAPHICAL ESSAY

A recent general introduction to the early and classical states of Southeast Asia is Taylor, Keith W., 1999, "The Early Kingdoms." Pp. 137–182 in *The Cambridge History of Southeast Asia*. Vol. 1: *From Early Times to Circa 1500*. Edited by Nicholas Tarling. Cambridge: Cambridge University Press. More akin to the main focus of the present book is Hall, Kenneth R., "Economic History of Early Southeast Asia." Pp. 183–275 in the same volume. A recent textbook on the period after 800 in Mainland Southeast Asia is Lieberman, Victor, 2003, *Strange Parallels: Southeast Asia in Global Context, c. 800–1830*. Vol. 1: *Integration on the Mainland*. Cambridge: Cambridge University Press.

A more detailed treatment is to be found in a number of older textbooks, but the reader then runs the risk of encountering somewhat outdated opinions: Cady, John F., 1964, *Southeast Asia: Its Historical Development*. New York: McGraw-Hill; Hall, D. G. E., 1968, *A History of Southeast Asia*. London: Macmillan/St. Martin's.

There is a rather voluminous body of publications written on the character and origins of these states. A classical text in this respect is Coedès, G., 1968, *The Indianized States of Southeast Asia*. Honolulu: University of Hawaii Press. Among other "classic" (but less antique) contributions I would like to mention Wolters, O. W., 1967, *Early Indonesian Commerce: A Study of the Origins of Srivijaya*. Ithaca, NY: Cornell University Press; Bronson, Bennet, 1977, "Exchange at the Upstream and Downstream Ends: Notes toward a Functional Model of the Coastal State in Southeast Asia." Pp. 39–52 in *Economic Exchange and Social Interaction in Southeast Asia: Perspectives from Prehistory, History and Ethnography*. Edited by Karl L. Hutterer. Ann Arbor: University of Michigan, Center for South and Southeast Asian Studies; Wheatly, Paul, 1983, *Nagara and Commandery: Origins of the Southeast Asian Urban Tradition*. Chicago: University of Chicago Department of Geography; Hall, Kenneth R., 1985, *Maritime Trade and State Development in Early Southeast Asia*. Honolulu: University of Hawaii Press; Bentley, G. Carter, 1986, "Indigenous States of Southeast Asia," *Annual Review of Anthropology* 15, pp. 275–305; Kulke, Hermann, 1986, "The Early and the Imperial Kingdom in Southeast Asian History." Pp. 1–22 in *Southeast Asia in the 9th to*

14th Centuries. Edited by David G. Marr and A. C. Milner. Singapore: Institute of Southeast Asian Studies/Research School of Pacific Studies Australian National University; Hagesteijn, Renée, 1989, *Circles of Kings: Political Dynamics in Early Continental Southeast Asia*. Dordrecht: Foris; Kulke, Hermann, 1991, "Epigraphical References to the 'City' and the 'State' in Early Indonesia," *Indonesia* 52, pp. 3–22; Christie, Jan Wisseman, 1991, "States without Cities: Demographic Trends in Early Java," *Indonesia* 52, pp. 23–45.

Studies on the early state in Southeast Asia dating from the last decade include Caldwell, Ian, 1995, "Power, State and Society among the Pre-Islamic Bugis," *Bijdragen tot de Taal-, Land- en Volkenkunde* 151, no. 3, pp. 394–421; Christie, Jan Wisseman, 1995, "State Formation in Early Maritime Southeast Asia: A Consideration of the Theories and the Data," *Bijdragen tot de Taal-, Land- en Volkenkunde* 151, no. 2, pp. 235–288; Pollock, Sheldon, 1996, "The Sanskrit Cosmopolis, 300–1300: Transculturation, Vernacularization, and the Question of Ideology." Pp. 197–248 in *Ideology and Status of Sanskrit: Contributions to the History of the Sanskrit Language*. Edited by Jan E. M. Houben. Leiden: Brill; Day, Tony, 2002, *Fluid Iron: State Formation in Southeast Asia*. Honolulu: University of Hawai'i Press; Manguin, Pierre-Yves, 2002, "The Amorphous Nature of Coastal Polities in Island Southeast Asia: Restricted Centres, Extended Peripheries," *Moussons* 5, pp. 73–99.

Special mention should be made of Leach, E. R., 1964, *Political Systems of Highland Burma: A Study of Kachin Social Structure*. London: Bell, which, although not referring specifically to early states, does discuss mechanisms that could be read as a model for centrifugal and centripetal sociopolitical tendencies.

Studies specifically referring to agriculture, irrigation, and other economic factors during the period under consideration are Dunn, F. L., 1975, *Rain-Forest Collectors and Traders: A Study of Resource Utilization in Modern and Ancient Malaya*. Kuala Lumpur: Malaysian Branch of the Royal Asiatic Society; Winzeler, Robert L., 1976, "Ecology, Culture, Social Organization, and State Formation in Southeast Asia," *Current Anthropology* 17, no. 4, pp. 623–640; Ishii, Yoneo, 1978, "History and Rice-Growing." Pp. 15–39 in *Thailand: A Rice-Growing Society*. Edited by Ishii. Honolulu: University Press of Hawai'i; Sedov, Leonid A., 1978, "Angkor: Society and State." Pp. 111–130 in *The Early State*. Edited by Henri J. M. Claessen and Peter Skalník. The Hague: Mouton; Jacob, J. M., 1979, "Pre-Angkor Cambodia: Evidence from the Inscriptions in Khmer concerning the Common People and Their Environment." Pp. 406–426 in *Early South East Asia: Essays in Archaeology, History and Historical Geography*. Edited by R. B. Smith and W. Watson. New York: Oxford University Press, and Ng, Ronald C. Y., 1979, "The Geographical Habitat of Historical Settlement in Mainland South East Asia." Pp. 262–272 in the same volume; Spriggs, Matthew,

1982, "Taro Cropping Systems in the Southeast Asian-Pacific Region: Archaeo-logical Evidence," *Archaeology in Oceania* 17, no. 1, pp. 7–15; Anderson, James N., and Walter T. Vorster, 1983, "Diversity and Interdependence in the Trade Hinterlands of Melaka." Pp. 439–457 in *Melaka: The Transformation of a Malay Capital c. 1400–1980.* Vol. 1. Edited by Kernial Singh Sanhu and Paul Wheatley. Kuala Lumpur: Oxford University Press; Lansing, J. Stephen, 1983, *The Three Worlds of Bali.* New York: Praeger; Dove, Michael R., 1985, "The Agroecological Mythology of the Javanese and the Political Economy of Indonesia," *Indonesia* 39, pp. 1–36; Stargardt, Janice, 1986, "Hydraulic Works and Southeast Asian Polities." Pp. 23–48 in *Southeast Asia in the 9th to 14th Centuries.* Edited by David G. Marr and A. C. Milner. Singapore: ISEAS/RSPacS, ANU, and Jacques, Claude, 1986, "Sources on Economic Activities in Khmer and Cham Lands." Pp. 327–334 in the same edited volume; Reid, Anthony, 1988, *Southeast Asia in the Age of Commerce 1450–1680;* Vol. 1: *The Lands below the Winds.* New Haven: Yale University Press; Boomgaard, Peter, 1989, "The Javanese Rice Economy, 800–1800." Pp. 317–344 in *Economic and Demographic Development in Rice Producing Societies: Some Aspects of East Asian Economic History, 1500–1900.* Edited by Akira Hayami and Yoshihiro Tsubouchi. Tokyo: Keio University; Aung-Thwin, Michael, 1990, *Irrigation in the Heartland of Burma: Foundations of the Pre-Colonial Burmese State.* DeKalb: Northern Illinois University, Center for Southeast Asian Studies; Miksic, John, 1990, "Settlement Patterns and Sub-regions in Southeast Asian History," *Review of Indonesian and Malaysian Affairs* 24, pp. 86–144; Rigg, Jonathan, ed., 1992, *The Gift of Water: Water Management, Cosmology and the State in South East Asia.* London: School of Oriental and African Studies; Wicks, Robert S., 1992, *Money, Markets, and Trade in Early Southeast Asia: The Development of Indigenous Monetary Systems to AD 1400.* Ithaca, NY: Cornell University South East Asia Program; Boomgaard, Peter, 1993, "Economic Growth in Indonesia, 500–1990." Pp. 195–216 in *Explaining Economic Growth: Essays in Honour of Angus Maddison.* Edited by A. Szirmai, B. van Ark, and D. Pilat. Amsterdam: North-Holland.

Of the literature on economic-agricultural aspects of the period published during the last decade, use has been made of O'Connor, Richard A., 1995, "Agricultural Change and Ethnic Succession in Southeast Asian States: A Case for Regional Anthropology," *Journal of Asian Studies* 54, no. 4, pp. 968–996; Reid, Anthony, 1995, "Humans and Forests in Pre-colonial Southeast Asia," *Environment and History* 1, no. 1, pp. 93–110; Glover, Ian C., and Charles F. W. Higham, 1996, "New Evidence for Early Rice Cultivation in South, Southeast and East Asia." Pp. 413–441 in *The Origins and Spread of Agriculture and Pastoralism in Eurasia.* Edited by David R. Harris. London: UCL Press; Christie, Jan Wisseman, 1998, "Javanese Markets and the Asian Sea Trade Boom of the

Tenth to Thirteenth Centuries A.D.," *Journal of the Economic and Social History of the Orient* 41, no. 3, pp. 344–381; Condominas, Georges, ed., 1998, *Formes Extrêmes de Dépendance; Contributions à l'Étude de l'Esclavage en Asia du Sud-Est*. Paris: EHESS; Junker, Laura Lee, 1999, *Raiding, Trading, and Feasting: The Political Economy of Philippine Chiefdoms*. Honolulu: University of Hawai'i Press; Allen, S. Jane, 2000, "In Support of Trade: Coastal Site Location and Environmental Transformation in Early Historical-Period Malaysia and Thailand," *Indo-Pacific Prehistory Association Bulletin* 20, pp. 62–78; Pottier, Christophe, 2000, "Some Evidence of an Interrelationship between Hydraulic Features and Rice Field Patterns at Angkor during Ancient Times," *Journal of Sophia Asian Studies* 18, pp. 99–120, and Goodman, James, 2000, "Reinterpreting Angkor; The Water, Environment and Engineering Context," in the same journal and issue; Brookfield, Harold, 2001, "Intensification, and Alternative Approaches to Agricultural Change," *Asia Pacific Viewpoint* 42, nos. 2/3, pp. 181–192; Morrison, Kathleen D., and Laura L. Junker, eds., 2002, *Forager-Traders in South and Southeast Asia*. Cambridge: Cambridge University Press; Chang, Te-Tzu, 2003, "Origin, Domestication, and Diversification." Pp. 3–26 in *Rice; Origin, History, Technology, and Production*. Edited by C. Wayne Smith and Robert H. Dilday. Hoboken, NJ: John Wiley and Sons; Boomgaard, Peter, 2003, "In the Shadow of Rice: Roots and Tubers in Indonesian History, 1500–1950," *Agricultural History* 77, no. 4, pp. 582–610; Boomgaard, Peter, 2003, "Bridewealth and Birth Control: Low Fertility in the Indonesian Archipelago, 1500–1900," *Population and Development Review* 29, no. 2, pp. 197–214; Dewar, Robert E., 2003, "Rainfall Variability and Subsistence Systems in Southeast Asia and the Western Pacific," *Current Anthropology* 44, no. 3, pp. 369–388; Christie, Jan Wisseman, 2004, "Food-crop Production and Animal Husbandry in Java and Bali before 1500." Pp. 47–68 in *Smallholders and Stockbreeders: Agriculture and Livestock in Southeast Asian History*. Edited by Peter Boomgaard and David Henley. Leiden: KITLV, and Hill, R. D., 2004, "Towards a Model of the History of 'Traditional' Agriculture in Southeast Asia." Pp. 19–46 in the same volume.

PART THREE

CLIMATIC CONDITIONS

4

CLIMATIC TRENDS AND FLUCTUATIONS

Before embarking upon a description and analysis of what is usually called the Early-Modern Period, the moment has come to interrupt the chronological narrative.

Only some forty years ago no historian would have known what an ENSO (El Niño–Southern Oscillation) event was, or would have heard of the Little Ice Age, let alone dedicate an entire chapter to such topics. Over the last decades, however, these and related notions have become indispensable tools of the trade, not only for environmental and economic historians but also for those of us who are specialists in political history. However, scholars with a meteorological background and an interest in forecasting (and therefore in long-term trends and fluctuations) have been quite familiar with such notions since the 1920s and 1930s.

Among historians, interest in and knowledge of climatic trends and cycles were initially strongly linked to the debate on the so-called seventeenth-century crisis in Europe. The early phase of the debate, held in the 1950s and 1960s, was conducted largely in the pages of the then recently established academic journal *Past and Present*. In the discussion, in which renowned historians such as Eric Hobsbawm, Hugh Trevor-Roper, and Pierre Goubert participated, the weather was rarely mentioned as a possible causal factor, if it was mentioned at all. Jan de Vries, who in 1976 published a study on the seventeenth-century crisis in the European economy, dedicated half a page to climate change as a possible cause. However, when Geoffrey Parker in 1979 published his study on the European seventeenth-century crisis, the first section of the first chapter was entitled "The Little Ice Age," a phenomenon purported to have reached its nadir in the seventeenth century, as will be shown presently. Climate history had arrived, and it was almost immediately admitted to the hallowed halls of mainstream history.

If the debates on both the seventeenth-century crisis and the Little Ice Age had so far been focused on Europe (and mainly on Northwestern and Central Europe, at that), around 1990, Asian history came into view. In that year the academic journal *Modern Asian Studies* published a series of articles on the question of whether there had been a seventeenth-century crisis in Asia; climatic factors figured perhaps not prominently, but at least they played a modest role. It was not a propitious start for those who believed in a climate-related seventeenth-century crisis in Asia, along the lines of the European one. Among the authors in this special issue, only Anthony Read finds a "crisis" during most of the seventeenth century (until ca. 1680?) in Southeast Asia, strongly related to the weather anomalies of the trough of the Little Ice Age. John Richards did not think there was a general crisis in India, and according to William Atwell there was no "general" or "long-term" crisis in East Asia (China, Korea, Japan) either, but only a period of crisis years in the 1630s and 1640s. Atwell argued that there may have been a link between the bad years and a series of volcanic eruptions. Niels Steensgaard, who tried to sum up the findings of the other authors, had to conclude that there was no "general Eurasian crisis" in the seventeenth century.

That was certainly not the end of the discussion, but before we can continue with this debate, we should have a closer look at the theoretical notions we are dealing with—the climatological terminology that dominates the writings of the climate historians, and those who have taken their cue from them.

THE LITTLE ICE AGE (SECULAR CLIMATE CHANGE)

In Chapter 1, glacial and interglacial periods were mentioned. In common parlance a glacial period is called an ice age, and as we have seen, the last one peaked around 21,000 BP. Since then deglaciation occurred, with ice caps retreating to the poles and sea levels rising, while around 10,000 BP, at the start of the Holocene, the present interglacial had begun. For a long time now the world has known the present sea level without experiencing fluctuations on such a scale.

However, evidence has been mounting over the last decades that during the last millennium we may have experienced a period of one or more centuries of unusually cold winters and wet or cold summers. Meteorologists coined the term "Little Ice Age" for that period. It is argued by an increasing number of scholars that this was a period during which adverse weather conditions led to harvest failures, sometimes famines, epidemics, higher mortality, stagnating or even decreasing population numbers, and general economic depression. Ultimately such adverse conditions may have been instrumental in bringing about political events of moderate to colossal importance, varying from local peasant unrest to the French Revolution.

The notion of a Little Ice Age is fraught with difficulties. In the first place, there is considerable difference of opinion as to how to date this phenomenon. There are those who appear to confine the term to the seventeenth century—albeit often a "long seventeenth century" (1550–1700). Other scholars have opted for a longer period, including the eighteenth and part of the nineteenth century

(1550–1850). Others still emphasize the horrors of the fourteenth century and want to include the difficult years at the end of the nineteenth century as well, which would yield a Little Ice Age starting after 1300 and ending around 1900. In fact, the earliest signs of a colder climate in Northern Europe were already to be found shortly after 1200, which implies that a legitimate claim could be made for a Little Ice Age beginning around 1200 and lasting until 1900.

Be that as it may, most climate historians would probably agree that the Little Ice Age was a period with many stretches of adverse weather, causing human suffering (at least in most of Europe) on a much wider scale than during the Medieval Warm Period (also called the Little Climatic Optimum), between 900 and 1250; in the twentieth century, of course, global warming came to the fore. Interest in global warming, by the way, explains why so many meteorologists are interested in climate history. They want to know if such warming has ever occurred before, and whether this twentieth-century global warming is largely or even exclusively due to an (anthropogenic) increase in so-called greenhouse gases.

It would also appear that there is a gradual, almost imperceptible, shift in the way the climatic variation during the Little Ice Age is represented. Until perhaps even ten years ago, the usual image conjured up by the climate historians was that of a gradual cooling between 1300 and the 1690s, and a gradual warming between 1700 and 1850. Both the downswing and the upswing phases witnessed short-term reversals to warmer and cooler periods, respectively, but the "central tendency" appeared in both cases to be unmistakable. In the recent literature a more complicated picture is emerging as more data are being included in the quantitative data sets used. Now five cold periods appear to be distinguished—the middle of the thirteenth century, the middle of the fourteenth century, the middle of the fifteenth century (perhaps the nadir of the entire period), the late seventeenth century, and most of the nineteenth century (although the middle of the century was slightly better).

The reader should also be aware of the spatial limitations of the notion of a Little Ice Age. The overwhelming majority of the publications deal with the Northern Hemisphere, and with the Temperate Zone at that, while within that area a large majority refer to North, West, and Central Europe. This state of affairs is, of course, related to the type of sources available. As the Little Ice Age theory seeks to establish a relationship between adverse weather conditions and human misery, historians are looking for climate-related data, such as temperatures, rainfall, flooding, snow, and frost, that then could somehow be linked to data on prices, wages, harvests, births, deaths, and marriages.

Daily climate data is available in various Western European countries from somewhere in the nineteenth century onward, but occasionally it dates back to the late seventeenth century. However, if such material is not available, we sometimes have proxy data at our disposal, such as the dates of the first harvest in the vineyards of France, or the number of days per month that the canals froze over in the Dutch Republic or the lakes of Switzerland. The farther back in time we go, the more we have to depend on such sources as the diaries of certain individuals who, for years on end, kept records of rough, often non-numerical observations of temperature and precipitation.

The same countries also tend to have records, going back centuries, reflecting economic activity on a day-to-day or month-to-month basis. Thus, long-term developments in prices, wages, the movements of ships, and data on exports and imports may be expressed as graphs in order to facilitate comparison with climate data. However, more information is coming in from other parts of the world, and not all of it confirms the patterns mentioned for Northwestern and Central Europe.

Data on Africa around the equator, West Africa, and Mexico are a mixed bag, apart from the fact that the term "Little Ice Age" must, of course, be regarded as a misnomer as soon as we are dealing with tropical areas. Some of the findings from those regions suggest similarities with the European Little Ice Age pattern, but others do not.

Students of Southeast Asian environmental history should also be interested in the climate history of Southeast Asia's neighbors. As regards China they are in luck, in the first place because the data are extraordinarily rich, and probably more detailed than the data set we have for Europe; in the second place, Chinese climate history has recently received quite some attention. The earliest temperature series, compiled by Zhu Kezhen, were published in 1972, but as the publication was in Chinese it did not attract the attention of European climate historians. From around 1990, however, English language publications on climate change in China became increasingly available.

There is ample evidence of a period for which the term "Little Ice Age" would be applicable, as between 1470 and 1850 the average temperature in China dropped by 1 degree centigrade. Within that period we find, according to one study, cool years between 1470 and 1560, warmer years between 1560 and 1620, colder ones from 1620 to 1740, warmer years between 1740 and 1830, and cooler years from 1830 to 1850. This pattern does not fit the European Little Ice Age model very well, although China shares part of the cold seventeenth century. However, in a later study the warm and cold periods appear to have shifted somewhat, more in the direction of the European model. Temperatures declined from around 1500 to the 1650s, then rose to warmer levels in the eighteenth century, after which cooler conditions were obtained again during the first half of the nineteenth century. Particularly for the period 1600 to 1960, the Chinese temperatures fitted the European ones quite well, although perhaps in both Chinese series the nineteenth century was colder than it was in Europe; it also may have been wetter.

As regards the other large neighbor, India, we are less fortunate. To my knowledge no series of temperatures or rainfall over four or more centuries have been published, and perhaps the data for such a reconstruction prior to, say, 1850 are just not there. What we do have are references to droughts and famines, which may or may not have any connection with Europe's Little Ice Age, between the late sixteenth and late eighteenth centuries. There is some evidence that these years of crisis were linked to ENSO events, but a possible link to the Little Ice Age at present appears to be rather tenuous.

In a recent article on the Pacific Islands it is suggested that there have been momentous changes in the climate around 1300, which could be interpreted as a shift from a Little Climatic Optimum pattern to a Little Ice Age. There are indications of rapid cooling and a fall in sea level; this could have amounted to 75 cm between 1270 and 1325 and to 40 cm between 1455 and 1475. There are also indications of increased precipitation levels, which might come as a surprise to those who, intuitively, would expect drier circumstances during a cooler phase. However, such a link applies only on a global scale, while regionally changes in precipitation are strongly linked to changes in atmospheric circulation and not necessarily to fluctuations in temperature. In tropical areas rainfall patterns are strongly influenced by the monsoons and the ENSO system.

Moving farther south, and firmly into the Southern Hemisphere, we also encounter evidence for long-term climate variation in Australia, based on a 342-year proxy record of runoff to the Great Barrier Reef, using annual coral growth data. This record shows, among other signals, evidence for the period 1650 to 1770 of consistently higher discharge punctuated by more large floods from the Burdekin River than at any time since. By contrast, the period 1770 to 1880 was much drier. Here the phases are clearly out of sync with the European Little Ice Age pattern.

Data from a 1,100-year tree ring chronology from a Tasmanian huon pine suggest warm intervals between 940 and 1000, 1100 and 1190, 1475 and 1490, as well as after 1965; Little Ice Age cooling was not apparent.

Finally, radiocarbon dating of abandoned penguin rookeries on the coastal area of Antarctica suggests that there was a mild phase between 1250 and 1450, and again between 1670 and 1840—again, out of sync with the Northern Hemisphere.

This overview strongly suggests that the whole Little Ice Age discussion is still very much in flux, even as regards Europe. The chronology of the climate swings seems to be known roughly, but the notion of a seventeenth-century nadir may no longer be tenable. It is important to remember, however, that the overwhelming majority of the literature on the cooler period between 1200 and 1900 refers to the Northern Hemisphere, and more in particular to Northern, Western, and Central Europe, an observation that also applies to the phases we now distinguish within this period. Literature on Africa around the equator and on China shows some similarities to the European Little Ice Age patterns, but there are also numerous observations that do not fit. Australia and the Antarctic also show phases that are not synchronic with the European ones. That was, of course, to be expected, as those areas do not have the same "normal" patterns of temperatures, precipitation, and prevailing winds as does Europe. Therefore, we would be well advised not to expect Southeast Asian patterns that fit the European ones, either. The following quotations, taken from recent publications, illustrate this point:

> "There is no evidence for a world-wide, synchronous and prolonged cold interval to which we can ascribe the term 'Little Ice Age.'" (Bradley and Jones 1995, 659)
>
> "…regionally strong but nonglobal centennial-scale episodes, such as the medieval Warm Period or the Little Ice Age." (McIntosh, Tainter, and McIntosh 2000, 19)
>
> "Early suggestions of globally synchronous cooling and warming, such as the Little Ice Age and Medieval Warm Period, have given way to a better-documented view that late Holocene climate variability is expressed as multi-decadal temperature anomalies of about 0.5 to 1.5°C that tend to be region-specific." (Dunbar in McIntosh et al. 2000, 45)

Before moving on to causation, the question should be asked whether there is agreement on the question of to what extent variations in harvests or prices can be explained by fluctuations in rainfall and temperature. The first scholar who, to my knowledge, attempted a rigorous test of this proposition was the U.S. economic historian Jan de Vries (in 1977). He correlated proxy values for climate fluctuations in Holland (number of days per year that the canals froze over) between 1634 and 1839 with six variables representing socioeconomic variables such as burials, grain shipments, and grain prices. The results were disappointing, and he had to conclude that the year-to-year variations did not appear to explain the variations in mortality and prices.

Ten years later, however, climate historian Christian Pfister did find high and significant correlations for his Swiss data, covering the period 1550–1850. It seems likely that the impact of the Little Ice Age was much greater and more direct on a landlocked and rather subsistence-oriented country like Switzerland than on coastal Holland, the hub of Northwestern Europe's trade network, where cereals and other foodstuffs could be easily imported in large quantities in years of bad harvest. Here, again, the message seems to be that local circumstances determine to what extent the Little Ice Age is a meaningful analytical and explanatory tool.

THE CAUSES OF SECULAR CLIMATE CHANGE CYCLES

As the term "Little Ice Age" does not have much significance for tropical regions, and has mainly historiographic relevance, it would be better to use a hemisphere and latitude-neutral description such as secular or long-term climate change. In this section we look at the mechanisms behind these cycles of climate change, to the extent, of course, that they have been understood.

Usually, three types of factors are distinguished: internal factors (variability of the climate system, of which ENSO is a good example), external factors (solar irradiance, volcanic activity), and global warming. In the literature of the last decade, two mechanisms—apart from global warming—appear to have received the most attention: cycles of solar irradiance and variations in volcanic activity. A third factor, variations in ENSO events, is also occasionally mentioned. It would appear, however, that theory formation in this respect is still in an incipient and speculative phase, at least as regards the influence of ENSO events on global and secular climate cycles. In contrast, the short-term and local effects of ENSO events are well established. Finally, global warming should be dealt with as a fourth factor, but one that has been in operation only since the late eighteenth century.

SOLAR RADIATION

Even a few decades ago, someone who would have argued that there was a relationship between long-term economic cycles and sunspots would have been regarded as a crackpot by most, except for a handful of scholars. Tradition has it that in 1611 the connection was made for the first time between the appearance of spots on the sun—which as such had been discovered and described much earlier—and the blocking of solar radiation. Observations increased in frequency after the invention of the telescope in Holland in 1607 had made sunspot observation much easier. Connected to the incidence of sunspots are the so-called northern or polar lights, or aurora borealis, which have been observed for a long time not only in Europe but also in China. Nineteenth-century scholars systematically studied the periodicity of sunspots and auroras in the past, and by the early twentieth century the following long-term periodicity of very low sunspot numbers had been developed:

Wolf Minimum ca. 1280–1340
Spörer Minimum ca. 1450–1545 (elsewhere: 1400–1510 and 1415–1535)
Maunder Minimum ca. 1645–1715
Dalton Minimum ca. 1795–1825

Comparison of these periods of sunspot minima with long-term data series on temperatures revealed since the late 1970s—thanks to the publications of John A. Eddy, though there had been earlier publications linking sunspot numbers with the Little Ice Age—that the sunspot troughs roughly coincided with periods of low temperatures in Europe and North America. Generally speaking, periods with very few or no sunspots are also periods of low solar irradiance (a "quiescent" sun) and therefore low temperatures on earth, or at least on many places on earth.

However, the fit between sunspot numbers and temperature fluctuations was far from perfect, and nowadays a much more complicated model is used, based on solar cycle length, cycle decay rate, mean level of solar activity, solar rotation rate, and fraction of penumbral spots. There is no point in going into detail as regards all of these variables. They have in common that they are measures of variations in solar irradiance, and the point made here is mainly that variations in solar radiation explain much of the long-term cycles of climate change, of which the Little Ice Age is an example.

A few words are in order regarding the notion of a solar cycle. Series of sunspot numbers reveal a cycle of around eleven years, also called a Schwabe cycle. That same cycle length is to be found in long-term series of tree ring measurements, carried out in many places on earth. These measurements show patterns of wet and dry years (a narrow tree ring being formed in a dry year and a broad one in a wet year) and can also be used to measure changes in relative carbon-14 content (another measure of solar activity). The most impressive series is one based on very long lived bristlecone pines in California (dating back to 3431 BCE); closer to Southeast Asia, however, study has been made of Formosan cypresses, of which one dates back to 865 CE. The only series to my knowledge available for Southeast Asia is the one published by H. P. Berlage, Jr., in 1931, for which teak trees from Java had been used, dating back to 1514 CE.

In some series of prices and climatic data there seem to be cycles of approximately fifty years that do not appear to correspond with any known natural phenomenon, although economic historians might associate this periodicity with the industrial (post-1800) Kondratieff cycle.

Many more types of cycles purported to reflect solar activity have been found in large numbers of proxy data series, but there is no point in discussing them here.

Volcanic Eruptions

Much of the long-term trend in hemispheric temperature variations, therefore, appears to be explained by a model combining various measures for the variation of the sun's radiation. However, as those measures are not always in agreement, some of the variation does not seem to be explained satisfactorily. It is now thought that variations over time in the number, strength, and composition of volcanic eruptions, expressed in one index, may explain some of the short-term climatic change patterns. There are currently at least three indices that have been applied to climate history research—the Dust Veil Index, the Volcanic Explosivity Index, and the Ice Core Volcanic Index.

Volcanic eruptions are held to have "forced" the normal climate patterns on many occasions in the past, because they produce large amounts of sulfur dioxide (SO_2), which remains in a gaseous state and is eventually converted into sulfuric acid, condensing into sulfate aerosols that deflect the sun's radiation back into space. However, they do not block outgoing earth radiation and thus are capable of causing considerable cooling, the more so as these dust veils can stay in orbit for two to four years.

The role of volcanic eruptions in climate forcing depends partly on the strength of the explosion and on the composition of the ejecta (the quantity of sulfur dioxide), the height of the dust plume, but also on the geographical position of the volcano. Volcanoes located between 20 degrees northern or southern latitude and the equator spread their dust veils within a few weeks over their own latitude, and within a few months over the entire globe. Volcanoes in the higher latitudes influence only their own hemisphere, and after a year or more the dust veil is concentrated in the higher latitudes.

This implies that the volcanoes of Southeast Asia make themselves felt all over the world, while the region is not troubled by ejecta from volcanoes outside the tropics. It stands to reason, therefore, that whereas some of the variation to be found in the Little Ice Age climate fluctuations in Europe has no doubt been produced by Indonesian and Philippine volcanoes, the fluctuations in climate change in Southeast Asia will not show the influence of eruptions of Eurasian volcanoes of 20 degrees latitude and above. That is another reason why the classical Little Ice Age pattern of the Temperate Zone of the Eurasian landmass may be expected to be different from Southeast Asia's secular climate change.

Examples of influential Southeast Asian volcanic eruptions are not difficult to find. The best-known example is, of course, that of Krakatoa in 1883, but the eruption of Mount Tambora (Sumbawa, Lesser Sunda Islands, Indonesia) in 1815 was even more forceful.

El Niño–Southern Oscillation (ENSO)

Nowadays, many students of Southeast Asia will have heard of El Niño, particularly since the drought that accompanied the 1997–1998 El Niño and that coincided with the onset of the financial and economic crisis of those years. Not

Illustration of the Indonesian volcano on the island of Krakatoa in the 29 September 1883 issue of Harper's Weekly. *The volcano exploded on 26 August 1883, triggering tsunamis that killed 36,000 Indonesians, leveled the volcano, submerged much of Krakatoa, and created several new islands. (Library of Congress)*

long ago, however, only people with a professional interest in climate change, including perhaps a handful of historians, were aware of this phenomenon and its repercussions.

The term *El Niño* comes from Peruvian fishermen, who used the term (which means "the Christ child") for the warming of the sea around Christmas once every few years, which put an end to their fishing activities. The term "Southern Oscillation" was designed by scholars, who thus described the seesaw between a situation of low pressure in and around Indonesia and high pressure over the Eastern Pacific in some years, and the reverse situation in other years, of which the sudden warming of the sea near Peru was a symptom. Nowadays the two terms are abbreviated to ENSO.

The El Niño stricto sensu (the warming of the Eastern Pacific) is now called an ENSO event, or, to be precise, a warm ENSO event, while the reverse development, known as La Niña (the cooling of the Eastern Pacific), is called a cold ENSO event. However, as a rule, the term "ENSO event" without qualification refers to the warm event.

The effect of an ENSO event in large parts of Southeast Asia and Australia is that the rainy season is unusually dry (droughts), while a cold ENSO event is normally accompanied by increased precipitation in the same areas. A warm ENSO event is often followed by a cold one, but occasionally there are two warm ENSO events in a row. It will not come as a surprise that these double ENSO events usually have devastating effects in areas like Indonesia and New Guinea and even in areas far removed from the Pacific, as was the case with the 1877–1878 El Niños, probably the biggest (double) ENSO event since statistics for the calculation of a Southern Oscillation Index—that is, after 1867—have been available.

An Indonesian villager drags his goats as he walks past a dry paddy field hit by drought in West Java, Indonesia. Droughts have been a common effect of El Niño in Southeast Asia for centuries. (Corbis)

However, not all ENSO events produce droughts, and not all droughts are the products of ENSO events. It has been estimated that around 90 percent of the droughts in Indonesia are associated with ENSO events, while 80 percent of the ENSO events are linked with droughts. If we look at moderate and strong events only, the average interval would be close to six years. It is clear, therefore, that in various areas of Southeast Asia, "normal" weather patterns are disturbed with sad regularity.

The ENSO signal is felt throughout Southeast Asia, in India, China, the Pacific, and Australia, but not everywhere to the same degree. Moreover, Southeast Asia can be divided into a part in which rainfall variations show a significant positive relationship with Indian monsoon rainfall (that is, are in phase with IMR), and another part that is out of phase. The areas in phase with IMR are (apart from central India and northern China) northwestern Thailand, Borneo, the southwestern Philippines, Sulawesi, and New Guinea. Out of phase are the regions surrounding the China Sea: the northwestern Philippines, southern Vietnam, Cambodia, Malaysia, and Sumatra. It is, in view of all this, unlikely that Southeast Asia would have reacted uniformly to ENSO disturbances.

GLOBAL WARMING

The term "global warming" is usually reserved for the observed increase in average temperature in most areas of the world during the nineteenth and particularly the twentieth centuries, according to current scholarly opinion largely the result of the increased emission of so-called greenhouse gases (carbon dioxide, methane, and nitrous oxide) since the late eighteenth century. The increased production of those gases was, in turn, linked to the Industrial Revolution and the process of industrialization and increased urbanization that followed in its wake.

In a recent publication using an energy-balance model to compare the combined effects of solar and volcanic factors together with anthropogenic climate-forcing factors such as greenhouse gases, with a decadal average temperature reconstruction for the Northern Hemisphere for the period 1600 to 1980, it was clear that solar and volcanic effects could explain a large part of the variation prior to 1800. It was also clear, however, that those factors could not explain the rapid warming of the last two centuries. The same publication suggested that not more than 25 percent of the warming of the past century could be explained by solar activity.

However, taking into account the anthropogenic factors just mentioned certainly does not solve all of the riddles posed by the astonishing increase in temperatures over the last century or so, perhaps particularly in the region that we are dealing with. If we look at the annual trend in air temperature exhibited by the mean monthly values of temperature data for Jakarta and Semarang (both Java) over the period 1866–1993, we find an increase of 1.64 degrees centigrade. For Manila (Philippines) we find a similar warming trend (2.2 degrees over a period of a hundred years). If we deduct 0.5 for global warming, and perhaps another 0.5 for the so-called city effect (Jakarta and Manila are much more strongly urbanized than they were a century ago, and large cities generate extra heat), there is still between 0.5 and 1.0 degree centigrade to be explained as an effect unique to the region.

Does this mean that climate fluctuations in Southeast Asia in the past were also stronger than they were in higher latitudes? Or is this phenomenon restricted to the warm phases? As a rule, the climate is supposed to vary more at higher latitudes than in tropical regions, and not, as these data suggest, the other way around. A possible explanation could be that aerosols (air pollution), which locally have a cooling effect, remain in the areas of emission. As Europe emitted until recently more air pollution than did Southeast Asia, part of the difference in the rise of average temperatures might be explained by differences in the emission of aerosols. If that is the case, this phenomenon would not have occurred during earlier warm phases.

The average global rise of sea level, linked to the global warming trend, has been estimated to be between 10 and 20 cm over the past century, a process that places cities like Jakarta and Bangkok at risk in the near future. Was there a comparable drop between, let us say, 1250 and 1600?

Finally, the question could be asked whether global warming leads to stronger ENSO events. The answer to that question has to be rather cautious: although recent ENSO events were evidently quite severe (1982–1983, 1997–1998), and more severe than any ENSO event after 1879, the severity of the 1877–1878 drought has not been surpassed. We will have to wait and see whether this record will hold.

Perhaps this entire section can be summed up by referring once again to the just mentioned model in which solar and volcanic effects are represented, in addition to serial data showing global warming. The three variables taken together show a significant (0.78) correlation with a series of decadal temperatures covering the period 1610–1980. Apart from the global warming proxy series, which was needed to explain the rapid rise in temperatures of the last two centuries, the volcanic series produced a better fit, generally speaking, with the temperature data than did the variable representing solar radiation. In general, the solar series was better at explaining long-term trends, while the volcanic series did better as regards shorter-term fluctuations. Be that as it may, the solar responses are small relative to the volcanic responses. However, addition of the solar data improved the correlation between the volcanic and temperature series considerably for the period prior to 1800. The model did not include a variable representing ENSO fluctuations, and for the time being attempts to relate ENSO to the secular climate change cycles remain inconclusive.

CONCLUSION

Reviewing the literature on the Little Climate Optimum/Little Ice Age, or, in the terminology used here, the secular "cycles" of climate change, the main points appear to be that such fluctuations have been reported from a great many areas all over the world, that there are clearly similarities in the timing of these cycles, particularly as regards the Northern Hemisphere; data from tropical areas and from the Southern Hemisphere, however, although they also show secular changes, often differ in periodicity from Northern Hemisphere cycles. Moreover, as more information is being collected and combined into new series of temperature data (often decadal averages), the timing of the ups and downs in the Northern Hemisphere climate cycles appears to be shifting away from one clear trend (downward from 1200 or 1300 to the seventeenth century, and then upward until the nineteenth century) to a graph with various peaks and troughs, among which the so-called nadir of the seventeenth century no longer holds pride of place.

It is, therefore, theoretically speaking unlikely that we will find a "classical" Little Ice Age pattern for Southeast Asia—that is, one in phase with the "traditional" European pattern with one trough in the seventeenth century.

Moreover, given the strength of the ENSO signal in Southeast Asia, it is theoretically possible, and even likely, that climate patterns in that part of the world were and are dominated by ENSO cycles, particularly in Indonesia, and that therefore the "Little Ice Age signal," in whatever shape it appeared, will have an ENSO signal superimposed.

Finally, it should be pointed out that it is unlikely that we will find one pattern of climatic change and fluctuations for Southeast Asia as a whole, in view of the fact that some areas of Southeast Asia are in phase with the Indian monsoon rainfall pattern, while others are out of phase.

Concluding this chapter, it could be suggested that climate patterns (particularly ENSO-related signals) and volcanic eruptions in Southeast Asia, especially those from Indonesia and the Philippines, appear to have (and to have had) more influence on global climate fluctuations than the other way around.

In the following chapters the chronological narrative is resumed, and we will look at the empirical evidence, such as it is, for secular climate change in Southeast Asia. It will be shown that some authors—Anthony Reid and Victor Lieberman in particular—have given the secular climate change fluctuations an important place in their description and analysis of the political and economic ups and downs of the region. Both authors also observe similarities between Europe and Southeast Asia. Reid argues, as we have seen earlier, that the seventeenth-century crisis we know so well from Europe is also to be found in Southeast Asia, and that it represents the nadir of the Little Ice Age. Lieberman, who avoids terms like "Little Ice Age" and "seventeenth-century crisis," argues that the "warm" and "wet" periods observed in Europe—mainly 950–1300, and less dramatically 1470–1590 and 1710/1720–1805—coincide entirely or partly with "wet" periods in Mainland Southeast Asia, periods that benefited agriculture and population growth. Conversely, climate in the fourteenth and middle to late seventeenth centuries was less favorable.

It is also argued that evidence for such secular climate fluctuations as good as that available for Europe or China is lacking for Southeast Asia. Evidence for Java, until now the only area in Southeast Asia for which more or less annual data on droughts, floods, harvest failures, and the like are available for the period 1600–1900, does not suggest that the seventeenth-century climate fluctuations and their effects on harvests and mortality differed much from comparable events in the eighteenth century. Nor is there any evidence that tree-ring measurements, ENSO events, or volcanic eruptions differed much in quantitative or qualitative respect between the two centuries (but there are, as is shown, difficulties with the registration of ENSO events). At least for Java, therefore, a seventeenth-century nadir of a European type Little Ice Age is not supported by the sources.

For all Southeast Asian areas outside Java it is too early to confirm or reject any of the suggestions presented by Lieberman, as we simply do not have the data required for either confirmation or rejection. Given the nature of the societies we are dealing with during the period under consideration, it would be wise not to be too optimistic about finding such data buried in the archives or libraries. Perhaps our best bet is that one day such data will be unearthed for Luzon in the Philippines, given the nature of the Spanish bureaucracy, both governmental and ecclesiastical.

BIBLIOGRAPHICAL ESSAY

The classic general text on climate history is Lamb, H. H., 1995, *Climate, History and the Modern World*. London: Routledge. Hubert Lamb, a scientist, started publishing on climate and history in 1968. Just one year earlier the historian Emmanuel Le Roy Ladurie published his *Histoire du climat depuis l'an mil* (Paris: Flammarion, 1967), translated as *Times of Feast, Times of Famine: A History of Climate since the Year 1000* (Garden City, NY: Doubleday, 1971). These two books can be regarded as representing the dawn of climate history as a discipline.

Many of the articles on the seventeenth-century crisis that appeared in *Past and Present* were reprinted in Aston, Trevor, ed., 1965, *Crisis in Europe 1560–1660*. London: Routledge and Kegan Paul; also mentioned in the text are De Vries, Jan, 1976, *The Economy of Europe in an Age of Crisis, 1600–1750*. Cambridge: Cambridge University Press; Parker, Geoffrey, 1979, *Europe in Crisis 1598–1648*. N.p.: Fontana. The articles of John F. Richards, Anthony Reid, William S. Atwell, and Niels Steensgaard on a possible Asian crisis in the seventeenth century are to be found in *Modern Asian Studies* 24, no. 4 (1990), pp. 625–698, and are reprinted in Parker, Geoffrey, and Lesley M. Smith, eds., *The General Crisis of the Seventeenth Century*. London: Routledge, [the first edition, dated 1978, does of course not contain these articles]. On climatic fluctuations in Mainland Southeast Asia, see Lieberman, Victor, 2003, *Strange Parallels: Southeast Asia in Global Context, c. 800–1830*. Vol. 1: *Integration on the Mainland*. Cambridge: Cambridge University Press.

There are two monographs on the Little Ice Age: Grove, Jean M., 1988, *The Little Ice Age*. London: Methuen; and Fagan, Brian, 2000, *The Little Ice Age: How Climate Made History 1300–1850*. New York: Basic. Important edited volumes on this topic are Bradley, Raymond S., and Philip D. Jones, eds., 1995, *Climate since A.D. 1500*. London: Routledge; McIntosh, Roderick J., et al., eds., 2000, *The Way the Wind Blows: Climate, History, and Human Action*. New York: Columbia University Press (particularly the introduction and an article by Robert D. Dunbar). Lamb's *Climate* (1995) deals extensively with the topic, and also with detailed information on areas outside Europe. A somewhat shorter but more recent summary is Richards, John F., 2003, *The Unending Frontier: An Environmental History of the Early Modern World*. Berkeley: University of California Press.

Recent detailed discussion of (graphic representation of) compound times series of temperatures over the last 600 to 1000 years may be found in Mann, Michael E., Raymond S. Bradley, and Malcolm K. Hughes, 1998, "Global-scale Temperature Patterns and Climate Forcing over the Past Six Centuries," *Nature* 392, pp. 779–787; Crowley, Thomas J., 2000, "Causes of Climate Change over the Past 1000 Years," *Science* 289, pp. 270–277.

Richards's *Unending Frontier* (2003) provides detailed information on the Little Ice Age signal in Africa and China, including references for further reading. On China, see also Marks, Robert B., 1998, *Tigers, Rice, Silk, and Silt: Environment and Economy in Late Imperial South China*. Cambridge: Cambridge University Press (particularly ch. 6). For Mexico, see O'Hara, Sarah L., and Sarah E. Metcalfe, 1995, "Reconstructing the Climate of Mexico from Historical Records," *Holocene* 5, no. 4, pp. 485–490. For India, see Grove, Richard H., and John Chappell, 2000, "El Niño Chronology and the History of Global Crises during the Little Ice Age." Pp. 4–34 in *El Niño: History and Crises*. Edited by Grove and Chappell. Cambridge: White Horse; Boomgaard, Peter, 2001, "Crisis Mortality in Seventeenth-Century Indonesia." Pp. 191–220 in *Asian Population History*. Edited by Ts'ui-jung Liu et al. Oxford: Oxford University Press (particularly pp. 196–197). On Australia, see Gagan, Michael K., and John Chappell, 2000, "Massive Corals: Grand Archives of ENSO." Pp. 35–50 in *El Niño: History and Crises*. Edited by Richard H. Grove and John Chappell. Cambridge: White Horse. On the Pacific, see Nunn, Patrick D., and James M. R. Britton, 2001, "Human-Environment Relationships in the Pacific Islands around A.D. 1300," *Environment and History* 7, no. 1, pp. 3–22.

For correlations between time series of climate data and socioeconomic variables in Europe, see De Vries, Jan, 1977, "Histoire du climat et économie: des faits nouveaux, une interprétation différente," *Annales, Économies, Sociétés, Civilisations* 32, pp. 198–226; Pfister, Christian, 1988, "Fluctuations climatiques et prix céréaliers en Europe du XVIe au XXe siècle," *Annales, Économies, Sociétés, Civilisations* 43, pp. 25–53.

A good and fairly recent study on climate change and the sun is Hoyt, Douglas V., and Kenneth H. Schatten, 1997, *The Role of the Sun in Climate Change*. New York: Oxford University Press. Eddy's seminal article, mentioned in the text, is Eddy, John A., 1977, "Climate and the Changing Sun," *Climatic Change* 1, pp. 173–190.

Berlage's original article (in Dutch) is Berlage, Jr., H. P., 1931, "Over het verband tusschen de dikte der jaarringen van djatiboomen (Tectona grandis L.f.) en den regenval op Java," *Tectona* 24, pp. 939–953. These data have been reanalyzed and expanded in Boer, H. J. de, 1951, "Treering Measurements and Weather Fluctuations in Java from A.D. 1514," *Proceedings Koninklijke Nederlandse Akademie van Wetenschappen* B 54, pp. 194–209; Murphy, J. O., and P. H. Whetton, 1989, "A Re-analysis of a Tree Ring Chronology from Java," *Proceedings Koninklijke Nederlandse Akademie van Wetenschappen* B 92, no. 3, pp. 241–257; Palmer, J. G., and J. O. Murphy, 1993, "An Extended Tree Ring Chronology (Teak) from Java," *Proceedings Koninklijke Nederlandse Akademie van Wetenschappen* 96, no. 1, pp. 27–41; D'Arrigo, Rosanne D., Gordon C. Jacoby, and Paul J. Krusic, 1994, "Progress in Dendroclimatic Studies in Indonesia," *TAO: Terrestrial, Atmospheric and Oceanic Sciences* 5, no. 3, pp. 349–363. On the Taiwan tree ring measurements, see Mori, Yukihiro, 1981, "Evidence of an 11-year Periodicity in Tree-Ring Series from Formosa related to the Sunspot Cycle," *Journal of Climatology* 1, pp. 345–353.

An informative and recent article on the influence of volcanic eruptions is Atwell, William S., 2001, "Volcanism and Short-Term Climatic Change in East Asian and World History, c. 1200–1699," *Journal of World History* 12, no. 1, pp. 29–98.

On the ENSO phenomenon, see Diaz, Henry F., and Vera Markgraf, eds., 1992, *El Niño: Historical and Paleoclimatic Aspects of the Southern Oscillation*. Cambridge: Cambridge University Press [see esp. the contributions by Roger Y. Anderson, David B. Enfield, and the longtime ENSO propagandist William H. Quinn.]; Fagan, Brian, 1999, *Floods, Famines, and Emperors: El Niño and the Fate of Civilizations*. New York: Basic; Grove, Richard H., and John Chappell, eds., 2000, *El Niño: History and Crises*. Cambridge: White Horse. For a study on the worldwide ramifications of one of the biggest ENSO events of the last 150 years, the double 1877–1878 ENSO, see Davis, Mike, 2001, *Late Victorian Holocausts: El Niño Famines and the Making of the Third World*. London: Verso.

More details are to be found in Quinn, William H., et al., 1978, "Historical Trends and Statistics of the Southern Oscillation, El Niño, and Indonesian Droughts," *Fishery Bulletin* 76, no. 3, pp. 663–678; Quinn, W. H., and V. T. Neal, 1995, "The Historical Record of El Niño Events." Pp. 623–648 in *Climate since A.D. 1500*. Edited by Raymond S. Bradley and Philip D. Jones. London: Routledge; Harger, J. R. E., 1995, "Air-Temperature Variations and ENSO Effects in Indonesia, the Philippines and El Salvador: ENSO Patterns and Changes from 1866–1993," *Atmospheric Environment* 29, no. 16, pp. 1919–1942; Kripalani, R. H., and Ashwini Kulkarni, 1997, "Rainfall Variability over South-East Asia— Connections with Indian Monsoon and ENSO Extremes: New Perspectives," *International Journal of Climatology* 17, pp. 1155–1168; Kane, R. P., 1999, "El Niño Timings and Rainfall Extremes in India, Southeast Asia and China," *International Journal of Climatology* 19, pp. 653–672.

On global warming, see Free, Melissa, and Alan Robock, 1999, "Global Warming in the Context of the Little Ice Age," *Journal of Geophysical Research* 104, D16, pp. 19,057–19,070. For a recent popular overview on global warming, see *National Geographic* 206 (2004), no. 3, pp. 2–75.

PART FOUR

THE EARLY-MODERN PERIOD

5

GENERAL OVERVIEW AND
POPULATION GROWTH

The term *Early-Modern* has been coined for Europe between the end of the Middle Ages and the time of the Industrial and French revolutions. Although it is unusual to use the term *Middle Ages* for Southeast Asia, there are various features of European history, customarily referred to as marking the end of the Medieval Period, that could be applied to Southeast Asian history as well. The following factors come to mind.

From the fifteenth century onward, many European states became more centralized, the bureaucracy expanded and turned more professional, and at the same time the power of the ruler increased vis-à-vis the nobility, a phenomenon indicated by the term *absolutist state*. It is also the period in which the Roman Catholic Church started to lose its religious monopoly and its very large influence on matters that in our view are the preserve of the state. It is at the same time the period of the European voyages of discovery to America and Asia, which boosted Europe's trade and that would increase in importance (both in absolute and relative terms) in other ways as well. Here terms like *merchant capitalism* and *mercantilism* come to mind.

As will be shown presently, many of these features can be applied to Southeast Asia as well, with one major qualification—whereas the Europeans were the "discoverers," subjects of the voyages of discovery, the Southeast Asians were the objects, the "discoverees."

At the tail end of the period no French Revolution–type event occurred in Southeast Asia (although it could be argued that it did in Java), while the Industrial Revolution was imported from outside and did not affect the lives of the people in any great way until much later. The first steam engines (in sugar factories) appeared in the 1820s, but that was a rather isolated phenomenon. It makes more sense to take the textbook starting point of the period of modern imperialism, conventionally pegged at 1870, as the end of the Early-Modern Period in Southeast Asia. This symbolizes the opening of the Suez Canal (1869), which boosted the growth of trade, the victory of steam over sail, and the spurt in colonial conquests, which brought most of the region under the sway of the

Western powers. In this book, therefore, the Early-Modern Period in Southeast Asia is defined as starting in the early fifteenth and lasting until the late nineteenth century.

A GENERAL OVERVIEW

The field of Southeast Asian history of the Early-Modern Period has undergone some rather drastic changes over the last fifteen years or so. We owe much of this to two scholars who have recently written their textbooks—Anthony Reid and Victor Lieberman. Of course, they based their findings on recent literature produced by other researchers in addition to their own, but it is nevertheless no exaggeration to say that they have reshaped Southeast Asian history.

Whereas earlier historians had depicted Early-Modern Southeast Asia—and often Asia as a whole—as stagnant, passive, and inward looking, Reid and Lieberman have emphasized Southeast Asian "agency," its internal dynamism, structural changes, and willingness, sometimes even eagerness, to participate in foreign trade and to learn from foreign examples. In this respect, the recent historiography of Southeast Asia is clearly influenced by, or at least "in phase" with, the so-called great divergence debate about similarities in development between Europe (particularly Northwestern Europe) on the one hand and East Asia (China, Japan) on the other. In this debate, historians of Japan and China—no doubt inspired by the incredible economic boom in Japan after World War II and the one in China starting in the 1990s—have argued that, around 1750 or 1800, the most developed areas in Japan and China were more or less on the same level in terms of production and consumption per capita as the most advanced areas of Europe, particularly England. Therefore—so the argument goes—the enormous gap between China and Europe around 1900 in terms of the standard of living and GDP per capita must have originated in the nineteenth century, and not earlier.

Something similar is now being postulated by Reid and Lieberman, although there are some differences between them. Reid argues that the levels between Europe and Southeast Asia around 1500 or 1600 were roughly similar, but that the latter area started to "underdevelop" in the seventeenth century; Lieberman emphasizes almost continuous growth in Southeast Asia, in political, economic, and cultural respects, between 1400 and 1850. Again, it is not difficult to see the almost uninterrupted economic growth in Southeast Asia between, say, 1970 and 1997 (the year of the financial crisis) as a factor contributing to these revisionist tendencies.

Both views are doubtlessly important corrections of the older image of Southeast Asia as a region of primitive and stagnant "hermit kingdoms," but the problem is that most evidence upon which these recent views are based is rather impressionistic. The statistical data needed for calculations of GDP per capita and similar measuring rods of income and production are simply not available for most areas prior to 1900 or even later. However, some scholars are now undertaking the reconstruction of pre-1900 "national" income figures for the more promising areas, and it is to be hoped that we will have, in the not too remote future, fairly reliable cross section data and time series at our disposal. Having said that, it must be admitted that some parts of the following sections are necessarily based on rather shaky foundations. Here I will present the ideas of the two scholars in broad outline.

Anthony Reid argued in his *Age of Commerce* that Southeast Asia's increased participation in international trade did not start, as is usually assumed, around 1500 with the arrival of the Portuguese, but about a century earlier, around 1400. This had been set in motion by developments in China, and the arrival of the Portuguese was the effect, not the cause, of growing maritime trade. Revenue from international trade led to the emergence of stronger and more centralized Southeast Asian kingdoms. Imported firearms contributed to the increased power of monarchs against their subjects, and thus Absolutist states could develop here as they did in Europe, also partly the result of a "military revolution."

From around the middle of the seventeenth century these upward trends were interrupted. Less silver arrived from America and Japan, the trade with Japan and China stopped, European traders shifted their interest to non–Southeast Asian products, and the prices of some Southeast Asian export products, such as pepper, were dropping. These factors, in turn, influenced political centralization processes negatively; there were many agricultural setbacks and local uprisings. In addition to all of that, the VOC (Verenigde Oostindische Compagnie, or Dutch East India Company) siphoned off much of the wealth that originally had been earned by indigenous rulers and merchants. Finally, Reid argues that the area was hit by the cooling process of the Little Ice Age (see previous chapter), which made itself felt through droughts. This led, so his argument goes, to large numbers of harvest failures and famines.

As a result of all these factors the positive developments of the fifteenth and sixteenth centuries were stopped; in many cases stagnation or even decline set in, spelling the end of once powerful states and causing a reversal to smaller, weaker and less centralized polities, with less prosperous populations. In fact, Reid argued, Southeast Asia would never recover from this setback, and the paths taken by Europe and Southeast Asia, which had run more or less parallel during the fifteenth and sixteenth centuries, started to diverge from the seventeenth century.

In a later publication, Reid had to admit that he had been too negative about Southeast Asia after the seventeenth century. He now postulated a "Chinese century" in Southeast Asia during the period 1740–1840. It was "Chinese" because the Chinese emperors finally rescinded their trade bans with Southeast Asia, thus stimulating not only Chinese maritime trade between China and the region but also the emigration of large numbers of Chinese. According to one estimate there were 1 million Chinese in the region around 1830: about 3 percent of the total population. That meant an enormous boost for the regional economy, leading to fairly high rates of economic diversification and growth.

Victor Lieberman's vision of Southeast Asia's early-modern history shares certain similarities with Reid's view, but he is also critical of a number of features of the latter. His main criticism is that Reid's view works well for Maritime Southeast Asia but not for the mainland. Particularly the seventeenth-century watershed would fit the mainland badly.

According to Lieberman, the mainland polities went through four more or less synchronous cycles of political consolidation between ca. 800 and 1830. The first—what he calls "charter" phase—ended with a general political and social crisis from the late thirteenth to the late fourteenth and early fifteenth centuries. A second phase, during which considerable political integration took place, started in the middle to late fifteenth century but came to an end between ca. 1540 and 1610. Reform in the early to mid-1600s inaugurated a third phase of consolidation, ending with a short late-eighteenth-century collapse (1752–1786). The fourth and final phase lasted until the European powers took over (except in Thailand). Between 1340 and 1820, the twenty-three independent Mainland Southeast Asian kingdoms that were present at the earliest date had been consolidated into three kingdoms—Burma, Siam, and Vietnam—at the end of the period. During each phase demographic and economic growth (of some description) and political and cultural integration and consolidation had taken place.

These synchronous phases may have been driven, in Lieberman's view, by climate fluctuations, in the sense that the agricultural and maritime cycles that characterized the phases of territorial consolidation may have depended on a third factor—climate change. I hasten to add that Lieberman stresses the hypothetical character of what he has to say about the role of climate change. I should also emphasize that he distinguishes many more factors that shaped these phases. In his own words: "The combination of accelerated political integration, firearms-based warfare, broader literacy, religious textuality, vernacular literatures, wider money use, and more complex international linkages (both cultural and material) marks the years between c. 1450 and 1800/1850 as a more or less coherent period...."

Nevertheless, climate change plays an important role in the views of Southeast Asian early-modern history of both Reid and Lieberman, though not the same role. In the eyes of Reid the "seventeenth-century crisis" was a point of no return, while Lieberman does not think that there was a seventeenth-century crisis—at least not on the mainland; in his view, the period 1600–1752 was a consolidation.

*An illustration of the flag of the Dutch East India Company. The Dutch had
headquarters in Batavia (now Jakarta) in Java. (Library of Congress)*

Finally we should turn to what Lieberman has to say about Maritime
Southeast Asia. Clearly this region had to contend with other forces than the
mainland: the European maritime expansion. Malacca (on the Malay Peninsula)
was conquered by the Portuguese in 1511; the Spanish conquest of what we now
call the Philippines started in 1565; and the Dutch established the headquarters
of the VOC in 1619 at a place they called Batavia (now Jakarta), in Java. The
Portuguese would be driven out by the Dutch after more than a century, but the

Spaniards and the Dutch were there to stay (at least during the period we are dealing with), a development not paralleled by anything on the mainland.

And yet, according to Lieberman, some parallels may be pointed out nevertheless. Prior to its conquest, Malacca had been developing as one of the important players during the consolidation phase that on the mainland lasted from 1350 to 1570. During the following mainland expansion and consolidation phase, 1600–1752, a number of coastal ports of trade showed promising developments until the Dutch ended that. Finally, during the final mainland consolidation phase (1760–1830/1840), the Dutch and the Spaniards consolidated their hold on Java and Luzon-Visayas, respectively, where they also introduced new methods of agrarian extraction, based on both old and new crops. During the same period a number of indigenous polities also went through a growth phase, stimulated by outside (America, China) demand for various commodities.

With these images in mind, as a broad outline, the following pages, dealing with a number of developments in more detail, should be read.

POPULATION GROWTH

Demographic development is one of the baselines for environmental historians. There is a strong link in all epochs between the growth of the numbers of people on the one hand, and land reclamation and resource use—and therefore environmental change—on the other: hence the importance of an assessment of population growth rates in Southeast Asia during the period under consideration. However, population data for Southeast Asia prior to 1850 are hard to come by and, if available, not very reliable; also, the farther back in time we go, the more difficult it gets to find meaningful figures. Lieberman has presented very rough estimates for Mainland Southeast Asia for the years 1400 and 1820, as shown in Table 5.1.

Table 5.1 Rough population estimates, in millions, for the main polities of Mainland Southeast Asia, 1400–1820, and average annual growth rates.

Area	Population (000,000)		Average annual growth rate (%)
	1400	1820	
Vietnam	2	7.0	0.30
Burma	2	4.5	0.19
Siam	2	4.0	0.17
Mainland SEA	6	15.5	0.23

Source: Lieberman, 2003, Strange Parallels: Southeast Asia in Global Context, c. 800–1830. Volume 1: Integration on the Mainland. Cambridge, Cambridge University Press.

There are no estimates for Maritime Southeast Asia around 1400. However, for 1600, Reid has presented figures, based on heroic assumptions, for all Southeast Asian areas, including those of Maritime Southeast Asia. For reasons to be explained presently, I believe that Reid's estimates err on the low side, and the figures given here are slightly higher. In Table 5.2 these figures are presented, together with some estimates for Europe, India, and China.

Table 5.2 Rough population estimates, in millions, for various Eurasian regions, 1600–1800, and average annual growth rates.

Area	Population (000,000)		Average annual growth rate (%)
	1600	1800	
China	180	320	0.30
Southeast Asia	25	35	0.17
India	120	160	0.14
Europe	100	180	0.30

Source: Reid, Anthony, 1988, Southeast Asia in the Age of Commerce 1450–1680. *New Haven: Yale University Press; Livi-Bacci, Masimo, 1992,* A Concise History of World Population. *Cambridge: Blackwell.*

Note: Reid's figures for Java are much too low and have been corrected for this table. Instead of 5 million in 1800, my own research suggests that it was closer to 7.5 million; the figure for 1600 has been adjusted accordingly.

It cannot be emphasized enough that all of these estimates are ballpark figures. Nevertheless, they are sufficiently robust for a comparison between regions. This comparison shows us clearly that the total number of people living in Southeast Asia was small compared with the total populations of China, India, and Europe.

That is true both in absolute terms and in relative ones; whereas Southeast Asia's population density around 1600 was approximately 6 persons per square kilometer (today it is 115), that figure was about 45 in India and China around the same time. However, there were also huge differences in population densities within Southeast Asia. High population densities—for the period under consideration—obtained in parts of Indonesia, such as Java (35) and Bali (80), and what today would be northern and central Vietnam (20). Very low figures were found in Borneo (1) and Mindanao-Sulu (Philippines) (1.5). Given these intra–Southeast Asian differences, the question should be asked whether it can be assumed that the entire area was characterized by one "demographic regime," as one expects high- and low-density areas to have shown different rates of increase.

According to tables 5.1 and 5.2, it would appear that Southeast Asia's population was growing faster (0.17–0.23 percent) than that of India (0.14 percent), but less rapidly than China's population (0.3 percent). However, we must again question the wisdom of postulating one growth rate for Southeast Asia as a whole, because the figures also suggest that population growth in Burma and Siam—the "Indic" part of Mainland Southeast Asia—was close to that of India, while the growth rate of Vietnam—the "Sinic" part of the area—was closer to that of China.

Interestingly enough, it has been argued that India has two demographic patterns, separated by a line running southwest to northeast. In the twentieth century, the demographic pattern to the south and east of this line was akin to patterns found in Southeast Asia, and it has been suggested that this demographic contrast is of long standing.

Unfortunately, we have very little reliable data on growth rates in Island Southeast Asia prior to 1800. The growth rate in the Philippines hinges on the choice one makes among the various estimates for the population at the time of the arrival of the Spaniards, in the 1560s. If we may trust a fairly recent estimate of between 1 and 1.25 million, we would arrive at a growth rate for the period 1565 to 1800 of around 0.14 percent, which is the lowest of all the Southeast Asian figures so far.

Data on Java, the area with the largest population of Maritime Southeast Asia, are not available prior to 1800, unless one is willing to convert the so-called *cacah* (pronounced CHACHAH) figures to population data, which I, for one, am not. The cacah figures are (usually) round numbers, indicating units of taxation, corvee labor, and potential soldiers in times of war, and the literature mentions any factor between 4 and 10 (and locally even more) as possible multipliers.

What we do know is that between 1601 and 1640 death rates in Java were unusually high owing to frequent wars, harvest failures, famines, and epidemics, which was also the case between 1657 and 1682. In fact, the entire period from 1676 to 1755, which has been called with only slight exaggeration the Javanese eighty-year war, was probably a period of very high mortality. Given the small margin of fertility over mortality in normal years, these periods must have shown stagnating population numbers at best. There are indications that, as a result of the absence of war, population growth between 1755 and 1800 was fairly high, although most modern authors fail to mention that West Java was hit by three or four years of high mortality owing to epidemics and harvest failures between 1756 and 1760, while East Java was the scene of a number of VOC military campaigns between 1767 and 1778. Be that as it may, it is unlikely that there was on balance much population growth between 1600 and 1800, and the 0.2 percent on average per year postulated by Reid may be too high.

MORTALITY

Although higher than the (hypothetical) growth rate we found for the millennium prior to the fifteenth century (0.5–1.0 percent per year on average; compare Chapter 3), the calculated population growth figures of the Southeast Asian areas during (part of) the Early-Modern Period have in common that they are low in comparison with twentieth-century growth rates, and in comparison with early-modern rates in China and Europe, but not in comparison with the rate obtained in India. The explanation for the former difference is rather straightforward—better medicine and higher income per capita now than in 1400 or 1600. It is perhaps less easily explained why during the Early-Modern Period growth rates in Southeast Asia (and India) were lower than those in Europe and China.

Population growth is by definition fertility minus mortality plus the migration balance. Assuming that the influence of migration was slight (possibly even positive, which makes it even more difficult to explain the lower growth rate), we should be looking for factors leading to higher mortality in Southeast Asia than in Europe and China, or to lower fertility.

The most obvious explanation, and one to be found often in the literature, is that tropical and subtropical areas are less healthful than regions in the temperate zone, at least partly because of the much higher parasite load. All of Southeast Asia and a large part of India are located in the intertropical zone, whereas Europe and China are not. Therefore, under similar circumstances, mortality would be higher in Southeast Asia than it would be in China and Europe purely for that reason alone.

This explanation has been challenged by Reid, who argued that Southeast Asians might have been at least as healthy as the Europeans who came to the region in the sixteenth and seventeenth centuries. He based this on the observation that various Europeans reported home about the tall stature and good health of assorted local peoples. However, that is a dubious argument. The quoted observations referred to adults, and it is precisely the absence of weak and unhealthy grownups that suggests high rates of particularly infant and child mortality.

Generally speaking, our knowledge of diseases in Southeast Asia during this period is somewhat better than that of the preceding period, but still it is far from satisfactory. It is generally assumed that diseases like malaria, dengue, and schistosomiasis had been present for a long time. However, that is based largely on backward projections, and in the case of dengue, for instance, the oldest description of the disease as present in the region dates from the late eighteenth century. We can also be sure about the ancient pedigree of smallpox and all kinds of gastro-enteritic diseases.

We are much less certain about the plague (of which the fourteenth-century variant is often called the Black Death), partly because sources easily use the term *pestilential* for severe epidemics in general, and in part because the sources contain contradictory information in this respect. We do know that the Black Death hit China and India around 1330 and Europe in the 1340s, but nevertheless there does not seem to exist convincing evidence for its spread to Southeast Asia in that period. In the seventeenth century, however, there are some signs that it had reached Maritime Southeast Asia. If both suppositions are true—a big if—this might suggest a pattern that can be applied to other diseases.

To understand this pattern the notion of the "civilized disease pool" must be introduced here. Ever since this idea was launched in the 1970s by Emanuel Le Roy Ladurie and William H. McNeill—to my knowledge independently from each other—most scholars have accepted the notion of a common Eurasian "disease pool," perhaps in existence as early as 1200 but certainly by the fourteenth century. The Black Death is a good example of this "microbial unification," which implied that a number of infectious diseases had, somewhere during the Medieval Period, reached at least both Europe and China (and parts of India, from which many diseases may have originated). After initially having suffered very high mortality rates from these "new" diseases, European and Chinese populations slowly built up a certain degree of immunity as the disease became endemic, or was at least reintroduced regularly. It is well known that such immunity was a boon, not only because mortality owing to these diseases dropped to lower levels but also because in hostile confrontations it gave the edge to those who were immune over a "virgin" population. The best known example is the conquest of Mexico and Peru by the Spaniards, who brought diseases such as smallpox and measles to an unsuspecting population that had never before come across those diseases, and therefore succumbed to them in great numbers.

Now the purported absence of the plague from Southeast Asia in the fourteenth century and its alleged presence in the seventeenth may suggest that Southeast Asia might have joined the common disease pool between those dates. It is a notion that can be defended on theoretical grounds as well, given the low population densities in the region, which would make transmission of infectious diseases and the transition from epidemicity to endemicity difficult, while the increase of trade and population between, say, 1300 and 1650 could very well have triggered the arrival of new disease patterns.

However, if Southeast Asia joined the club of plague-stricken areas in or just prior to the seventeenth century, it did not necessarily at the same time sign up for all of the other diseases from the pool as well. For instance, smallpox was present in the Indonesian and Philippine archipelagoes as early as 1550, and possibly much earlier, but measles may have been rare prior to the nineteenth century.

Nor is it likely that all Southeast Asian areas joined the pool at the same time. From detailed studies of smallpox in Indonesia prior to vaccination we know that inland areas of Borneo and Sumatra were hit by smallpox once every thirty years or less, while parts the Moluccas and Java were visited every seven to eight years in the seventeenth century. Around 1800, smallpox hit Java's north coast every three years. It is plausible to assume that similar differences existed in Mainland Southeast Asia between the fairly well populated valleys and the relatively isolated uplands, with their far smaller numbers of people. This also gave the lowlands—almost always the core areas of the Southeast Asian states—an edge versus the uplands, the dwelling places of the various tribal, nonstate peoples.

Upland people were well aware of the dangers of contact with people from the coastal lowlands, and the former were quite circumspect in their contacts with the latter. In a number of instances the highlanders avoided all contact, trading with the lowlanders through what has been called "silent barter"—that is, trading at a distance, or through an intermediary. Such groups also frequently recognized a goddess or (female) spirit of the sea whose coming to the uplands invariably heralded the arrival of epidemics and put the population to flight (thereby spreading the disease). In the same areas people also attempted to bar access to their areas—sometimes symbolically—when epidemics were afoot. If it is true that many diseases were relatively new to the region during the Early-Modern Period, that state of affairs would be part of the explanation for higher mortality rates in Southeast Asia than in China and Europe.

Finally it should be mentioned that the Europeans, precisely because they shared a disease pool with at least the Chinese, probably did not introduce many diseases that were new to Asia as a whole—a situation, therefore, that differed from "first contact" in America. The only certain exception to this rule is syphilis, of which the arrival in Asia more or less reads like the itinerary of the first Portuguese, around 1500. Most scholars now accept that syphilis came back with the Spaniards from the Americas, and if one reads the descriptions of the spread of the disease over Europe in the late fifteenth century, when the armies of Charles VIII of France were disbanded after the sack of Naples (1494), it is evidently a new and terrible disease to the stricken Europeans. From Europe syphilis was soon introduced into Asia, and it may be assumed that its arrival was as terrible as it had been in Europe. It is a general rule that a disease that is new to an area often kills large numbers of people before settling down as a much less lethal affliction. We might expect, therefore, that during the sixteenth century, Asians who came into sexual contact with syphilitic European visitors must have paid dearly.

There are also reasons to assume that yaws (*Framboesia tropica*) was new to the area around 1500, and although it is supposed to have come from Africa originally, it was probably introduced by Spaniards and Portuguese who brought along African slaves. The Dutch called it the Amboina pox, because Ambon, in the Moluccas, was the area where the disease was to be found around 1600—a plausible Asian yaws connection, as the Moluccas had received quite some attention from the Spaniards and the Portuguese in the sixteenth century. Later on, the disease evolved into a nonlethal but potentially very bothersome affliction, but it could have been lethal directly after first contact. It is no coincidence that both diseases had been recently introduced into Europe from outside, and therefore were not yet part of the "civilized disease pool."

With this expose of disease patterns we have not exhausted the environmental explanations for higher mortality in Southeast Asia than in China and Europe in the Early-Modern Period. Other environmental factors would be the influence of ENSO events, and the role of volcanic eruptions, earthquakes, cyclones, and the like.

ENSO

As we have seen in Chapter 4, ENSO (El Niño–Southern Oscillation) events strongly influence weather patterns in Asia. It is now well known that during an El Niño event Indonesia (particularly central and eastern Indonesia) experiences drought conditions. The same is true for large parts of India and Southeast Asia outside Indonesia, although it appears that the effects are strongest in the Indonesian Archipelago. This pattern has also been found in the past. As we know that in an agrarian society rainfall is by far the best predictor of a good harvest, as is the lack of rainfall for harvest failures, it stands to reason that during the Early-Modern Period, when survival depended almost entirely on the agricultural sector, droughts were literally a matter of life and death.

For both the seventeenth and eighteenth centuries, Quinn and Neal registered seventeen ENSO events that could be categorized as Very Strong, Strong, or between Moderate and Strong, while twenty such events were found for the nineteenth century. It should be realized that the effects of an ENSO drought were often felt during more than one year (and after a double ENSO event, of course, for more than two years), as people had been physically weakened, had eaten the portion of the (rice) harvest that should have been kept as seed, or had borrowed money or food that had to be paid back. This implies that one should double the number of serious ENSO years in order to arrive at an impression of its impact on daily life in parts of India and Southeast Asia. Droughts thus influenced the main foodcrop harvest (and often other harvests as well) in thirty-five to forty years per century because of ENSO events alone.

However, the relationship between registered El Niños and droughts is not unproblematic, if we look at the well-documented example of Java. For the seventeenth century, for instance, we have the series of El Niños as registered by Quinn and Neal; there were a series of dry years according to Berlage's tree ring measurements, in addition to a series of dry years, crop failures, high rice prices, and famines as found in the historical records and compiled by Boomgaard. Berlage found twenty-one years or clusters of years with very dry conditions (according to tree ring measurements) for seventeenth-century Java, but only nine of the seventeen serious El Niños coincide with those dry years or groups of years. Of the eleven years or clusters of years described as dry by historical sources, five were at least partly also El Niños, and of the twelve years and groups of years during which (according to the historical record) crop failures and high rice prices had occurred, seven were also El Niño years.

These findings underscore recent criticism of the lists of El Niño years made by Quinn and collaborators. Many of the registered El Niños from these lists, and particularly the earlier, sixteenth- and seventeenth-century ones, have been called into question. It seems sensible, therefore, to wait for new lists to be drawn up before we confront them with local data. Until that moment, the local data on droughts, harvest failures, high rice prices, and famines might be better proxy variables for "real" ENSO events than the lists compiled by Quinn et al., based on data from Latin America.

Nevertheless, it should not be forgotten that, according to the recent literature (and based on good data from recent years), dry years could and did occur outside "serious" ENSO years, while not all ENSO years were dry. This implies that, historically, dry years, either according to tree ring measurements or as found in the historical record, will never coincide entirely with El Niños.

Be that as it may, it is clear that historically many areas in Southeast Asia were hugely troubled by recurring droughts, most of which were strongly related to ENSO events, which put many parts of Southeast Asia (and India) at a disadvantage in comparison with China and Europe, and for that reason alone would have caused higher mortality than in the latter regions.

And then we have not even talked about floods and excessive rains, features that might be expected during La Niña (the return El Niño, so to speak) years. The historical record contains far less data on heavy rain and flooding in Java, which might suggest that the effect of excessive precipitation on harvests is less destructive than that of droughts, at least in that area.

VOLCANIC ERUPTIONS

As we have also seen in Chapter 4, weather patterns throughout the world are (and were) influenced by volcanic eruptions. The more forceful ones among them produce a dust veil in the atmosphere that will stay there for a number of years.

This dust veil, provided it is sufficiently dense, shields the earth from the rays of the sun, thus leading to lower yields per acre in agriculture. If the volcanic eruptions are sufficiently frequent and large, the dust veil can be the cause of a series of harvest failures.

Seen from a Southeast Asian perspective, the most important eruptions are the ones produced by volcanoes located between 20 degrees northern and 20 degrees southern latitude. For this area, Simkin and Siebert found 300 eruptions during the seventeenth century and 444 for the eighteenth. That was not a real increase, however. We are aware that the number of volcanoes known to Western scholars and observers increased as well between 1600 and 1800. In the nineteenth and twentieth centuries the number of known volcanoes as well as that of the registered number of eruptions would increase even further, which implies that the further back we go in time the larger is the under-registration. However, for our purpose, that is not necessarily a problem. What we want to know is whether we find more years with harvest failures during and shortly after years with above-average numbers of strong eruptions. Such averages are relative averages, established by looking at the surrounding decades.

There are a number of cases for the island of Java—at the moment the only Southeast Asian area for which there is fairly reliable year-by-year historical evidence for the presence or absence of droughts, harvest failures, rice shortages, and famine, including data on rice prices—for which the evidence for such a link is rather convincing. The most obvious case is that of the 1670s, during which most years showed very high rates of volcanic activity, a series of dry to very dry years, and an almost uninterrupted series of poor harvests, rice shortages, and famines. On a somewhat more modest scale, a long series of above-average eruption years in the 1790s also led to at least five years of harvest failures and high rice prices, perhaps particularly between 1796 and 1800, which was also a period of drought. Interestingly enough, the 1780s had seen a low incidence of eruptions and hardly any mention of harvest failure. Shorter series of bad years that may have been caused by above-average volcanic dust veils are the early 1660s, the years between 1715 and 1724, a number of years in the 1730s, and between 1742 and 1747. A series of very bad years (with peak mortality in Western Java) starting in 1757 may have been prolonged by three years of many eruptions, 1759–1761, and a series of above-average eruption years in the early 1770s partly coincided with dry years and (local) harvest problems.

However, there are also anomalies. For instance, the period between 1632 and 1646 showed a lot of volcanic activity—leading to a series of bad years in East Asia—without doing much damage in Java, although during those years there may have been under-reporting. The opposite occurred between 1703 and 1710, mainly years with indifferent rice crops and high rice prices but with no volcanic eruptions to speak of. As this was also a period of war (1704–1708), and—at least according to Quinn et al.—a moderate to strong El Niño (1707–1709), this lack of coincidence is not difficult to explain.

Another anomaly occurred in the 1690s, the period with the highest total volcanic eruptive index of the seventeenth century in the tropical zone. During that same period two drought years were mentioned in Java (1695, 1698), but the historical record does not speak of harvest failures, let alone famine. And yet, this was the nadir of the so-called Little Ice Age in Europe, with severe winters, terrible harvest failures, and famines across the board. Under-reporting as regards Java is not likely in this period. Generally speaking, however, the link between a series of strong volcanic eruption years and a number of bad harvests in a row seems well established for Java between 1600 and 1800.

As was shown in Chapter 4, there are good reasons to be wary of the application of the notion of a Little Ice Age, designed for the temperate zone of the Northern Hemisphere, to Southeast Asian history. Increased frost does not mean much in most of the region—it occurs only in the most elevated mountain areas, such as in New Guinea—and it is unlikely on theoretical and empirical grounds that Southeast Asia would experience spells of good and bad weather during the same periods as did Europe (or China).

However, as we have just seen, it makes good sense to investigate the link between clusters of strong volcanic eruptions, one of the forcing mechanisms behind climatic fluctuations (of which the Little Ice Age is just one specific, regionally defined case) in both Southeast Asia and Europe, on the one hand, and periods of bad harvests and increased mortality on the other. Even then, we should be aware of the fact that synchronicity between Southeast Asia and Europe regarding good or bad periods as a result of volcanic activity is not always to be expected. The dust veils of volcanic eruptions outside the zone between 20 degrees northern and 20 degrees southern latitude have little influence outside their own hemisphere, whereas the eruptions from within the tropical zone after some time spread over the entire globe. Therefore, Southeast Asia was not much troubled by the eruptions of Etna, Vesuvius, or Stromboli, but Europe did feel the effects of Krakatoa and Tambora.

If we compare the number and strength of the volcanic eruptions in the tropical zone in the seventeenth century with those registered for the eighteenth century, we find, as has been said, higher numbers for the latter, which is probably an artifact of the historical reconstruction. However, there are no good reasons to assume that the seventeenth century was worse than the eighteenth, something that is implicit in Reid's notion of a seventeenth-century crisis in

Southeast Asia, which represented the nadir of the Little Ice Age. We can go no further than the conclusion that in both centuries, clusters of strong volcanic eruptions indirectly led to harvest failure and high mortality in many years.

As a footnote, it should be pointed out that we have been talking only about the effects of the volcanic dust veils. We have not discussed the direct local effects of people being killed by the eruptions themselves or the indirect local effects of agricultural lands with standing crops being covered by lahar streams or layers of ash and rocks, thereby causing scarcity and in some cases famine. The two best known examples are the eruption of Tambora (Sumbawa, one of the Lesser Sunda Islands, Indonesia) in 1815, and that of Krakatoa (between Sumatra and Java) in 1883. Tambora may have killed some 10,000 people directly and about 82,000 indirectly, through starvation and disease, while Krakatoa is reported to have caused the death of more than 36,000 people, largely because of the tsunami that was the result of the eruption. The total death toll for the two volcanoes, therefore, would have been just under 130,000, or 65,000 per volcano on average.

However, it would be entirely unjustified if we applied this figure to the total number of known eruptions in order to get at some estimate of numbers of direct and indirect local deaths owing to volcanic eruptions. Fortunately, Simkin and Siebert have listed eighty-two Indonesian volcano eruptions of which the number of victims is known. If we divide the total number of victims (182,749) by 82, we get an average figure of 2,230 deaths per eruption. As we found some sixty-five eruptions in seventeenth-century Indonesia, and a few more in the eighteenth century, the total death toll per century between 1600 and 1800 would have been something like 150,000, or 1,500 annually—not a significant number by any stretch of the imagination.

We will briefly discuss the category of hazards as a whole, including earthquakes, tropical cyclones, typhoons, floods, and tsunamis. Is there any way to establish whether Southeast Asia was hit proportionally harder than other areas by such natural disasters, which then might partially explain higher mortality than in Europe or China?

A tentative answer is suggested by Table 5.3, which reproduces data taken from a recent publication by Greg Bankoff, giving an impression of the relative "weight" per unit of surface area of hazards in various Asian regions.

Clearly, Southeast Asia and South Asia have a much higher rate of hazards per unit of surface area than does East Asia, at least if we look at twentieth-century data. That is in keeping with the lower population growth rates for Southeast Asia and India during the Early-Modern Period, although two caveats must be entered.

In the first place, one always has to be careful with backward projections, and it may not be assumed that the proportions to be found for the twentieth century were also obtained for the Early-Modern Period. Earlier data are, however, not available.

Table 5.3 Comparison of hazards to land mass by region in parts of Asia, 1900–1997.

Region	Percent of all hazards, 1900–1997	Percent of total land mass
Southeast Asia	12.1	3.0
East Asia	12.7	7.9
South Asia	12.2	3.0

Source: Bankoff, Greg, 2003, Cultures of Disaster: Society and Natural Hazard in the Philippines. *London: RoutledgeCurzon.*

In the second place, we should have some impression of the scope of the influence of natural disasters on the death rate. One of the most impressive disasters in recent times was the 26 December 2004 underwater earthquake (9.3 on the Richter scale) followed by the tsunami that hit South and Southeast Asia. With an estimated 225,000 dead, including, in Sumatra alone, probably some 170,000 deaths, in addition to 500,000 people left homeless, and a million fishing vessels destroyed, we might be inclined to assume that the influence of natural hazards in general was huge. However, it is likely that, as we have seen when looking at the number of victims of volcanic eruptions, this was an exceptional occurrence, and that the average number of direct victims of all hazards is rather modest. However, if we look at the number of people affected by such disasters because they lost their house, crops, boats, or livestock, it must be concluded that the influence must have been considerable. It may be assumed, then, that indirect mortality owing to hazards—as the result of poverty, disease, or starvation, particularly by way of infant and child mortality—may have made a notable contribution to the overall death rate. Although rare, tsunamis such as the 2004 one were not entirely unheard of. In 1833 and 1861 Sumatra had been hit by similar, though probably somewhat smaller, tsunamis.

WAR AND CONFLICT

In addition, it has been argued that Southeast Asia was an area with constant, although often low-level, warfare. There were, of course, wars between neighboring Southeast Asian states throughout the period under consideration. Perhaps equally important were the attempts by provincial governors or regional princes and aristocrats to break away from the "galactic polity" of which they were a part, to oust the incumbent on the throne of the core area, or even attempts by regional centers to become the new supraregional core (compare Chapter 3). Wars of succession were frequent as well, as in these societies of bilateral kin reckoning and rulers with very large harems, there were almost always large numbers of contestants to the throne and no clear rules of succession. There were also wars between Chinese, Indians, Portuguese, Spaniards, Dutch, and English, on the one hand, and indigenous rulers and peoples on the other, and also occasionally between the Europeans (and their indigenous allies) among themselves.

It has been suggested that in Southeast Asian warfare people were not killed in battle in great numbers. Recent studies have shown that there is also evidence for many a bloody battle, and that the numbers of casualties resulting from war may have increased during the period under consideration because of the growing use of firearms. Be that as it may, the side effects of war were considerable. In wars between and within Southeast Asian states, it was the peasantry who were conscripted in large numbers to fight the battles of their leaders, as there was no professional standing or mercenary army (although occasionally elite mercenaries with imported firearms were involved). This meant that many agricultural tasks could not be carried out, leading to lower returns or even to total harvest failures if mobilization occurred during harvest time. As fighting almost invariably took place during the dry season, because roads were impassable and seas too choppy during the wet monsoon, harvesting was often jeopardized. Even if the harvesting was carried out by women—which was often the case—the men were needed to stand guard over the harvesters in order to protect them against wild animals (tigers, leopards, wild boar, elephants) and various types of raiders. In the wet-rice areas, neglect of irrigation channels and dikes, the upkeep of which was a typically male task, led to crops withering in the fields when the men had been conscripted.

Another cause of high death rates was the inability of most Southeast Asian war leaders to feed their armies and to prevent the outbreak of epidemics, at beleaguered cities or among armies in the field. The logistics of feeding large armies appear to have been beyond the capacity of many rulers and their generals, or it was simply assumed that the army would live off the land. Thus, high rates of mortality obtained in the early-modern Southeast Asian armies, probably not mainly because of casualties in combat but owing to malnourishment and epidemics.

At the same time it was not unusual to set fire to the standing crops, villages, and towns encountered by a Southeast Asian army on the move, and occasionally one reads that even forests were set on fire. Water running through a beleaguered city was deliberately polluted. There are also examples of large numbers of ships and boats belonging to the enemy being burned. In other words, people were deliberately deprived of their health, their food, and their livelihood by their opponents. Recovery from war must have been long and arduous, and it stands to reason that mortality figures would be high for a considerable period of time after a war, while fertility figures would be low.

It should be mentioned that rulers routinely rounded up large numbers of war captives from the conquered areas in order to increase the population of the core areas, where they were settled as a semiservile labor force. Southeast Asia was, as we have seen, sparsely populated, and rulers were always on the lookout for more producers, taxpayers, and potential soldiers. Sometimes the war captives were left to clear and farm an area not too far from the capital, but often they were given specific tasks to perform, such as brick making and other artisanal activities that could be better carried out by specialists than as a sideline by peasant-agriculturists. One can imagine how many human lives were lost when large groups of vanquished people were forcefully resettled in areas far away from their country of origin, probably without an adequate infrastructure to take care of the newcomers. Regular raids of lowland states against bordering tribal uplands, often at least partly in order to acquire slaves, may be regarded as a variant of this theme.

Nor should we underestimate local conflicts between nonstate adversaries: tribal groups or villages, and raiding parties by pirates. We can think of the large groups of Tai speakers that were cut adrift and were looking for new areas to settle in the thirteenth and fourteenth centuries after the Mongols had displaced them. Or of the large-scale movements of the Dayaks in nineteenth-century Borneo. Complete wars were fought over disputed territories, not once but repeatedly.

One important aspect of life in tribal societies, chiefdoms, and early (small) states in Southeast Asia was the quest for status. That, in turn, was linked to raiding, competitive feasting (which could easily lead to bloody battles when people felt slighted), and headhunting. Chiefs and rulers had to demonstrate their prowess in battle and had to add constantly to their collection of heads and women. Headhunting raids were also associated with marriage—and therefore can be seen as rites of passage—and with the fertility of women, animals, and the soil. Raids, undertaken either to acquire heads or to bring back captives that could be employed as slaves or sold (or both), were also related to status seeking and the acquisition of fame and fortune. Both activities, often undertaken at the fringes of a state, invited retaliation by that state, and often bloody attempts to suppress both headhunting and slave raiding. Thus an upward spiral of violence existed until the (colonial) state finally succeeded in suppressing both activities, usually not before the nineteenth century. Nevertheless, it would be wrong to suggest that these mechanisms caused high death rates in all stateless societies in Southeast Asia during this period. We should consider the possibility that these types of violence were of importance only locally and temporarily.

Slavery was, as was shown earlier, an important "mode of production" in Southeast Asia during the Pre-Modern Period. As that was also true during the Early-Modern Period, there was a constant demand for slaves and therefore for slave raids. As slaves were often transported to slave markets under conditions that left much to be desired, and from there to their final destinations, mortality among slaves was high during their early captivity. In their new surroundings it often remained high, as they were now confronted with other environmental circumstances, including disease. In addition, men encountered by the slave raiders were often—but it is not clear how often—killed immediately, and only the women and children were kept alive to be sold.

Finally, one should take into account feuds that resulted from unpaid debts, and that in a society in which indebtedness was more or less ubiquitous, and had no doubt increased with the expansion of monetization during the period under discussion.

War and other forms of violent conflict, therefore, often led (in combination with the other factors mentioned) to high levels of mortality. It is likely that there were peaks and troughs in the death rate caused by war and conflict, as is suggested by Lieberman's notion of three periods of crisis between 800 and 1830. His chronology of crises suggests that we may expect mortality peaks (owing to wars) during the late thirteenth, the fourteenth, and the early fifteenth centuries, during the period from 1540 to 1610, and between 1752 and 1786. However, between those political crises there were often rebellions, or wars between states. Therefore, what we really need is a much more detailed chronology per area—like the one given above for Java between 1600 and 1800—before we will be able to be more specific about the long-term fluctuations of the Southeast Asian death rate owing to wars and other violent conflicts.

Finally, among the causes of high mortality rates in Southeast Asia, infanticide should be mentioned. Technically, infanticide should be dealt with under mortality, although demographers regard it generally as postponed abortion, a topic to be dealt with under the rubric of fertility.

Information on infanticide in Southeast Asia is not abundantly available for the period we are dealing with, but there is no doubt that it was resorted to. As this was also the case in India and China, infanticide is not the factor we are looking for to explain higher mortality in Southeast Asia than in China. However, as it was virtually absent in Europe, its practice in Southeast Asia could explain part of the difference in mortality rates in the Early-Modern Period between those two areas.

Summing up our findings regarding mortality in Southeast Asia during the Early-Modern Period, it can be said that mortality rates in (much of) the region may have been higher than in Europe and China because of the role of ENSO events, the joining of the civilized disease pool, and high levels of war and violent conflict. Infanticide did make a difference in comparison with Europe, but not with China. Clusters of volcanic eruptions can explain the timing of series of harvest failures, which may and often will differ from the timing in Europe, because we are talking about different volcanoes, but it is probably unlikely that they led to higher mortality rates in Southeast Asia than in Europe or China.

Many of these factors are evidently related to environmental factors. Even the high level of violent conflict may be perceived as being partly explained by the highland-lowland dichotomy that is so typical of much of Southeast Asia's history. It could also be argued that the many wars fought between the cores and the peripheries of the "galactic polities" that were equally typical of the area until quite late had an environmental background—that is, if we can agree that the persistence of such polities was related to geophysical factors, such as the difficulties involved in travel from one center to another, and the problems that this entailed in the disciplining of the subaltern centers by the core.

Final proof of higher mortality rates in Southeast Asia than in China or Europe rests in the end, of course, on the availability of quantitative data, and in that respect there is not much on offer regarding Southeast Asia. For early nineteenth-century Java a case can be made for a life expectancy at birth for men of around twenty-three years and for women of twenty-five years. Figures for Europe and China for men around 1800 were between thirty and forty. If Java is representative of Southeast Asia in this respect, and I cannot see many reasons why that should not be the case, the higher Southeast Asian mortality rates during the Early-Modern Period (at least for its tail end) postulated here appear to be confirmed.

FERTILITY

If population growth during the Early-Modern Period was lower in Southeast Asia (and India) than it was in China and Europe, that may also have been caused at least partly by a lower rate of fertility. As a rough generalization it can be said that Western Europe was characterized during this period by a pattern of high marital fertility and restricted marriage, while China had low marital fertility but early and universal marriage. On balance there may not have been much of a difference between the two regions in overall levels of fertility.

The received wisdom about Southeast Asia is that marriage was early and universal, but there are reasons to assume that to be too easy a generalization. For Indonesia it has been demonstrated that prior to 1900 women married late or not at all in many areas outside Java, particularly in regions with patrilineal kin groups, and to a lesser extent among the groups with bilateral systems. These were often also areas in which extended families and longhouse-type dwellings were frequent, if not dominant. The reason for the low rate of nuptiality was the high brideprice payments that were customary in these regions. Generally speaking, population growth rates were low in these areas, which was attributed to low overall fertility owing to late marriages and the fact that marriage was not universal. During the nineteenth century, a combination of urbanization and further monetization and commercialization, as well as attempts by the colonial state and Islamic religious authorities to rid Indonesian societies of both bridewealth payments and complex households, brought about the breakdown of this system (but note that some commercialization had originally been conducive to higher brideprice payments).

Information on these points regarding the mainland is rare, but there is at least one publication, on Upper Burma, in which elements of this complex are to be encountered. The low growth rate to be found in this area between 1920 and 1950 was partly explained by a late age at marriage among females and a high percentage of bachelors and spinsters. There is no mention of a connection with the level of the brideprice, however. Of course, this publication refers to a much later period than the one we are dealing with here, but there is no reason to assume that this state of affairs was a recent introduction.

The link between the longhouse-type dwelling, usually home to complex families, and low fertility rates could perhaps be partly explained by the possibility that adoption of children or "child sharing" by members of the household is easier than in the case of nuclear families. Thus the (economic) need for many children of one's own would be less pressing.

In the higher status households there were also often slaves to be found, and slaves were not all that different from adopted children. There are indeed some indications that fertility was lower among slave-keeping groups, partly because the slaves themselves did not have many children (who would, in any case, be the possessions of their masters), and in part because the slave-holding women had fewer children.

THE TWO-TRACK DEMOGRAPHIC PATTERN

On the other hand, it is also clear that there were some areas, of which Java is a good example, in which nuclear families and early and universal marriage predominated, and where marital fertility (and therefore also overall fertility) was high. This might suggest a two-track demographic pattern for early-modern Southeast Asia—a fast track in the wet-rice producing lowlands (and mid-altitudes) and a slow track in the hunter-gatherer and shifting cultivation uplands. In the wet-rice–producing valleys one would find early and universal marriage and high

The Padang Highlands in Sumatra in 1903 were still a rugged and isolated region in Southeast Asia. Upland regions tended to develop much more slowly than well-connected and commercialized lowlands. (National Oceanic and Atmospheric Association)

fertility (and therefore relatively high rates of population growth under conditions of "normal" mortality), while in the upland areas marriages would be late, celibacy relatively high, and fertility rather low (and rates of natural increase low). Generally speaking, the uplands were rather isolated, while the lowlands were commercialized and well connected to regional and international markets.

Not only do we find this dichotomy in pre-1900 Indonesia, but there is also a theoretical justification for it. Wet-rice production is a highly labor intensive activity, with a high degree of female participation. It has been argued (the so-called demand-for-labor hypothesis) that this gives particularly women an incentive to bear more children, as young children, who participate in agricultural activities at quite an early age, will lighten the woman's burden. There is some evidence for this effect from the Indonesian Archipelago. In slash-and-burn areas labor requirements are lower, while in the case of hunter-gatherers, young children are quite a burden on the foraging mothers. Such societies, therefore, would have incentives to keep the number of children at modest levels.

Moreover, as a rule the upland populations were often animists, or at least only marginally influenced by the so-called world religions (Buddhism, Hinduism, Islam, Christianity), while the wet-rice–producing lowlands, usually the core areas of the Southeast Asian states and centered around the big cities, had been converted to those religions either before or during the period we are dealing with. As the position of women in Southeast Asia is generally assumed to have been more autonomous prior to the arrival of the more patriarchal world religions, and as it is also often supposed that if women could do what they wanted they would have fewer children than when men could impose their wishes, we would have an additional explanation for lower fertility in the upland areas.

This two-track demographic pattern was reinforced by slave raiding from the lowlands in the uplands and by the deliberate transplantation of large groups of war captives, carried out by victorious rulers of the core areas, from the highlands to the lowlands. The upland areas were also the headhunting regions, and although there is no proof that headhunting meant a considerable loss of life, the idea that young men who wanted to get married were supposed to present their prospective brides with a hunted head may have been sufficiently widely spread to make a marginal difference, thus reinforcing the two-track pattern.

It was also reinforced by the differential sensitivity of the two types of areas for epidemics. As time went by the relatively densely populated lowlands became part of the "civilized disease pool," which meant that diseases became less virulent, and the population acquired immunity against many of them; the isolated and sparsely populated uplands, on the other hand, were visited by epidemics only occasionally, but when that occurred the results could be devastating.

Spontaneous migration from the uplands to the lowlands must have played a role as well, particularly when people in the uplands ran into economic difficulties. However, in troubled times in the lowlands we would expect temporarily reversed flows, particularly with the introduction of new crops suited to the uplands (maize, tobacco), a topic dealt with in the next chapter.

As the population in the rice bowls was growing more rapidly than that of the uplands, it may be supposed that at the beginning of the Early-Modern Period—say, ca. 1400—the proportion of the population living in the slash-and-burn and forager areas was larger than it would be in the nineteenth century. This implies that the low-fertility areas had a much larger weight at the beginning than at the end of the period, which, in turn, implies that—other things being equal—the birth rate would be gradually increasing during the period under consideration.

PRUDENTIAL CHECKS

A number of additional points can be made that should render the idea of low fertility regimes at least in some areas and periods more plausible. The first point to be made is that of abortion. It is quite clear that abortion was known among many Southeast Asian peoples and that it was practiced on a wide scale. Of course, we do not have statistics on this point, but there is ample qualitative evidence that women could and did use abortifacients successfully. A related issue is that of the prevention of conceptions in an era prior to the invention and widespread introduction of modern methods of birth control. There are references to the practice of tilting the uterus (*retroflexio uteri*) for various areas in Southeast Asia, which was carried out by traditional female healers. This would render the woman temporarily infertile, and could be reversed when she wanted to conceive.

Women, therefore, could and did manipulate their fertility in the era prior to the birth-control pill. This notion is reinforced when we recall that infanticide was also widely practiced; its effects on population growth rates were tantamount to those of abortion.

This is an important conclusion, as people have often assumed that Asian population fluctuations were entirely mortality driven, a situation to which the term *Malthusian* is usually applied. According to Malthus, the Chinese peasantry lived in misery because of unchecked population growth. That view has recently been contested, partly based on the observation that infanticide, which was widely practiced in China in the Early-Modern Period, should be regarded as a method of family planning. This view can be seen as yet another aspect of the recent "divergence" debate, in which the revisionists argue that China and Europe were much closer to each other around 1800 in terms of income and production per capita than has been assumed generally. It was already accepted orthodoxy that (Northwestern) Europe had avoided "Malthusian situations" (population growth outstripping economic growth) by so-called prudential or preventive checks. Those terms refer to late marriages and high rates of celibacy in periods of economic distress, with lower ages at first marriage and a larger proportion marrying during better times. (The term *positive checks* was used by Malthus for the situation in which higher mortality restored the balance between population numbers and resources.) Now it is argued that China had a similar mechanism of preventive checks whereby contraception, abortion, and infanticide lowered the rate of natural increase.

It is suggested in the present study that similar mechanisms were in operation in Southeast Asia as well. In times of economic distress, women had a choice of means to opt for a smaller family at their disposal, and there are nineteenth-century data for Java to suggest that they did just that. Nineteenth-century data for Java and the Philippines also suggest that the age at marriage and the proportion of people marrying were influenced by the state of the economy, particularly by the average size of landholdings. Again, data for the mainland are regrettably rare, but the notion espoused here that demographic behavior in Pre-Modern Southeast Asia was less Malthusian than many might have thought may be hoped to set the agenda for future research in this region.

One of the links that should be researched in more detail is that between out-migration and the mechanisms of family planning mentioned above. This link has been established for the Ilocos Coast in the nineteenth-century Philippines and for the twentieth-century Iban of Sarawak, where out-migration of men could be observed in times of (relatively) high population pressure in a situation of limited resources. Prolonged absence of men led to lower rates of marriage and higher rates of celibacy. It could be argued that something similar occurred in the Upper Burman Dry Zone in the early twentieth century.

Strangely enough, a source on southern Sumatra in the early nineteenth century suggests that the causal link could have been the other way around: because brideprices were high, young men of restricted means left the area, in search of less demanding potential parents-in-law. If sufficient numbers of young men had done so, the number of women remaining single would have gone up in the area of origin. The effects would be the same as in the Ilocos Coast area: out-migration of men in combination with higher rates of female celibacy must have led to lower natural growth rates in the areas from which the migrants came, and higher rates in the places where they settled.

CELIBACY AS A CALLING

Finally, the demographic effect should be gauged of the existence of Buddhist monasteries, hermits, Beguines, and the possibly religion-induced taboo on remarriage of widows and divorcees. A source dated around 1500 referring to Java, when it was still largely a Hindu-Buddhist area, alludes to hermits "who do not know women," to high proportions of women in nunneries, and to widows who, instead of joining the funeral pyre of their husbands, became Beguines (that is, lay nuns). Many areas of Mainland Southeast Asia were (and are still) Buddhist, and large numbers of monasteries and monks were typical of the area. However, to be a (Theravada) Buddhist monk today is largely a life-cycle phenomenon, and it is doubtful that it influences the birth rate even marginally. This was different in the past, when being a monk was a lifelong calling, which, if the numbers were large enough, may have influenced the average age at first marriage and the proportion of people never marrying. Another point is that

Two Laotian Buddhist monks stand by a temple. The existence of large numbers of Buddhist monks leading celibate lifestyles in the past may have influenced population growth rates. (Corel)

under Theravada Buddhism nunneries are a rarity and have been so for a long time, although in fifteenth-century Burma fully ordained women were still to be found. In Mahayana Buddhist areas (Java, Cambodia, Vietnam), nunneries were less rare.

Nevertheless, it appears likely that between 1400 and 1870 the percentages of lifelong monks and fully ordained nuns in the total population gradually dropped, partly because Theravada Buddhism took over and partly because of the arrival of Islam. On that score, therefore, one would expect a gradually dropping age at first marriage and a decreasing proportion of people remaining celibate, which could imply a gradually increasing birth rate. It is entirely unclear how these proportions compare with those in China and Europe, where there were also monasteries and nunneries, although not in the Protestant areas of Europe.

We would also like to know how strong the objections were against divorcees and widows remarrying. Remarriage of widows was discouraged in Early-Modern Vietnam, although young childless widows probably remarried. In northern Burma during the early twentieth century such remarriages were taboo, but in Islamic nineteenth-century Java they were not. Such taboos were reportedly also strong—but perhaps not everywhere or always—in Early-Modern India and China, but not, to my knowledge, in most areas of Europe. Again, we must conclude that the importance of this factor in Southeast Asia is unclear.

Summing up, it can be said that there are rather firm indications that marriage was not everywhere early and universal (as it was in China). As a rule, such a situation is conducive to lower fertility than in regions where marriage is early and universal. This applies particularly when mortality rates are high (and therefore every year of postponement of a marriage increases the risk of the death of one of the partners) and remarriage is taboo.

We found indications of a two-track demographic pattern for this period, with low fertility in slash-and-burn and hunter-gatherer upland areas, and high fertility in the wet-rice–producing lowlands and mid-altitudes. The differences between these two types of areas were reinforced by slave raids, deportations, migration, and a different impact of epidemics. As the relative weight of the upland areas diminished over time, the downward pressure exerted by the highlands on the average rate of fertility must have decreased gradually. Theoretically, therefore, fertility in Southeast Asia should have been closer to the Chinese pattern in the nineteenth century than it was around 1400, when we might have expected the Southeast Asian rate to have been lower than that of the Chinese.

We also found that women could and did manipulate their fertility. They appear to have done so particularly in situations of great resource pressure. It would seem, therefore, that the usual supposition that Asian population development was mortality driven does not apply (fully) to Southeast Asia. As has been recently suggested for China, it is now possible that there was no "Malthusian situation" in (parts of) Southeast Asia, either. However, as mortality in Southeast Asia appears to have been higher than in China, this is certainly a point that needs to be clarified further and awaits much more research.

CONCLUSION

There is a good deal of evidence that mortality was higher and fertility lower in Southeast Asia than in Europe or China, which tallies nicely with the calculated differences in rates of population increase, which would appear to have been lower in Southeast Asia than in the other regions. It is possible that at the tail end of our period the differences became less great. Fertility may have increased in Southeast Asia as the relative weight of the rice bowls increased, while mortality may have dropped because of the same reason: fewer lethal effects of epidemics as the growing rice bowls were increasingly drawn into the sphere of influence of the civilized disease pools. Throughout this chapter we have seen how often environmental factors may have played a role in the differences in demographic regimes of the regions we have been comparing.

A final footnote is due on migration. For the Early-Modern Period there is no evidence of out-migration from Southeast Asia to places abroad. There is ample proof, however, for immigration, particularly of Chinese people who came from overseas as merchants, artisans, and manufacturers, but also as peddlers and coolies. There is an estimate of nearly a million Chinese in the region around 1830, which would have been some 3 percent of the total population. At the same time, tribal non–Han Chinese groups from north of the border between China and Southeast Asia continued to migrate to Southeast Asia, escaping from oppressive rule. This went on even in the twentieth century, as witness the case of the Hmong who ended up in Laos. In sum, then, people immigrated to Southeast Asia, but there was no out-migration. Therefore, if average population growth in Southeast Asia had been lower than in China and Europe, the average rate of natural increase (that is, births minus deaths) had been even lower than that, and as China and Europe probably experienced more out-migration than immigration, the rates of natural increase in those regions would have been higher, which makes the contrast with Southeast Asia even starker.

BIBLIOGRAPHICAL ESSAY

The most important authors on this period are Anthony Reid and Victor Lieberman, who, between them, have revolutionized our idea of Southeast Asia during the Early-Modern Period. Reid's magnum opus is *Southeast Asia in the*

Age of Commerce 1450–1680. 2 vols. New Haven: Yale University Press, 1988, 1993; see also Reid, Anthony, 1999, "Economic and Social Change, c. 1400–1800." Pp. 116–163 in *The Cambridge History of Southeast Asia*. Vol. 1, Part 2: *From c. 1500 to c. 1800.* Edited by Nicholas Tarling. Cambridge: Cambridge University Press; and Reid, Anthony, 1997, "A New Phase of Commercial Expansion in Southeast Asia, 1760–1840." Pp. 57–82 in *The Last Stand of Asian Autonomies: Responses to Modernity in the Diverse States of Southeast Asia and Korea, 1750–1900.* Edited by Anthony Reid. Houndmills: Macmillan; New York: St. Martin's.

Lieberman's most important book (from the point of view of a general Southeast Asian history textbook writer) is *Strange Parallels: Southeast Asia in Global Context, c. 800–1830.* Vol. 1: *Integration on the Mainland.* Cambridge: Cambridge University Press, 2003; the second volume, which will deal mainly with areas other than Southeast Asia, had not yet been published at the moment of writing. For an earlier article in which the main points of the book had already been made, see Lieberman, Victor, 1993, "Local Integration and Eurasian Analogies: Structuring Southeast Asian History, c. 1350–c. 1830," *Modern Asian Studies* 27, no. 3, pp. 475–572.

For the "divergence" discussion, see the special issue of the journal *Itinerario* 24, nos. 3/4 (2000), with contributions by Chris Bayly, Janet Hunter, Patrick O'Brien, Sevket Pamuk, Kenneth Pomeranz, and Jan Luiten van Zanden. See also Frank, Andre Gunder, 1998, *ReOrient: Global Economy in the Asian Age.* Berkeley: University of California Press; Pomeranz, Kenneth, 2000, *The Great Divergence: China, Europe, and the Making of the Modern World Economy.* Princeton: Princeton University Press.

On demographic history during this period, see the books and articles by Reid and Lieberman cited above. Additional data and discussions are to be found in Zelinski, Wilbur, 1950, "The Indochinese Peninsula: A Demographic Anomaly," *Far Eastern Quarterly* 9, no. 2, pp. 115–145; Nash, June and Manning Nash, 1963, "Marriage, Family, and Population Growth in Upper Burma," *Southwestern Journal of Anthropology* 19, pp. 251–266; Padoch, Christine, 1982, *Migration and Its Alternatives among the Iban of Sarawak.* The Hague: Nijhoff; Boomgaard, Peter, 1987, "Multiplying Masses: Nineteenth-Century Population Growth in India and Indonesia," *Itinerario* 11, no. 1, pp. 135–147; Dyson, Tim, 1989, "An Overview of South Asian Historical Demography in Southeast Asian Perspective," unpublished paper, Washington; Boomgaard, Peter, 1989, *Children of the Colonial State: Population Growth and Economic Development in Java, 1795–1880.* Amsterdam: Free University Press; Yu, Insun, 1990, *Law and Society in Seventeenth and Eighteenth-Century Vietnam.* Seoul: Asiatic Research Center, Korea University; Knaap, Gerrit, 1995, "The Demography of Ambon in the

Seventeenth Century: Evidence from Colonial Proto-Censuses," *Journal of Southeast Asian Studies* 26, no. 2, pp. 227–241; Corpuz, O. D., 1997, *An Economic History of the Philippines*. Quezon City: University of the Philippines Press; Xenos, Peter, 1998, "The Ilocos Coast since 1800: Population Pressure, the Ilocano Diaspora, and Multiphasic Response." Pp. 39–70 in *Population and History: The Demographic Origins of the Modern Philippines*. Edited by Daniel F. Doeppers and Peter Xenos. Madison: Center for Southeast Asian Studies, University of Wisconsin; Lavely, William, and R. Bin Wong, 1998, "Revising the Malthusian Narrative: The Comparative Study of Population Dynamics in Late Imperial China," *Journal of Asian Studies* 57, no. 3, pp. 714–748; Knapen, Han, 2001, *Forests of Fortune?: The Environmental History of Southeast Borneo, 1600–1880*. Leiden: KITLV; Boomgaard, Peter, 2001, "Crisis Mortality in Seventeenth-Century Indonesia." Pp. 191–220 in *Asian Population History*. Edited by Ts'ui-jung Liu et al. Oxford: Oxford University Press; Boomgaard, Peter, 2003, "Bridewealth and Birth Control in the Indonesian Archipelago, 1500–1900," *Population and Development Review* 19, no. 2, pp. 197–214; Henley, David, 2005, *Fertility, Food, and Fever: Population, Economy and Environment in North and Central Sulawesi c. 1600–1930*. Leiden: KITLV; Henley, David, 2006, "Rising Birth Rates in Colonial Sulawesi (Indonesia): The First Fertility Transition and Its Causes," *Population Studies* (forthcoming).

On the unhealthful character of tropical areas, see Burnet, Macfarlane, and David O. White, 1972, *Natural History of Infectious Disease*. Cambridge: Cambridge University Press; McNeill, William H., 1979, *Plagues and Peoples*. London: Scientific Book Club; Jones, Eric L., 1981, *The European Miracle: Environments, Economies, and Geopolitics in the History of Europe and Asia*. Cambridge: Cambridge University Press; Cohen, Mark Nathan, 1989, *Health and the Rise of Civilization*. New Haven: Yale University Press.

On diseases, see the titles mentioned above regarding demographic history and the disease patterns of tropical areas. In addition, the reader is referred to Le Roy Ladurie, Emmanuel, 1973, "Un Concept: L'Unification Microbienne du Monde (XIVe–XVIIe Siècles)," *Revue Suisse d'Histoire* 23, no. 4, pp. 627–696; Owen, Norman G., ed., 1987, *Death and Disease in Southeast Asia: Explorations in Social, Medical and Demographic History*. Singapore: Oxford University Press (see esp. contributions by Peter Boomgaard and Barend Jan Terwiel); Boomgaard, Peter, 1993, "The Development of Colonial Health Care in Java: An Exploratory Introduction," *Bijdragen tot de Taal-, Land- en Volkenkunde* 149, no. 1, pp. 77–93; Newson, Linda A., 1998, "Old World Diseases in the Early Colonial Philippines and Spanish America." Pp. 17–36 in *Population and History: The Demographic Origins of the Modern Philippines*. Edited by Daniel F. Doeppers and Peter Xenos. Madison: Center for Southeast Asian Studies, University of

Wisconsin; Boomgaard, Peter, 2003, "Smallpox, Vaccination, and the Pax Neerlandica: Indonesia, 1550–1930," *Bijdragen tot de Taal-, Land- en Volkenkunde* 159, no. 4, pp. 590–617; Boomgaard, Peter, "Harvest Failures, Famine, Epidemics, War and Weather in 18th-Century Java, Indonesia," unpublished paper prepared for the ICAS4 conference, Shanghai.

On ENSO, see the bibliography of Chapter 4, and particularly Berlage, Jr., H. P., 1931, "Over het verband tusschen de dikte der jaarringen van djatiboomen (Tectona grandis L.f.) en den regenval op Java," *Tectona* 24, pp. 939–953; Quinn, W. H., and V. T. Neal, 1995, "The Historical Record of El Niño Events." Pp. 623–648 in *Climate since A.D. 1500*. Edited by Raymond S. Bradley and Philip D. Jones. London: Routledge. Very critical of the ENSO lists of Quinn et al. is Ortlieb, Luc, 2001, "The Documented Historical Record of El Niño Events in Peru: An Update of the Quinn Records (Sixteenth through Nineteenth Centuries)." Pp. 207–295 in *El Niño and the Southern Oscillation: Multiscale Variability and Global and Regional Impacts*. Edited by Henry F. Diaz and Vera Markgraf. Cambridge: Cambridge University Press.

On volcanic eruptions and other natural hazards, see Lamb, H. H., 1970, "Volcanic Dust in the Atmosphere: With a Chronology and Assessment of Its Meteorological Significance," *Philosophical Transactions of the Royal Society (London)* Series A 266, pp. 425–533; Simkin, Tom, and Lee Siebert, 1994, *Volcanoes of the World: A Regional Directory, Gazetteer, and Chronology of Volcanism during the Last 10,000 Years*. Tucson: Geoscience Press; Atwell, William S., 2001, "Volcanism and Short-Term Climatic Change in East Asian and World History, c. 1200–1699," *Journal of World History* 12, no. 1, pp. 29–98; Bankoff, Greg, 2003, *Cultures of Disaster: Society and Natural Hazard in the Philippines*. London: RoutledgeCurzon.

On the effects of war, including the deportation of war captives on a large scale, see Graaf, H. J. de, 1958, *De regering van Sultan Agung, Vorst van Mataram 1613–1645 en die van zijn voorganger Panembahan Seda-ing-Krapjak 1601–1613*. 's-Gravenhage: Nijhoff; Terwiel, B. J., 1989, *Through Travellers' Eyes: An Approach to Early Nineteenth-century Thai History*. Bangkok: Duang Kamol; Wyatt, David K., 1997, "Southeast Asia 'Inside Out,' 1300–1800: A Perspective from the Interior," *Modern Asian Studies* 31, no. 3, pp. 689–709; Charney, Michael W., 2004, *Southeast Asian Warfare 1300–1900*. Leiden: Brill; Grabowsky, Volker, 2004, *Bevölkerung und Staat in Lan Na: Ein Beitrag zur Bevölkerungsgeschichte Südostasiens*. Wiesbaden: Harrasowitz; see also a special issue of *South East Asia Research* 12, no. 1 (2004).

On feasting, feuding, raiding, slaving, and debts, see Rutter, Owen, 1986, *The Pirate Wind: Tales of the Sea-Robbers of Malaya*. Singapore: Oxford University Press; Leach, E. R., 1964, *Political Systems of Highland Burma: A Study of*

Kachin Social Structure. London: Bell; Freeman, J. D., 1955, *Iban Agriculture: A Report on the Shifting Cultivation of Hill Rice by the Iban of Sarawak.* London: Her Majesty's Stationary Office; Warren, James Francis, 1981, *The Sulu Zone 1768–1898: The Dynamics of External Trade, Slavery, and Ethnicity in the Transformation of a Southeast Asian Maritime State.* Singapore: Singapore University Press; Reid, Anthony, ed., 1983, *Slavery, Bondage and Dependency in Southeast Asia.* St Lucia: University of Queensland Press; Volkman, Toby Alice, 1985, *Feasts of Honor: Ritual and Change in the Toraja Highlands.* Urbana: University of Illinois Press; Hoskins, Janet, ed., 1996, *Headhunting and the Social Imagination in Southeast Asia.* Stanford: Stanford University Press; Junker, Laura Lee, 1999, *Raiding, Trading, and Feasting: The Political Economy of Philippine Chiefdoms.* Honolulu: University of Hawai'i Press; Knapen, Han, 2001, *Forests of Fortune?: The Environmental History of Southeast Borneo 1600–1880.* Leiden: KITLV; Boomgaard, Peter, 2003, "Human Capital: Slavery and Low Rates of Economic and Population Growth in Indonesia, 1600–1910," *Slavery and Abolition* 24, no. 2, pp. 83–96; Andaya, Barbara Watson, 2004, "History, Headhunting and Gender in Monsoon Asia: Comparative and Longitudinal Views," *Southeast Asia Research* 12, no. 1, pp. 13–52.

On the role of women, see Andaya, Barbara Watson, 2000, "Delineating Female Space: Seclusion and the State in Pre-Modern Island Southeast Asia." Pp. 231–253 in *Other Pasts: Women, Gender and History in Early Modern Southeast Asia.* Edited by Barbara Watson Andaya. Honolulu: Center for Southeast Asian Studies, University of Hawai'i at Manoa; Andaya, Barbara Watson, 2002, "Localising the Universal: Women, Motherhood and the Appeal of Early Theravada Buddhism," *Journal of Southeast Asian Studies* 33, no. 1, pp. 1–30.

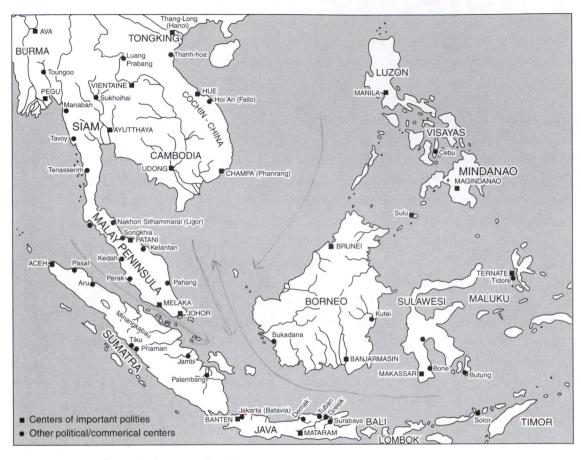

Southeast Asia, around 1600

6

STATE AND ECONOMY

INTERNATIONAL TRADE

Environmental change is caused by internal and external factors. In the period under consideration, population growth, the factor we dealt with in the last chapter, was the internal motor of change, while the external motor was trade, mainly international or long-distance trade. In fact, those factors are interdependent, and while population growth stimulated trade, trade, in turn, stimulated population growth. The term *international trade* refers here predominantly to trade between Southeast Asian polities (and some areas that hardly could be called polities) on the one hand, and European and Chinese (and to a lesser extent Indian) merchants on the other. Japan played a role as well, but during a large part of this period not by way of its own ships and nationals, who were forbidden to sail the inter-regional seas from the early seventeenth century.

Is it possible to decide which factor, the internal or the external, was responsible for the largest part of the environmental changes that took place in Southeast Asia during the Early-Modern Period? If we accept that environmental change in this period was largely the result of economic changes, the question would then be whether economic change was largely caused by internal or external factors.

Of one thing we can be fairly sure: the external influence (trade) was less important before the Early-Modern Period and more important after it. If we may assume that economic and environmental changes in Southeast Asia prior to the Early-Modern Period were set in motion almost entirely by internal developments (mainly climate fluctuations, disease patterns, war, and population growth), and if we also may assume that after approximately 1870 external factors (international trade and colonialism) became about as important as the internal ones in shaping the economy and natural environment, then external influences during the Early-Modern Period must have been modest but notable. Nevertheless, we should consider the possibility that internal factors might have shaped economic and environmental developments to a higher degree than did external ones.

Ideally we should be in a position to express at least the share of exports in GDP in figures, but as we are dealing with the prestatistical—or, at best, protostatistical—

period of Southeast Asia's history, that is extremely difficult. We do have a few pointers, though. It has been estimated that Burmese rice exports during this period never constituted more than 0.5 percent of the rice harvest. In Java during the same period, that figure cannot have been much lower than 2 percent but not much higher than 4 percent, and figures for Siam around 1750 must have been some 2 to 3 percent of the crop. For the Minahassa (Sulawesi, Indonesia) it has been calculated that at least 10 percent of the rice crop was exported during the seventeenth century, and possibly much more. Returning to Burma, we find that around 1800 about 3 to 4 percent of the population depended primarily on foreign trade for their livelihood, but it seems likely that the comparable proportions for Siam, Vietnam, and Java were higher, perhaps as high as 10 percent.

We are also quite confident about total growth for Southeast Asia's international trade as a whole between, say, 1400 and 1870. Again, precise figures are lacking, but it seems safe to say that there was an almost constant increase—although, as we have seen, punctuated by "crises" like that of the seventeenth century—in maritime and overland trade between Europe, India, and China on the one hand and Southeast Asia on the other. This could be regarded as the second globalization wave, the sequel to the one that occurred during the Pre-Modern Period, as was argued in Chapter 3. In the fifteenth century it was demand from China and Europe (via India and the Red Sea or the Persian Gulf). In the sixteenth century Portuguese and Spanish shipping, backed by silver from the Americas, joined the fray, and in the seventeenth century the Dutch took over part of this network, adding their own trade links in the process. The French and particularly the British also expanded the maritime contacts between Europe and Southeast Asia in the seventeenth and eighteenth centuries, and in the eighteenth century the Chinese returned in large numbers.

If, for example, we look at all European shipping with Asia from 1581–1590 to 1721–1730, we find a ninefold increase in the number of ships and an almost sevenfold increase in tonnage. Now, this refers to Asia as a whole, and it has been argued that after 1680 there was a shift away from Southeast Asian products to products from India and China, but part of the increase was certainly also applicable to Southeast Asia.

It should be mentioned here that European shipping at least partly displaced indigenous maritime transport, perhaps particularly after 1511 and between 1650 and 1750. Nevertheless, it is generally assumed that, measured over the period as a whole and across the entire region, European shipping grew more than indigenous shipping dropped.

Another example of long-term growth is the export of pepper, and in this case we do have figures for Southeast Asia alone. Between 1500 and 1800 there was a sixfold increase (in volume) in Southeast Asian pepper exports. That implies a

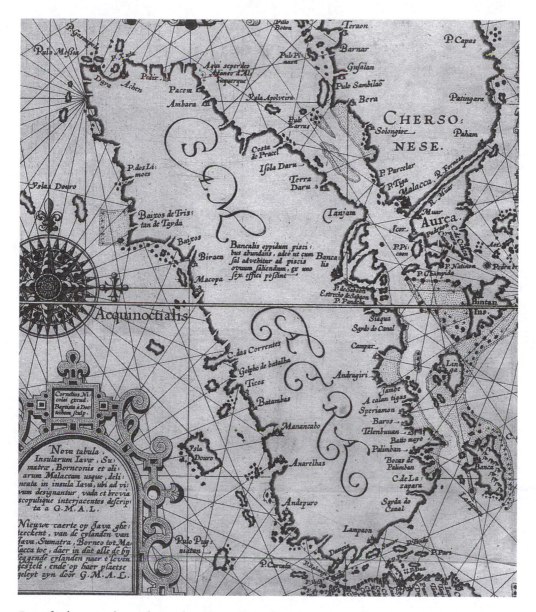

Detail of a map from the 16th century depicting Sumatra and regions of present-day Aceh Province in Indonesia. International trade throughout the region appears to have been growing during the period. (James Ford Bell Library, University of Minnesota)

growth rate of 0.67 percent per year, compared with a population growth rate of 0.2 percent or less. The increase in Southeast Asian sugar exports between 1640 and 1830 (on the eve of the so-called Cultivation System in Java) was fourfold, or 0.82 percent per year on average. Of course, not all export products were doing so well, but it seems likely nevertheless that exports were growing at a higher rate than the population, which implies that the share of exports of total GDP was

Table 6.1 Export products of Southeast Asia, by category and "country," seventeenth century.

Products	Burma	Malaysian Peninsula	Indonesia	Siam	Cambodia/ Laos	Vietnam	Philippines
Forest							
Benzoin			X	X	X	X	
Camphor			X				
Cardamom	X				X		
Cutch	X						
Deer hides	X			X	X		
Eaglewood			X	X		X	
Ebony			X			X	X
Elephants/ Ivory	X	X	X	X	X	X	
Lacquer	X	X	X	X	X		X
Mats		X					
Musk	X				X	X	
Rattan			X				X
Resins	X		X			X	X
Rhino horn			X		X	X	
Sago	X	X					
Sandalwood			X				X
Sappanwood	X		X	X	X		
Timber	X	X	X				X
Wax/Honey	X		X		X		X
Agriculture							
Abaca							X
Cloves			X				
Coconut oil			X	X			
Cotton	X		X		X		X
Ginger				X			
Long pepper	X						
Nutmeg			X				
Pepper		X	X	X			
Pulses	X		X				
Rice	X		X	X	X	X	
Silk				X	X		
Sugar			X	X			

Table 6.1 continued

Products	Burma	Malaysian Peninsula	Indonesia	Siam	Cambodia/ Laos	Vietnam	Philippines
Livestock							
Hides (cattle, buffalo)				X			
Mining							
Gems	X		X				
Gold	X		X		X	X	
Iron		X				X	
Lead	X						
Petroleum	X						
Saltpeter	X						
Silver	X		X				
Tin	X	X		X			
Zinc	X						
Sea							
Pearls			X				X
Ray skins				X			
Salt	X		X				
Tortoiseshell			X				X

Sources: Stapel, F. W., ed., 1927–1954, Pieter van Dam: Beschryvinge van de Oostindische Compagnie. *5 vols. in 7 parts. 's-Gravenhage: Nijhoff; Glamann, Kristof, 1958,* Dutch-Asiatic Trade, 1620–1740. *'s-Gravenhage: Nijhoff; Wheatley, Paul, 1959, "Geographical Notes on Some Commodities Involved in Sung Maritime Trade,"* Journal of the Malayan Branch, Royal Asiatic Society 32, no. 2, pp. 5–140; Smith, George Vinal, 1977, The Dutch in Seventeenth-Century Thailand. *DeKalb: Northern Illinois University; Bruijn, J. R., F. S. Gaastra, and I. Schöffer, 1987,* Dutch-Asiatic Shipping in the 17th and 18th Centuries. *Vol. 1: Introductory Volume. The Hague: Nijhoff; Ptak, Roderich, and Dietmar Rothermund, eds., 1991,* Emporia, Commodities and Entrepreneurs in Asian Maritime Trade, c. 1400–1750. *Stuttgart: Franz Steiner; Knaap, Gerrit J., 1996,* Shallow Waters, Rising Tide: Shipping and Trade in Java around 1775. *Leiden: KITLV; Nagtegaal, Luc, 1996,* Riding the Dutch Tiger: The Dutch East Indies Company and the Northeast Coast of Java, 1680–1743. *Leiden: KITLV; Corpuz, O. D., 1997,* An Economic History of the Philippines. *Quezon City: University of the Philippines Press; Kersten, Carool, 2003,* Strange Events in the Kingdoms of Cambodia and Laos (1635–1644). *Bangkok: White Lotus; Dijk, Willy Olga, 2004, "Seventeenth Century Burma and the Dutch East India Company 1634–1680," unpublished Ph.D. diss., Leiden University.*

increasing during the Early-Modern Period.

International trade, therefore, appears to have been growing throughout most of the period under discussion—with temporary downswings, no doubt—not only in absolute but also in relative terms, as a percentage of GDP. It must also be supposed—and we look into this question in more detail presently—that this growth of international trade was reflected by the increasing exploitation of natural resources (in addition to the increase resulting from population growth). Clearly, products obtained from the exploitation of forests, mines, the sea, and agriculture, including livestock keeping, played a predominant role in the exports of Southeast Asia, as is shown in Table 6.1.

In other words, Southeast Asia exported primary products (raw materials), while importing mainly manufactured goods (cotton and silk textiles, ceramics, and opium), in addition to bullion and specie (nonminted and minted gold, silver, and copper).

A few comments on Table 6.1 are in order. We do not have complete trade statistics for any Southeast Asian country for any year prior to the nineteenth century, let alone series of statistics that could show developments over time. Entries in the table are based partly on Dutch East India Company (VOC) trade data and partly on other (often also European) observations. This could be a source of some distortion—for instance, because the reports used might under-represent overland trade between Mainland Southeast Asian countries and India and China. Another point is that the boundary between agricultural and forest products is blurred. Cardamom, for instance, was collected in the forest in some areas, but cultivated in others. Finally, this table under-represents the contribution of the "industrial" (or rather, artisanal) sector, because industrial products have been lumped together with the raw material used for their production. Thus, cotton textiles have been classed as "cotton" under "Agriculture," together with cotton yarn and raw cotton, and the same applies to silk. Mats and coconut oil should also be considered artisanal products. Generally speaking, however, the industrial component of the items listed here was small. Locally, the export of cotton cloth may have been important, but as a rule Southeast Asian countries were importing cotton textiles (from India), not exporting them.

SHARE OF ECONOMIC SECTORS IN EXPORTS

Do we have any idea in what proportions the four sectors (forests, agriculture, mining, and fishing) were represented in the exports of the countries concerned? We cannot produce figures for the whole region and for the entire period under consideration, but for two regions in the seventeenth century we do have some notion of proportions, as witness Table 6.2. This table presents a breakdown as to sector for the ten or so most important commodities exported by the VOC from Burma and Thailand during the seventeenth century.

Table 6.2 demonstrates that it is difficult to generalize for the whole of Southeast Asia, as there was evidently quite some variation between states regarding the share of the economic sectors in the total international (maritime)

Table 6.2 Proportions (%) in terms of value of the most important VOC export commodities from Burma and Thailand during the seventeenth century, by sector of the economy.

Sector	Burma	Thailand
Mining	80	8
Forests	18	64
Agriculture	2	26
Fishing	—	2

Sources: Smith, George Vinal, 1977, The Dutch in Seventeenth-Century Thailand. DeKalb: Northern Illinois University; Dijk, Willy Olga, 2004, "Seventeenth Century Burma and the Dutch East India Company 1634–1680," unpublished Ph.D. diss., Leiden University.

export package (assuming that the VOC exports may be held to be representative). We cannot say much more than that, in the two areas concerned, mining and forestry between them were responsible for the majority of exports (in terms of value), while fishing was the least important sector of the four, with agriculture taking up a position in between.

For the part of the Indonesian Archipelago under Dutch rule, and that of the Philippines under the Spaniards, we have some export data for around 1800, as is shown in Table 6.3.

Again, the variation is enormous, but in both cases the agricultural sector is clearly much larger than it was in the seventeenth century in Burma and Thailand (Table 6.2). The data for Indonesia, dating from 1797, and representing exports from Batavia (present-day Jakarta), are possibly distorted by the fact that at that moment there was a war on in Europe—and therefore also in the Asian territories under European dominance. The war had severed the usual trade connections, which led to bad times for Dutch shipping, and to a shift to neutral (U.S. and Danish) ships. However, the very large share accruing to agriculture in Table 6.3 does reflect the normal situation well, as we know from comparable figures dating from the 1740s and from 1795. The products constituting this sector were, first and foremost, coffee, followed by sugar, cloves, pepper, and nutmeg. Of these products coffee was entirely new—that is, introduced in Java after 1700.

Table 6.3 Proportions (%) of the most important export commodities from VOC Indonesia and the Spanish Philippines ca. 1800, by sector of the economy.

Sector	Philippines	Indonesia
Mining	1	7
Forests	46	1
Agriculture	40	92
Fishing	10	—

Sources: *De Jonge, Jhr. J.K.J., 1888,* De Opkomst van het Nederlandsch Gezag in Oost-Indië, *Vol. 12. 's-Gravenhage: Nijhoff; Corpuz, O.D., 1997,* An Economic History of the Philippines. *Quezon City: University of the Philippines Press.*

As coffee represented about half of all exports mentioned here, it can be imagined that the proportions were totally different around 1700. The large share of agriculture in the Philippines is largely due to indigo and sugar, two products that were insignificant prior to 1700. In all likelihood, therefore, the share of agriculture in the exports of the two areas (the western-dominated parts of the Indonesian and Philippine archipelagoes) had been much smaller in the seventeenth century, which would be in keeping with the findings for that period for Burma and Thailand, as reflected in Table 6.2.

In the eighteenth century the importance of the fishery sector may have increased, mainly because of growing interest by the Chinese markets for products of the sea such as trepang (sea cucumber) and tortoiseshell. That would explain the fairly large share of the fisheries sector in the Philippines around 1800, a sector that would probably have been much smaller in 1700, which is in keeping with the data for Burma and Thailand in Table 6.2. The absence of data for this sector in the Indonesian figures is caused by the fact that the figures in the table represent only the VOC exports, while, for instance, trepang and tortoiseshell were being exported in fairly large quantities to China, but not by the VOC. This is also the explanation for the very small share of the forestry sector in the figures of the Indonesian VOC exports, as the export of edible birds' nests by private traders alone was of considerable importance.

Summing up these findings, it could be said that the importance of the (often large) share of forest products in Southeast Asian exports may have declined somewhat between 1600 and 1800, while the shares of fishing and particularly agriculture increased. Mining was very important in Burma, but elsewhere its share in export production appears to have been rather modest.

Table 6.1, therefore, would also have looked different if data from the eighteenth century had been included. Particularly during the latter half of the

Workers pick coffee beans along irrigation ditches in Singapore. An introduced crop, coffee quickly became a major export. (Library of Congress)

century—and in many cases, concentrated in the last quarter of the century—we see the expansion of the production and export of new products, introduced from outside, often at a much earlier date, in addition to the spread, sometimes to other areas, of more traditional products from the region. Examples of the first category are coffee and tobacco, introduced from Arabia and America, respectively. Expansion of old crops refers to rice, indigo, cotton, gambier, sugar, and pepper. We also witness an increase in the exports of tin, timber, and ships.

Generally speaking, there was a gradual shift between 1500 and 1800 away from expensive luxury goods characterized by high value for small volume (spices, valuable woods, resins, musk) to more bulky wares with low value per volume (rice, tin, timber, rattan, cutch, sugar, pepper, coffee, kapok, and sappan). This does not mean that the products with high value per unit became unimportant, but that their share of all exports diminished. It seems likely that the growth of the share of bulk products in the expanding world market led to greater environmental impact in Southeast Asia than would have been the case if the same growth, in terms of value, had been in products with high value per volume. In other words, the increased exports of, for instance, timber and tin led to more environmental "damage" than would have been the case if the export of spices had increased for the same amount of money.

Exports of products of the sea appear to have increased as well, as did their share in total exports, but there does not seem to be a similar tendency for bulkier products with low value per unit for this category of products: trepang had a high value per unit, and tortoiseshell a very high value.

THE STATE, THE RULER, AND THE ARISTOCRACY

So far the environmental impact—in very general terms—has been mentioned in regard to the increased production of commodities for the growing export trade from Southeast Asia during the Early-Modern Period. As economic exploitation is "mediated" by the state, some reflection on the role of the ruler and his apparatus is in order.

The most important point to be made here is that a large share of the long-distance trade in the area was "managed" trade. The state, in many instances, was the only buyer, or rather "receiver," of commodities for the international market, and the only seller as well. At the very least the prices for which these commodities could be sold or bought were set by the state. The notion of free trade may have existed regarding products of purely local importance, but internationally traded, and therefore rather easily taxable, goods that were being exported in large quantities (or at least in sufficiently large amounts to generate a notable source of income) usually had to be channeled through state institutions, which were often monopolists and monopsonists at the same time. However, particularly in the Malay world, there were some free-trading ports, such as Malacca before the arrival of the Portuguese and Makassar before VOC dominance. They may be regarded as collection points for stateless areas. Although, again, statistics have not yet been collected in such a way that comparison is feasible, it would appear that these ports, in terms of the value of cargo traded, could not compete with the value of managed trade of the larger political entities.

It is well known that the Dutch VOC in the Indonesian Archipelago and the Spanish in the Philippine Archipelago attempted to establish monopolies in the most important export (and import) commodities. This was successful in some but not all cases. The VOC acquired a monopoly in cloves, nutmeg, and mace (products to be found only in the Moluccas) but did not succeed in monopolizing products such as pepper, sappan, and tin, although they did acquire a large share of total exports of those products. The latter commodities were produced in a fairly restricted area, but still too large for the VOC to control entirely.

Another famous example of a monopoly was the so-called Manila Galleon. Every year the Spanish Crown transported products (mainly produced in China, not in the Philippines) from Manila to Acapulco in Mexico, and exported silver from Mexico to the Philippines in order to pay for them. In the eighteenth century, the cultivation and export of tobacco was established as a state monopoly by the Spaniards in the Philippines.

In both the Dutch- and the Spanish-ruled areas, a range of products had to be produced for the state, either for free (as tax in kind) or for a price far below the market value, set by the state itself. These commodities were mainly export products, but they included timber, needed for the construction of urban buildings and ships.

It is perhaps less well known that indigenous monarchs were, as a rule, also monopolists as regards the most important export commodities. In the case of Thailand, for instance, where the export sector was relatively large, a large array of export commodities was produced by the villagers in lieu of tax. Examples were cotton, cardamom (which was collected from forests), sappanwood, eaglewood, iron, silver, and gold. This also applied to products needed for the activities of the state itself, such as brick and lime. In the case of sappanwood, we know that the people who had to deliver this tax-in-kind were expected to serve two months in the year, during which time they had to produce fifty pieces of sappan of about 2 meters by 30 centimeters. Other export products were taxed at the source, which was the case with palm sugar, pepper, gamboge (a resin), and tobacco.

In this way, it could be argued, both the colonial and the indigenous states stimulated export production, while at the same time appropriating what was probably the larger share of the surplus generated by increased production. However, because corruption (as we would call it today) was rife in all of these societies, we can be sure that an unknown proportion of such earnings did not reach the court and remained in the region, albeit in all likelihood not in the hands of the producers. This was called smuggling by its detractors. In later chapters it is shown that such illegal production—in the eyes of the central state—would remain an important feature for local power holders, with equally important environmental consequences, right up to the present.

It was not only the ruler who profited from increased exports, but also the nobility, which was more or less identical with the bureaucracy, perhaps with the exception of the lowest rung of the ladder, the village heads (they also profited, but they were, as a rule, not regarded as aristocrats). These noblemen-cum-bureaucrats were often involved in trade and in the various monopolies just mentioned. A good example of this is the group of people in the Malay world who were called collectively *orang kaya*. This term, which means "rich people," is the usual expression for the merchant-nobility of the ports-of-trade-cum-sultanates in the Malay Peninsula and the Indonesian Archipelago. Unfortunately, we do not have as much information about these people as we might wish for, as in most other societies the well-to-do merchants who were either aspiring to noble status or at least involved in the administration of the realm as members of parliament, tax farmers, or as members of city councils were one of the most dynamic elements of society.

However, the impression one gets from this group in Southeast Asia is that they were not so much risk-takers and (direct) investors as rent-seekers (portfolio investment). As a rule, rent-seeking is not an engine of growth. It stimulates people who have been appointed by the ruler in a certain function, often for a limited period of time, to make the most of it in terms of income during their tenure. As they will not remain in office for long, there is no point in investing; they would not reap the rewards of that investment. These short-term office-holders, therefore, had every incentive to wring their subjects dry, which they could do with impunity, and without ill effects on their own income, because they would no longer be around when the impoverished people of their region might fail to pay the required taxes or rise up in rebellion. Eric Wolf has used the term *prebendal*—a word loosely based on the writings of Max Weber—for this type of behavior, as opposed to the "feudal" arrangements in European societies, where the aristocracy, including those who had bought their titles, were long-term stakeholders in the areas they administered.

This is not to say that there was no indigenous investment at all in foreign trade. We do know, for instance, that in the mid-seventeenth century the sultan of Banten (Java) owned, equipped, and loaded ships that he sent on trading expeditions, all financed by himself. Similar ventures were undertaken by the rulers of Malacca, Acheh, and Siam, often investing large amounts of money together with family members and other aristocrats, a type of arrangement that is usually called *commenda* by economic historians. However, direct investment on such a scale by indigenous capital owners seems to have remained fairly exceptional—more exceptional, it would appear, than in China and India, and certainly more so than in Europe.

FOREIGN MERCHANTS

Instead of a prosperous and indigenous (non-noble) merchant class that made money by risk-taking, much of the long-distance trade was often in the hands of foreigners. Of course, there were the chartered European trading companies and the European crown-sponsored and -financed trading ventures, but there were also many Armenian, Arab, Persian, Indian, and particularly Chinese merchants who occupied a profitable niche, one that had not been filled by local traders. It has been suggested that this was a conscious policy of Southeast Asian rulers in order to avoid the growth of a truly local, wealthy (and non-noble) indigenous mercantile elite that could have become a threat to the ruler's authority. It was easier, so the argument goes, to control and discipline foreign traders who did not have many links to the local society.

Discipline in many ports of trade and other towns and cities included confiscation by the ruler of the possessions of a (foreign) merchant or a nobleman after his death, thereby also getting rid of possible IOUs signed by the ruler to the benefit of the deceased. The legal term for such a procedure is *escheat*. Certainly, the laws in most of the polities dealt with here during the Early-Modern Period contained clauses safeguarding property against theft, fraud, and the like, but the ruler was above the law and could fairly well do as he pleased, at least in the larger, more centralized states; in the smaller polities there was sometimes more of a balance of power. There were many clauses in Southeast Asian laws stating that all kinds of rules applied under normal conditions, but not in the case of natural disasters or acts of the ruler. In Siam, a law issued in 1635 stated unequivocally that the king could appropriate all property belonging to higher officials. In addition, there are many instances in seventeenth- and eighteenth-century Siam of noblemen being summarily executed by a displeased ruler. There is a seventeenth-century VOC source explicitly stating that there are no rich indigenous merchants in Banjarmasin (Borneo), because the sultan would fleece everyone with capital.

The ruler ruled absolutely, and his behavior was often erratic and despotic, although there were doubtlessly many rulers who had the best interest of their subjects at heart. The rule of law applied only as long as it pleased the ruler. There were no civic institutions to counterbalance the rule of the monarch, which made it difficult for a class of rich, indigenous merchants to flourish; that, in turn, must have limited economic growth. Even the indigenous nobility was, as was shown above, at the mercy of the ruler.

Foreign residents were well aware of what could happen if they incurred the displeasure of the ruler, and many of them maintained their networks—or even their headquarters, and part of their families (or another family)—in their country of origin. They also made sure that a sizable portion of their earnings was remitted to the home country, one more factor—and not one often mentioned—that must have had a negative impact on economic growth in Southeast Asia. It goes without saying that the drain of Portuguese, Spanish, Dutch, English, and French earnings would also have had a negative impact on the regional economy, a point that has been made by many scholars, past and present.

It may be added as a short footnote here that not too much should perhaps be made of the difference between indigenous nobility and foreign merchants, given the fact that successful foreign merchants in many cases became bureaucrats in the employ of the rulers; if they played their cards well, they were able to blend into the aristocracy-cum-bureaucracy. They sometimes married into the aristocracy and could end up as parents or parents-in-law of rulers-to-be, particularly in Siam. Be that as it may, the absence of the rule of law regarding the ruler's claims to the property of the rich, and the lack of civic institutions that could counterbalance the power of the monarch, must have limited the structural development of any class of independent well-to-do merchant-capitalists.

Perhaps geography played a role as well in all of this: while Canton was far away from Peking, and Surat and Bengal were far away from Agra and Delhi, ports-of-trade in Southeast Asia were either also capital cities of the polity or too close for comfort. As a result, everything took place under the watchful eyes of the monarch, while merchants in India and China benefited from the distance between them and the ruler. As we have seen so often in the previous chapters, scale does matter.

It seems to me that Victor Lieberman, in his eagerness to show how similar the political and economic development of Southeast Asia was to that of China and Europe, has not paid sufficient attention to these features. He does acknowledge that some rulers displayed behavior detrimental to the economy, but he does not appear to regard the absence of the rule of law in this respect—in addition to the absence of a civil society, the prevalence of rent-seeking, and the remittance of earnings to other regions—as structural factors that must have severely curtailed economic growth. This is in keeping with the fact that the notion of the "Oriental despot" has not been very fashionable during the last decades, partly because Edward Said's book *Orientalism* rendered this type of concept suspect, and partly because the Hydraulic Society hypothesis of Karl Wittfogel (see Chapter 3) has been largely discredited. Nevertheless, whatever term we use for this phenomenon, the monarch of the Early-Modern Period in Southeast Asia was evidently not bound by the rule of law, or kept in check by a parliament; rich people—both foreign and particularly indigenous—had always the threat hanging over them of confiscation of their possessions or even execution on trumped up charges (such as sorcery); therefore a strong and independent indigenous class of well-to-do merchants that was not identical with the aristocracy (and therefore not really independent) had not come into existence.

Strangely enough, it looks as if at least in Java a fairly strong indigenous class of non-noble merchants had existed prior to 1500, but around 1600 they seem to have disappeared. The growth of international trade, therefore, does not appear to have been conducive to the growth of such a class. On the contrary, international trade apparently strengthened the position of monarchs, aristocrats, and a number of foreign merchants. This, by the way, is remarkably similar to what would happen after the 1960s, as is shown in later chapters.

The foreign merchants evidently played a role in economic growth in the region. Particularly the Chinese merchants were not only heavily involved in international maritime trade but also were often the middlemen between the peasantry and the port of trade, providing transportation and also investing in backward linkages: agriculture, mining, fishing, and industry, topics dealt with presently. They also gathered large fortunes as moneylenders and tax-farmers.

In their latter capacity, they made huge sums of money available to the rulers, including the European ones. It has been argued that tax-farming was introduced by the Dutch in Southeast Asia. They did so in seventeenth-century Batavia, with the Chinese merchants residing there as the major tax-farmers. In the eighteenth century, and particularly in the second half, tax-farming spread to many other areas in Southeast Asia, together with the Chinese entrepreneurs who settled there in increasing numbers. For Southeast Asian states, it was a convenient system of revenue collection. Instead of having to set up an entire tax-levying apparatus, the state could auction off the right to collect specific taxes to the highest bidder, more often than not a Chinese capitalist.

Important sources of income were the sale of opium, tobacco, and alcohol, in addition to gambling and cockfighting—vices, in other words. In fact, everything that was marketable could be taxed and the right to do so farmed out. Import and export duties, the sale of rice, fish, livestock, and meat, as well as textiles—all were at the basis of revenue farms. In addition to commodities that were taxed when they were marketed, other products were taxed at the source—that is, the place where they were collected or produced, with salt and edible birds' nests as prime examples.

The largest single tax farm was usually the opium farm. The amounts of money involved were so large that, as a rule, no single merchant could pay the money for it up front: a group of Chinese capitalists—a *kongsi*—had to cooperate in order to bring together such an enormous sum. The same capitalists were also the ones who exploited tin mines or sugar plantations, which were operated with Chinese immigrant labor, often indentured labor, indebted to the same capitalist, which goes to show that certainly not all Chinese inhabitants of Southeast Asia were rich. As these laborers were often also addicted to opium or alcohol, and spent much of their money on gambling, the Chinese revenue farmers made good on their initial outlay many times over, thus accumulating more capital than indigenous entrepreneurs might have done. As a class they may not have done much for improving the livelihood of their laborers, but they certainly were engines of economic growth and props of the indigenous and colonial states.

THE STATE, ECONOMIC GROWTH, AND THE WORLD ECONOMY

As we have seen, the proportion of GDP consisting of foreign trade was modest. However, the influence of foreign trade on state income—or rather, the income of the ruler and his aristocracy—was much larger than its share in the total economy. In a number of cases income from foreign trade in the eighteenth century was probably the largest single source of Crown revenue. From the sixteenth century onward, this was partly translated into modern firearms and absolutism, and these features arguably led to more and more destructive wars and to a more effective repression of the state's own subjects, not unlike the way in which income from primary resources is spent today in many African states. The environmental effects of this are difficult to judge. On the one hand it must have led to the wanton destruction of means of production and stocks of food, and may thus be regarded as environmentally unsound behavior. On the other hand, it probably led to lower rates of economic and population growth, which might have been beneficial to the natural environment.

However, waging war and oppressing their citizens were not the only activities of rulers. The state—both colonial and indigenous—also attempted to stimulate economic growth, pursuing developmental policies avant la lettre. This was partly a question of economic incentives: individuals, often foreigners, would be given land, sometimes for free, and licensed to start a sugar plantation or a tin mine. Occasionally the state might even have given credit to prospective collectors or producers of export commodities. However, things were partly, and perhaps largely, arranged as in a command economy: local rulers or lower officials were simply told to produce so much coffee, pepper, or cotton yarn. A final incentive was that the state either itself imported certain commodities or permitted others to do so, thus allowing market forces to do their work by making people produce export commodities in order to acquire desirable (prestige) items, even if the import of the commodities in question, such as Indian textiles, was a government monopoly.

As a rule, the state fully supported the foreign—usually Chinese—tax-farmers-cum-merchants. However, occasionally relations between the Chinese on the one hand and the state or local population on the other broke down, as happened during the Chinese massacres in Manila in 1603, in Batavia in 1740, and in Siam in 1848. Anti-Chinese popular discontent, however, was not restricted to those conspicuous examples. Nor were all riots anti-Chinese or antiforeigner. The anti-Chinese movements are the ones that have impressed historians most—and they certainly appear to be a recurring phenomenon in post-1600 Southeast Asian history—but people obviously rebelled against many instances of a failing "moral economy," of whatever origin.

Production for the international markets could, in principle, be profitable for the producers, and there are instances in which that was, indeed, the case, as witness the rapid spread of coffee in western Java in the early eighteenth century, before it became a compulsory crop. What appears to be typical, however, for Southeast Asia in the Early-Modern Period, is that all too often commodities for for-

A young Javan man wearing the garb of the royal court, circa 1807. Indigenous rulers often worked with foreign merchants, playing a crucial role in the exporting of goods. (Hulton Archive/Getty Images)

eign markets were produced or collected by order of the ruler, who would either receive them in lieu of tax or would pay a price far below the going market rate. Looking, for instance, at pepper in seventeenth- and eighteenth-century Java and Sumatra, it is evident that the VOC paid a fairly low price to the rulers of Banten,

Cirebon, Central Java, Palembang, Jambi, and Sumatra's West Coast. They, in turn, paid even less to the middlemen or officials who delivered the pepper to them, while it goes without saying that the pepper growers received an even lower payment still.

Part of the difference in prices at the farm gate and at the monarch's warehouse could be accounted for as transportation costs. That is a factor that has not received much attention in the historical literature, but one should not underestimate the cost of transport as a factor in the total price structure, given the inaccessibility of many primary producing hinterlands in Southeast Asia. This factor in itself should be regarded as an obstacle to economic growth (and, as we have seen, state formation), and one that would be only gradually removed at a much later date, even though some improvements, such as the canals dug by the Siamese state during the eighteenth and early nineteenth centuries, came into being during the period under consideration.

Both features taken together—the high cost of transportation and the low prices paid by those higher up in the hierarchy—might explain why the local sedentary and semisedentary peasants, as well as the foragers, many of whom were participating in production or collection for the world market, did not benefit much from such participation.

And then we have not yet mentioned the many slaves, serfs, and people in debt-bondage who had to work for free or at least for not more than their upkeep, in addition to all of those who produced commodities in statute labor for the ruler. Although figures are lacking, these types of bonded labor appear to have been more important than they have been given credit for. Although bonded labor is very old in Southeast Asia, as was shown in Chapter 3, it would seem plausible that its expansion was stimulated by the growth of long-distance trade. Bonded labor may be expected to increase numerically under a number of mutually supportive conditions—growing demand for commodities, to be met in a region with low population densities, where natural and manmade hazards led to insecurity and indebtedness, while raids and wars produced a constant supply of bonded people. Therefore, the same quest for commodities that might have been a stimulus to somewhat higher rates of population growth (see Chapter 5) was also instrumental in enlarging or at least maintaining a bonded labor force, partly because population growth was too slow.

It could be argued that indigenous rulers, bureaucrats, aristocrats, and some foreign merchants thus shielded the indigenous peasantry from too much influence from the world market, although the latter did obviously produce for it in addition to being (limited) consumers of some of its products. In many cases, there was hardly any contact between European and other foreign exporters on the one hand and the producers on the other. For instance, the VOC never came into contact with the pepper producers in the Palembang uplands because the pepper was collected by the sultan of Palembang's men, who delivered the produce they had collected into the warehouses of the sultan or those of the VOC in the port of Palembang.

Perhaps this state of affairs has led scholars like Immanuel Wallerstein to suggest that Asia until 1750 remained outside the "Modern World-System" or the "European World-Economy"—that in contrast to Latin America, which was regarded as the periphery of this system. Quite apart from the questionability of a "European World-Economy" prior to 1800, as witness recent historiography (Andre Gunder Frank, Kenneth Pomeranz), it might be argued that Wallerstein's view of an Asia outside such a world economy does not do justice to the penetration of world market demand into even the remotest corners of Southeast Asia, and the growth of bonded labor that accompanied it. Granted that the impact of the world market on several areas of Latin America may have been larger than that regarding most of Southeast Asia, and granted that this impact increased after 1750, far-reaching world market influences prior to that date, even as early as the sixteenth century, nevertheless cannot be denied.

This, in turn, may have led scholars to underestimate the environmental impact of production for the world market in Early-Modern Southeast Asia.

BIBLIOGRAPHICAL ESSAY

The most important authors on this period are, as we saw in the last chapter, Anthony Reid and Victor Lieberman. That applies to the topics dealt with in this chapter as well: Reid, Anthony, 1988 and 1993, *Southeast Asia in the Age of Commerce 1450–1680*. 2 vols. New Haven: Yale University Press; Reid, Anthony, 1999, "Economic and Social Change, c. 1400–1800." Pp. 116–163 in *The Cambridge History of Southeast Asia*. Vol. 1, Part 2: *From c. 1500 to c. 1800*. Edited by Nicholas Tarling. Cambridge: Cambridge University Press; and Reid, Anthony, 1997, "A New Phase of Commercial Expansion in Southeast Asia, 1760–1840." Pp. 57–82 in *The Last Stand of Asian Autonomies: Responses to Modernity in the Diverse States of Southeast Asia and Korea, 1750–1900*. Edited by Anthony Reid. Houndmills: Macmillan; New York: St. Martin's; Lieberman, Victor, 2003, *Strange Parallels: Southeast Asia in Global Context, c. 800–1830*. Vol. 1: *Integration on the Mainland*. Cambridge: Cambridge University Press.

Data on export products can be found in Stapel, F. W., ed., 1927–1954, *Pieter van Dam: Beschryvinge van de Oostindische Compagnie*. 5 vols. in 7 parts. 's-Gravenhage: Nijhoff; Cortesão, A., ed., 1944, *The Summa Oriental of Tomé Pires: An Account of the East from the Red Sea to Japan, Written in Malacca and India in 1512–1515*. London: Hakluyt Society; Glamann, Kristof, 1958, *Dutch-*

Asiatic Trade, 1620–1740. 's-Gravenhage: Nijhoff; Wheatley, Paul, 1959, "Geographical Notes on Some Commodities Involved in Sung Maritime Trade," *Journal of the Malayan Branch, Royal Asiatic Society* 32, no. 2, pp. 5–140; Smith, George Vinal, 1977, *The Dutch in Seventeenth-Century Thailand*. DeKalb: Northern Illinois University; Viraphol, Sarasin, 1977, *Tribute and Profit: Sino-Siamese Trade, 1652–1853*. Cambridge, MA: Council on East Asian Studies, Harvard University; Bruijn, J. R., F. S. Gaastra, and I. Schöffer, 1987, *Dutch-Asiatic Shipping in the 17th and 18th Centuries*. Vol. 1: *Introductory Volume*. The Hague: Nijhoff; Boomgaard, Peter, 1989, "The Javanese Rice Economy, 800–1800." Pp. 317–344 in *Economic and Demographic Development in Rice Producing Societies: Some Aspects of East Asian Economic History*. Edited by Akira Hayami and Yoshihiro Tsubouchi. Tokyo: Keio University; Terwiel, B. J., 1989, *Through Travellers' Eyes: An Approach to Early Nineteenth-Century Thai History*. Bangkok: Duang Kamol; Ptak, Roderich, and Dietmar Rothermund, eds., 1991, *Emporia, Commodities and Entrepreneurs in Asian Maritime Trade, c. 1400–1750*. Stuttgart: Franz Steiner; Knaap, Gerrit J., 1996, *Shallow Waters, Rising Tide: Shipping and Trade in Java around 1775*. Leiden: KITLV; Nagtegaal, Luc, 1996, *Riding the Dutch Tiger: The Dutch East Indies Company and the Northeast Coast of Java, 1680–1743*. Leiden: KITLV; Corpuz, O. D., 1997, *An Economic History of the Philippines*. Quezon City: University of the Philippines Press; Bulbeck, David, et al., 1998, *Southeast Asian Exports since the 14th Century: Cloves, Pepper, Coffee, and Sugar*. Singapore: ISEAS/KITLV; Junker, Laura Lee, 1999, *Raiding, Trading, and Feasting: The Political Economy of Philippine Chiefdoms*. Honolulu: University of Hawai'i Press; Kersten, Carool, 2003, *Strange Events in the Kingdoms of Cambodia and Laos (1635–1644)*. Bangkok: White Lotus; Ariyasajsiskul, Supaporn, 2004, "The So-called Tin Monopoly in Ligor: The Limits of VOC Power vis à vis a Southern Thai Trading Polity," *Itinerario* 28, no. 3, pp. 89–106; Dijk, Willy Olga, 2004, "Seventeenth Century Burma and the Dutch East India Company 1634–1680," unpublished Ph.D. diss., Leiden University; Knaap, Gerrit, and Heather Sutherland, 2004, *Monsoon Traders: Ships, Skippers and Commodities in Eighteenth-Century Makassar*. Leiden: KITLV.

On the role of indigenous versus foreign merchants, including the lack of security even for the rich and powerful under Southeast Asian absolutist rulers, see Reid, Anthony, 1975, "Trade and the Problem of Royal Power in Aceh. Three Stages: c. 1550–1700." Pp. 45–55 in *Pre-Colonial State Systems in Southeast Asia: The Malay Peninsula, Sumatra, Bali-Lombok, South Celebes*. Edited by Anthony Reid and Lance Castles. Kuala Lumpur: Council of the Malaysian Branch of the Royal Asiatic Society; Lombard, Denys, and Jean Aubin, eds., 1988, *Marchands et Hommes d'Affaires Asiatiques dans l'Océan Indien et la Mer de*

Chine 13e–20e Siècles. Paris: EHESS; Colombijn, Freek, 1989, "Foreign Influence on the State of Banten, 1596–1682," *Indonesia Circle* 50, pp. 19–30; Dhiravat Na Pombejra, 1990, "Crown Trade and Court Politics in Ayutthaya during the Reign of King Narai (1656–88)." Pp. 127–142 in *The Southeast Asian Port and Polity: Rise and Demise*. Edited by J. Kathirithamby-Wells and John Villiers. Singapore: Singapore University Press; Kathirithamby-Wells, Jeyamalar, 1993, "Restraints on the Development of Merchant Capitalism in Southeast Asia before c. 1800." Pp. 123–148 in *Southeast Asia in the Early Modern Era: Trade, Power, and Belief*. Edited by Anthony Reid. Ithaca: Cornell University Press; Butcher, John, and Howard Dick, eds., 1993, *The Rise and Fall of Revenue Farming: Business Elites and the Emergence of the Modern State in Southeast Asia*. Houndsmill: Macmillan; Boomgaard, Peter, 1996, "Geld, Krediet, Rente en Europeanen in Zuid- en Zuidoost-Azië in de Zeventiende Eeuw." Pp. 483–510 in *Kapitaal, Ondernemerschap en Beleid: Studies over Economie en Politiek in Nederland, Europa en Azië van 1500 tot Heden*. Edited by C. A. Davids, W. Fritschy, and L. A. van der Valk. Amsterdam: NEHA; Trocki, Carl A., 1997, "Chinese Pioneering in Eighteenth-Century Southeast Asia." Pp. 83–102 in *The Last Stand of Asian Autonomies: Responses to Modernity in the Diverse States of Southeast Asia and Korea, 1750–1900*. Edited by Anthony Reid. Houndmills: Macmillan; New York: St. Martin's; Dhiravat Na Pombejra, 1998, "Princes, Pretenders, and the Chinese Phrakhlang: An Analysis of the Dutch Evidence concerning Siamese Court Politics, 1699–1734." Pp. 107–130 in *On the Eighteenth Century as a Category of Asian History: Van Leur in Retrospect*. Edited by Leonard Blussé and Femme Gaastra. Aldershot: Ashgate; Trocki, Carl A., 2002, "Opium and the Beginnings of Chinese Capitalism in Southeast Asia," *Journal of Southeast Asian Studies* 33, no. 2, pp. 297–314; Cooke, Nola, and Li Tana, eds., 2004, *Water Frontier: Commerce and the Chinese in the Lower Mekong Region, 1750–1880*. Lanham, MD: Rowman and Littlefield; Kwee Hui Kian, 2006, *The Political Economy of Java's Northeast Coast c. 1740–1800*. Leiden: Brill.

On the terms "prebendal" and "feudal" as used in the text, see Wolf, Eric, 1966, *Peasants*. Englewood Cliffs, NJ: Prentice-Hall. For Southeast Asian law during this period, see Hooker, M. B., 1986, *Laws of Southeast Asia*. Vol. 1: *The Pre-Modern Texts*. Singapore: Butterworth. Said's book mentioned in the text is Said, Edward W., 1979, *Orientalism*. New York: Vintage. On the notion of "moral economy" (originally designed by E. P. Thompson) as applied to Southeast Asia, see Scott, James C., 1976, *The Moral Economy of the Peasant: Rebellion and Subsistence in Southeast Asia*. New Haven: Yale University Press.

On the "Modern World-Economy" discussion, see Wallerstein, Immanuel, 1974, *The Modern World-System: Capitalist Agriculture and the Origins of the*

European World-Economy in the Sixteenth Century. New York: Academic;
Frank, Andre Gunder, 1998, *ReOrient: Global Economy in the Asian Age*.
Berkeley: University of California Press; Pomeranz, Kenneth, 2000, *The Great
Divergence: China, Europe, and the Making of the Modern World Economy*.
Princeton: Princeton University Press.

7

ENVIRONMENTAL CAUSES AND EFFECTS

FORESTS

Having dealt with the two major (human-induced) forces that led to environmental change—population growth and growing demand generated by international trade—it is now time to look at these changes themselves. The drop in forest cover was one of them. By 1400 perhaps as much as 85 to 95 percent of Southeast Asia might still have been under forest cover. We must assume that some areas were too wet, too dry, or too steep for forests to grow there, while human habitation (villages, towns, and cities), the clearing of fields for agriculture, and livestock grazing must also have made inroads on the existing tree cover, but such places were tiny islands in an arboreal sea.

As usual for this period, exact figures are absent or rare, at least for the early years. We have an estimate for the Philippines, dated 1565, which puts the forest cover figure at 92 percent, but one hesitates to see that as much more than a wild guess. By 1874, that figure had dropped to between 65 and 70 percent. Around 1840, Java's forest cover was some 50 percent. However, areas like Borneo, the Malay Peninsula, and New Guinea might, even by 1870, have had a forest cover close to between 80 and 90 percent, which makes it, again, difficult to generalize across Southeast Asia.

Nevertheless, it is quite clear that Southeast Asia was much more thickly forested than China and Europe, and almost certainly also more thickly forested than India. Around 1940, when much better statistics had become available, Southeast Asia had an estimated forest cover of between 60 and 65 percent, compared with 10 percent for China, 20 percent for India and Pakistan, and 30 percent for Europe.

The all–Southeast Asia average for around 1850 or 1870 must have been somewhere between 70 and 75 percent. Hidden behind that average figure, the one for Java (50 percent in 1840) probably constitutes the lowest value for a large area (small areas like Bali may have had even lower figures), while those for New Guinea or Borneo (perhaps 80 to 90 percent) should be regarded as maximum values. Southeast Asia, therefore, was still a heavily forested area, but it must have lost some of its tree cover during the Early-Modern Period.

Such losses could, indeed, be considerable. Between 1776 and 1840, central Java lost 40 percent of its teak forest cover, and another 30 percent between 1840 and 1870. Unfortunately, we have no figures prior to 1840 regarding the nonteak forests, so we cannot present figures for central Java's loss of total forest cover. Given the total lack of forest cover data for any other Southeast Asian area during this period, we will have to be grateful for what we do have. But even without other figures the message is clear: for selected types of wood, the reduction of forest cover could be very rapid indeed. However, we should not assume that mixed forests (the majority of all forests in Southeast Asia) would always fare better. The following paragraphs deal with the factors that played a role in deforestation.

The most important factor was no doubt population growth. As we have seen in chapter 5, the rate of growth was low according to later standards, probably not more than 0.2 percent annually, but it may have been higher than it used to be prior to 1400. In a predominantly agrarian society population growth usually equals clearing of land, and in Southeast Asia that meant, generally speaking, deforestation. It should be pointed out that this was not a process of uninterrupted forest clearing. In part becasue many agriculturists practiced shifting cultivation. Also partly because of the recurring crises dealt with in previous chapters (wars, epidemics, harvest failures and other natural hazards, and famines), which led to temporary demographic setbacks. Under a system of shifting cultivation, peasants cultivated a plot for one or two years, after which it reverted to secondary forest. This might turn again into a climax vegetation (old-growth forest) if left fallow long enough, but that was often not the case. And the growth of livestock, probably at a higher rate than that of the population, must be regarded as still another significant factor influencing forest cover.

The expansion of commercial crops for the international market also often took place on lands that had been previously covered with wood. This was particularly damaging in upland areas, as was the case with pepper and coffee.

TIMBER AND FIREWOOD

Another major factor was the need for timber. As we have seen earlier, almost all nonreligious buildings constructed by indigenous Southeast Asians were made either of timber or—as was the case with the majority of the rural dwellings—of bamboo (technically a grass, and usually not regarded as timber). Even palaces were often wooden constructions, a phenomenon also encountered in Japan. In the Malay Peninsula and in the Indonesian and Philippine archipelagoes, many

A fleet of fishing boats sits just offshore in Vietnam. Shipbuilding in Southeast Asia generated a major demand for timber. (Corel)

religious buildings (mosques and smaller prayer houses [*langgar*]) were also wooden structures.

With the arrival of the Europeans, that did not change immediately. The Spaniards, for instance, often built wooden churches, even cathedrals. Although the Dutch and the Spaniards soon issued regulations forbidding constructions of wood or bamboo in the inner cities of Batavia and Manila, where fire hazards were ubiquitous, fires in towns and cities were a perpetual threat to the urban populations throughout Southeast Asia. We have the diary of an Englishman who stayed in the city of Banten, western Java, as the representative of the East India Company around 1600, where he lived in constant fear for his life and goods because of the ever-present threat of fire. Hardly a week went by without a fire in some quarter of the town.

Urbanization as such must be regarded as a source of growing timber use, because the number of wooden houses—as opposed to the ones made of bamboo—increased at a higher rate than that of the total population. In addition, the larger cities witnessed the growth of the number of warehouses and "industrial" establishments (albeit in our eyes rather primitive ones), all made at least partly of timber. In addition there was the need for construction and regular repair or even replacement of landing stages, bridges, and fortifications, also often wooden constructions. Timber for urban construction was, therefore, in constant demand, not only because of slowly expanding urban populations but also because buildings burned down with sad regularity.

The other major demand for timber was generated by the shipbuilding industry. Southeast Asia had always constructed its own ships and boats, and indigenous shipbuilding continued to play a role, though locally a diminished one, in the Early-Modern Period. Areas like Burma, Siam, and Java, with their seemingly unlimited supplies of teak wood, regionally the best species for the construction of ships available, were admirably suited for shipbuilding. But also nonteak areas, such as Borneo, the Southern Moluccas, and the Philippines, were historically dotted with indigenous ship wharves. From the sixteenth century onward, the construction and exploitation by Southeast Asians of ships of large tonnage dwindled to insignificance for a variety of reasons, but smaller charters were still being built. Shipbuilders and merchants from India (Gujarat) and China (Fujian) took over the construction and use of the larger ships in the sixteenth century, while in the seventeenth century European companies played a similar role. At least in one case indigenous political development played a role as well. The expansion of the inland state of Mataram, Central Java, led to the demise of a number of trading polities (Jepara, Tuban, Surabaya) situated on Java's north coast, and therefore to less local shipbuilding.

Soon the European and Chinese newcomers to the region started to build their own boats and ships, particularly in and around Ayudhya (later near Bangkok), Luzon, and Rembang, a few hundred kilometers to the east of Batavia. A small island near Batavia's roadstead, Onrust, specialized in ship repair work. The Chinese in Ayudhya and Bangkok built Chinese-type junks, and the Spaniards built galleons in addition to smaller vessels; the Dutch built both lateen (European) and square-rigged (indigenous) vessels. As we have seen, there was a reduction in teak cover in Java between 1775 and 1840 of some 40 percent. That meant a drop in teak forest cover of 0.8 percent per year, much higher than the rate of population growth. We have no way of knowing whether similar percentages applied before 1776 in Java, or in Burma and Siam during or prior to the same years, but it seems safe to assume that the rates of deforestation of the teak forests were higher than the average growth rates of the population.

·Usually the literature emphasizes the construction of larger ships (junks, galleons, and so forth), but it should not be forgotten that the demand for small craft (canoes and the like) was increasing as well as a result of the expansion of trade. Locally, the activities of small-scale boatbuilders did lead occasionally to the disappearance of those tree species that were much sought after for such construction.

Up to now we have been discussing growing demand for timber as a major source of deforestation, but of course we must look at the supply side as well. In this respect technological innovation as a result of foreign influence should be mentioned. Talking specifically about forest exploitation, reference should be made to the use of the saw in seventeenth-century Java. The saw was almost certainly an innovation introduced by the Dutch, and one assumes that something similar happened in the Philippines. Perhaps even more interesting—and certainly less well known—is the introduction by the Dutch, also in seventeenth-century Java, of wind- and water-driven sawmills. Saw and sawmills speeded up and standardized production of beams, boards, and other types of processed timber.

Not much is known about possible concerns by indigenous rulers about depletion of forest, particularly timber, resources, but we do know that the VOC around 1775 tried to regulate the exploitation of the teak forests of Java with a view on sustainable production, even if a well-organized forest service would not arrive until much later.

The existence of sacred groves and forbidden forests in some areas of Southeast Asia was perhaps not consciously meant as a means of preserving forests, but it had very much the same effect. Such forests were often located on mountains and in watersheds, and their preservation was frequently vital for water retention and irrigation. Some of those groves have survived until today, and it has been suggested that they should be incorporated in present-day forest conservation planning.

Another potential threat to the forests is the collection of firewood, either for direct use or for the production of charcoal. Firewood and charcoal were the only fuels for a long time. Oil ("earth oil," or petroleum) was found in Burma and Java but does not appear to have been used as fuel; in any case, it was available only in small quantities. Coal does not appear to have been used by indigenous Southeast Asian people either, although the Chinese had been large-scale coal producers and consumers—particularly for their iron industry—for a long time. As northern China had been largely deforested at an early stage, the use of coal had more or less been forced upon them, as happened in England at a somewhat later date. Given the enormous forests that the people of Southeast Asia still had at their disposal, even at the end of the period, the need for coal had never arisen. The VOC imported coal from Europe in order to use it for ship repair and the production and repair of iron and steel implements, and it seems likely that this was also done by the other European powers. It was expensive, however, and not used on a large scale. Coal mining in Indonesia did not start until around 1850.

Finally, it should perhaps be mentioned that there are no signs of the use of dried cowpats as fuel, something that was already in evidence in India in the seventeenth century and that is today a well-known sight in dry and deforested areas. As was the case with coal, the need for the use of cowpats did not arise.

Firewood and charcoal had, of course, also been in demand prior to the period under discussion, as all households had to cook. On a modest scale, firewood and charcoal had also been used by artisans for "industrial" purposes (blacksmiths, goldsmiths, silversmiths, potteries, and producers of bricks and lime) before 1400.

In the Early-Modern Period both types of firewood and charcoal use increased, the first in proportion to the population but the second at a higher rate, because increased international trade led to a more than proportional increase of "industrial" production: the preparation of products like gambier, sugar, coffee, tobacco, and indigo required some form of heating.

With the expanding cities, the growth of the number of industrial establishments increased as well, and iron foundries, gun foundries, distilleries of arrack and other alcoholic beverages, brick and tile works, lime kilns, and for the European element bakeries all needed fuel in ever-increasing quantities.

NON-TIMBER FOREST PRODUCTS

Finally we arrive at the search for and exploitation of so-called Non-Timber Forest Products (NTFPs). Bamboo may have been the most widely used NTFP in Southeast Asia. It was (and partly still is) one of the defining characteristics of the region, together with rice paddies and buffaloes. It has already been mentioned in reference to house construction, but it was also used in huge amounts for temporary irrigation dams—that had to be renewed annually because they were swept away at high water—and for bridges. We may assume that its use was roughly proportional to the size of the population, and its production, therefore, must have increased with demographic growth. It was also often used for mat weaving and for the production of packaging materials. It was widely available in forests, often to be encountered in large almost pure stands, usually a sign of earlier forest disturbance. However, it was also grown around villages—often as a hedge—and in home gardens, which, technically, made it an agricultural product. It is tempting to speculate that its agricultural production had become attractive when the forests retreated, which occurred locally—for instance, in some densely settled areas such as the Red River delta in Vietnam and central Java—at an early epoch.

Bamboo was a bulk product with a low value per unit weight, and there are various other commodities belonging to that category, such as sappan and rattan, of which the value per unit was also small. International merchants often used such products to ballast their ships. However, in many cases NTFPs were high value for low-volume products, as was the case with various valuable woods for cabinetmaking, commodities with medicinal properties, aromatics (including incense), aphrodisiacs, raw materials for lacquers, and various kinds of resins and gums. Examples are beeswax, benzoin, birds' nests, camphor, copal, dammar, eaglewood, ebony, gutta-percha, jelutung, and sandalwood.

One of the interesting features of NTFPs is that they constituted a link between the world market and "tribal" groups—hunter-gatherers and shifting cultivators—living in remote inland and upland areas, usually far away from ports of trade. This underscores the remark made earlier that world market forces did penetrate into the most remote corners of Southeast Asia a long time prior to 1750.

Many of these commodities were quite rare and difficult to find, and only experienced woodsman were able to locate them and take care of their collection. Some commodities could not be found outside a restricted number of locations, which implied that the risk of overexploitation was always present, even at such an early date. For instance, sandalwood was to be found—outside Malabar in India—only in Java, and some of the Lesser Sunda Islands, mainly Sumba (also called the Sandalwood Island) and Timor.

The three areas—Java, Sumba, and Timor—represent three different ways of how indigenous polities reacted to world market demand. Java had been mentioned as an exporter of sandal prior to 1500, but in the seventeenth and eighteenth centuries exports had dwindled to very small quantities, which were exported only incidentally, perhaps during years of harvest failure, when the local peasantry was forced to look for alternative sources of income. In the nineteenth century foresters could no longer find the sites where it had been growing, which were rediscovered only after 1900, presumably after some regrowth had taken place. For a long time Sumba refused to cut sandalwood for the foreign merchants. Attempts by the VOC undertaken in the eighteenth century to get one of the rulers of Sumba to have it collected and to sell it to the company failed, because the ruler claimed that their ancestral spirits dwelled in the sandalwood forests, which made cutting the trees taboo. Only much later, probably after 1880, did the Sumbanese start exporting their sandalwood, and then in large quantities, because by 1920 it was all gone. Timor appears to have produced, or rather collected, sandal in a sustainable way for a long time. It was known as a sandalwood-exporting region to the Chinese around 1300, and it was selling considerable quantities on a regular basis to the VOC and the Portuguese between 1600 and 1800; it continued collecting it for the Dutch and the Portuguese during the nineteenth century. By 1915, however, it appears to have overexploited its sandalwood forests, and the colonial government imposed restrictions.

These three cases may be seen as representing three models—the sell-out model (Java), the abstinence model (Sumba), and the durable-production model (Timor). Explaining the differences is much more difficult than thinking up the models, but I suspect that scale has something to do with it. Java was a big place, and one product less did not really matter. Timor was small, however, and depended strongly on it. The Timorese rajas did not have many other sources of money income and therefore had to husband their sandalwood. In Sumba, the presence of alternative resources (horses, slaves) made it possible to keep the VOC at a distance. In addition, there may have been differences between the states in the way that they dealt with the exploitation of this resource. In Timor it was clearly a monopoly by the ruler; in Java that does not appear to have been the case. If sandalwood was a "managed" commodity, it was in the interest of a ruler (who had security of tenure) to exploit it in a sustainable way, whereas in a situation in which sandal was regarded as an open access resource or as common property, the institutions that could have taken care of sustainable management appear to have been lacking.

This point brings us to the question of the management of scarce resources. The well-known "tragedy of the commons" discussion of the last decades may appear to suggest (at least some of the participants do) that resources which are regarded as common property are easily overexploited. The Timor model suggests that management by the monarch may have been a solution to this problem, although it is unlikely that this notion—the ruler or the state managing a resource sustainably—would be universally valid. There are other examples, though, one of them being the management by the state of natural stands of sago in northern Sulawesi (Indonesia) and on the islands of Ambon (Moluccas), of sappanwood in Sumbawa (Lesser Sundas), and of cassia trees in the Batak area (northern Sumatra), being in the hands of a local raja. In the case of Ambon it was the VOC that undertook the management of the natural sago stands. If states are larger than in most of the cases mentioned, the ruler might no longer be in a position to control scarce natural resources as the lines of communication become longer, and the possibilities of illegal exploitation and smuggling greater. In smaller states, however, rulers may have managed these resources with the consent of the people, in which case social control would almost be sufficient to control possible free-riders. On the other hand, large states do have the means to police such resources.

However, there are other possibilities of sustainable foraging of scarce NTFPs. One of them questions the assumption of NTFPs as common resources. In the eyes of the local population, some trees, even if they were not domesticated but found in forests, far away from towns or villages, could be claimed by the person who had discovered them. He had to mark the tree in a way that could be recognized by his neighbors (exploitation of trees was usually a male occupation), who would then regard him as the owner. This applies to fruit trees like the durian, but also to trees yielding other NTFPs.

Another option was to start the cultivation of a valuable tree species that might be in the process of becoming overexploited or was to be found only far away from the villages. That is what happened with the trees (*Styrax spp.*) that yielded benzoin, also called gum benjamin. Around 1770 they were still found in the wild in North Sumatra, but the tree was also cultivated on some scale.

Finally, mention should be made of a substance like camphor, a product that requires the cutting of a large number of trees because even the specialists cannot see from the outside whether the tree will contain the camphor they are after. Such trees were indeed open access resources or common property—although the local ruler would no doubt claim his 10 percent tax—as they could not be marked or monopolized, because the tree had to be cut down first before it could be established whether it contained the valuable camphor.

Sometimes lack of proper management led to at least local extinctions, as witness a deforested hill in Sumatra, mentioned around 1800, called *kayu manis* hill, which points at cassia or wild cinnamon trees that must have stood there.

GAME, TRAPPING, AND HUNTING

A special category of Non-Timber Forest Products is formed by products derived from wild animals and by the animals themselves, although today the latter are no longer items of legitimate trade. It must be admitted, though, that elephants apart, live game was not an important category of commerce. There was a lively exchange of rare animals between the VOC headquarters in Batavia and the many rulers they were dealing with—the VOC had to establish a small zoo in order to house the animals they had been regaled with—and such presents were highly valued, but the numbers involved were small.

Elephants, however, were indeed traded regularly. Elephants occur naturally in most of the region, although not in the Philippines, the eastern Indonesian islands, or New Guinea. They were exported from Sumatra, and traded between Mainland Southeast Asia and India, where the rulers were always interested in expanding their already large collections. The VOC annually exported a few dozen of these animals to India from Sri Lanka as well. The rulers of Mainland Southeast Asia also had large numbers of elephants at their disposal, partly for pomp and circumstance, but certainly also because they were employed in their wars. The sultan of Acheh possessed large numbers of elephants, and around 1500 even some Javanese rulers had a fair number of them. Elephants were regularly employed for traveling and for work in the forests.

An elephant hauls and stacks lumber in Burma. There was a growing need for timber in Southeast Asia in the Early-Modern Period, helping lead to deforestation. (Library of Congress)

Products from wild animals were also traded. A well-known example is the trade of rhinoceros horn from Southeast Asia—both Maritime and Mainland—to China, where it was (and still is) considered to be an aphrodisiac. Ivory was an internationally traded commodity as well. Another example was the trade in deer hides from Siam (and Taiwan) to Japan, which annually imported thousands of them during the seventeenth and eighteenth centuries. Edible birds' nests were collected in places such as Borneo and Java's southern coast and then exported to China. It would appear that the collection of these nests was done sustainably, but the trade in deer hides and rhino horn might have set in motion a downward movement of the numbers of deer and rhino locally.

However, it would be an exaggeration to assume that all wild animals had entered a downward-sloping curve by the end of the period being studied. On the contrary, it might be argued that in many areas of Southeast Asia during the Early-Modern Period some animals were doing rather well. That applied particularly to the densely forested areas of the ever-humid tropics, where, as we have seen, people were late in coming, and where population densities were lower than elsewhere in the region. Such areas (Sumatra, Malay Peninsula, Borneo) were not very attractive to ground-dwelling herbivorous mammals like deer and wild boar, because there was not much food to be found on the forest floor. Therefore, ground-dwelling carnivorous and omnivorous mammals were also rather scarce. During the period under consideration, the population of these areas grew, albeit at a modest rate, which led to the creation of arable lands, grasslands, savannas, and secondary forests. They thus created ecotones—transitional zones between major biotic communities—that are particularly rich in plant life and also attract a highly diverse number of animals, which were often also interested in the planted crops. These forest fringes were the preferred habitat of, among other animals, tiger, deer, wild boar, and various monkeys and birds.

However, in certain areas several kinds of game had become rare or had vanished entirely by 1800. That was particularly the case near big cities like Bangkok and Batavia. To the south of Bangkok, for instance, crocodiles, still rather common in 1778, are no longer mentioned in travel accounts dating from the first half of the nineteenth century; even monkeys, also present in large numbers at earlier dates, seem to have disappeared. Elsewhere in Siam tigers and crocodiles were still a threat to travelers, although those traveling by elephant would not have been troubled much by those predators. The disappearance of game near some of the bigger cities was partly the result of hunting pressure—in both Siam and Java stimulated by the promise of bounties—but, of course, also of habitat loss owing to land reclamation.

Hunting was practiced by rulers and aristocrats, although not on such a scale as in India. It was also undertaken by the groups of hunters and gatherers still to be found in many areas of Southeast Asia, who combined it with trapping. Shifting cultivators were usually also hunters and trappers, and even sedentary peasants occasionally hunted as a sideline, although in their case trapping was probably more important. Hunting with firearms, often by Europeans, is mentioned as early as around 1600, but with the primitive matchlocks then available they were not much of a threat to wildlife. That situation "improved" somewhat with the introduction of flintlocks, but it was not until the arrival of the breechloading rifles of the nineteenth century that a more dependable weapon became available, which coincided with the arrival of larger numbers of Europeans in the second half of the century.

AGRICULTURE

As we have seen, population growth and increasing demand from the world market led to the clearing of "waste" land for purposes of agriculture. A number of developments are typical for the period, among which the most spectacular—at least in hindsight—was surely the so-called Columbian exchange. The term, taken from a classic study by Alfred Crosby, refers to the exchange, after the discovery by Columbus of the "New World" (the Americas, in 1492), of people, animals, diseases, and crops between the New World and the Old. Crosby deals extensively with the role of American crops in Europe; he is also quite informative about China, but is rather brief about Southeast Asia.

The Columbian exchange of crops is part of an ongoing process, already mentioned in Chapter 3, of introductions of alien or exotic species. As we have seen, this process started long before 1500. The most important American crops newly introduced in Asia after 1500 were no doubt maize and sweet potatoes. The white or "Irish" potato (also an American crop) came later to Asia, and was successful only in the higher and therefore cooler regions. Cassava (also called manioc or tapioca, although the latter term ought to be reserved for cassava flour) was also introduced fairly late in our period, and its success dates in fact from after 1870. It is originally an American crop as well, but it may have been introduced from Africa, where it had been transplanted by the Portuguese. Pulses, mainly from America and other places in Asia, should be mentioned, of which the peanut is arguably the most important example. In the early seventeenth century Chinese farmers were cultivating peanuts near Batavia, and it is likely that they were the ones who introduced this crop from China, probably after other Chinese farmers had become acquainted with it near Manila, where the galleons from Acapulco had brought many American crops. Peas were probably introduced both by Europeans and by the Chinese. Finally, chili peppers—not to be confused with "normal" (black or white) pepper—and yam beans should be mentioned, two other American crops with at least local importance in Southeast Asia around 1800.

What most of these alien introductions had in common was that they were subsistence crops—particularly maize and the roots and tubers (sweet potato, Irish potato, cassava, yam bean). This implies that they were grown for the cultivators' own consumption. These crops, therefore, revolutionized not only the existing cropping patterns but also the diet of a great many people. In addition to these staple crops there were various American fruit species that came to enrich an already impressive array of native ones.

But there were also introductions of crops grown entirely or largely for the international market. Among them were two very important commercial crops—tobacco and coffee, the first from America, the latter originally from Africa, but introduced in Asia from Arabia. Between the two of them they changed the face of the earth in many areas. Tobacco became important in China at an early stage, and somewhat later also in Southeast Asia, mainly in the Indonesian and Philippine archipelagoes, from where the crop was exported. Coffee arrived around 1700 in Java and soon became an enormous success; after a few decades Java was exporting large quantities to Europe. By 1800, coffee had spread to other Southeast Asian areas, such as Siam and the Philippines.

There are a few general points to be made about these new crops. In the first place, the sheer number of newly introduced alien crops—of which only the most important have been mentioned—must come as a surprise to those who thought that Southeast Asia prior to the age of modern imperialism (ca. 1870) was an isolated place. It was once rather generally assumed that the world market did not "penetrate" into these areas until the nineteenth century.

One of the most important points regarding our theme—environmental history—is that the arrival of these crops made for a rather radical change of cropping patterns that led in turn to changes not only in landscapes but also in agroecosystems. New plant communities came into existence that must have attracted a new mix of insects, birds, snakes, and other animals. They also led to increasing agro-diversity, which is a good thing for various reasons, one being ecological and another concerning risk spreading.

It could even be argued that the new species were less risky than the old ones, after the cultivators had overcome the teething problems of the unfamiliar crops. Crops often have specialized predators in their region of origin, which they lose when they are transplanted to other regions. There is a well-known story about rubber plantations in Brazil that failed because the trees came from Brazil, which made the plantations susceptible to insects specializing in rubber. Therefore, the first successful rubber plantations were established in Malaysia and the Indonesian Archipelago, where rubber had no natural enemies. Coffee, introduced in Java around 1700, had no serious plagues or pests until the 1880s. Less is known about the other crops, but in theory they might all have benefited from the principle just outlined, although they will have had problems with generalist predators. We do know that by 1940 the four alien roots and tubers to be found in Java (sweet potato, Irish potato, cassava, and yam beans) were all doing better in terms of surface area covered than were the roots and tubers of local origin (yams, taro); although there are several other explanations for this phenomenon, a lack of natural enemies (biologists call this "ecological release") surely must play a part.

In fact, many successful export products appear to have been foreign introductions, including pepper, although that was admittedly an early one (it came late, however, to areas like the Philippines). The only clear exceptions to this rule appear to be cloves, nutmeg, rice, and cane sugar. It will be shown in the next chapter that the link between introduction from outside and success as export crops would be valid after 1870 as well.

Another point is that many of these crops could be grown, or even had to be grown, in areas where rice could not be economically cultivated, because they were too dry, too high, too cold, too steep, or (in the case of cassava) too infertile.

Many of these conditions were, of course, found mainly in mountainous areas. The new crops, therefore, enabled people to go and live in areas where agriculture had hardly been possible before. This may have been a relief for poor peasants who were farming in marginal lowland areas; environmentally speaking it was a mixed blessing. If people took good care to keep the soil covered in these upland areas, erosion could be kept to a minimum, but that is a big if. We also know that the VOC wanted the coffee gardens in Java, which were laid out mainly on mountain slopes, to be well weeded, which must have been a source of erosion. Already by 1800 a number of central Javanese mountains had been stripped of their forest cover by peasant agriculturists, leaving only grasslands, a process that had led to water supply problems.

A second major trend during this period was that of the ongoing "sawahization," to use a term from Indonesian historiography. As was shown in Chapter 3, there is good evidence that irrigated wet-rice cultivation at an early stage became an important—locally, even the most important—farming system, for various reasons that are dealt with above.

It is possible, and even likely, that the spread of (better) irrigation methods went hand in hand with shifts in the types of rice that were used. What today is called upland rice (because it is grown almost exclusively in the uplands) used to be grown in the lowlands as well. By 1500, when irrigation had spread over large sections of the lowlands, it remained only in marginal regions and was replaced by lowland types. Among the lowland types of rice, the so-called slender-type rice, a high-yielding, more water intensive, late-maturing nonglutinous *indica*-type grain from Bengal, increasingly replaced the round-type rice, a relatively low-yielding, early-maturing, glutinous *japonica*-type rice from China.

The foregoing refers to the mainland, and information on Island Southeast Asia on this point is not as detailed. Nevertheless, after around 1650 sawahization took place in Java as well during this period, probably stimulated largely by urban demand, mainly from Batavia; it would appear that in the Philippines some of the friar estates constructed small-scale irrigation works with the same results. Construction of irrigation works and canals on the initiative of the state is mentioned in various Southeast Asian areas such as Burma, Thailand, Banten, Batavia, and Makassar.

It is often assumed that the spread of irrigation has increasingly led to monocrop rice cultivation, thus diminishing the older agro-diversity. That is undoubtedly true for the areas where, already around 1400, two (consecutive) rice crops a year were grown (on the same fields), as was the case in the Red River delta in northern Vietnam, in the Brantas delta of eastern Java, and in Bali. Given the character of those locations, it would appear likely that trade—and perhaps particularly international trade—was conducive to the double-cropping of rice.

To a somewhat lesser extent the increase of monocrop rice cultivation also may have occurred in the other irrigated wet-rice areas—where only one rice crop per year was grown—but that will have to be established empirically, case by case. In Java, for instance, most rice-cultivating peasants had home gardens as well, where a variety of nonrice crops were grown, while perhaps some 10 percent of all irrigated rice-cultivating peasants also grew "dry" (that is, nonirrigated) crops on their sawahs during the dry season. Around a number of villages—but we do not know what proportion of villages—small wooded lots were present, individually worked and probably also individually owned, with either a mix of annuals and perennials (bamboo, fruit trees) or just perennials.

It would appear, however, that Java was much better endowed in this respect than most Southeast Asian areas, although some parts of Thailand were also doing fine as regards garden culture. It has been suggested that regions where tenancy predominated were scoring badly as regards home gardens, while one expects that densely populated rural areas would not have had much room for orchard-type land use. Therefore, on balance, agro-diversity must have suffered sizable losses in the irrigated rice bowls and deltas of Southeast Asia.

Thus we observe two diametrically opposed trends: the oldest trend was the one just mentioned of increased sawahization and growing monoculture of rice; the other trend, which set in somewhat later, was the one of increasing agro-diversity caused by the Columbian exchange and other foreign introductions. Much of the latter crops were grown in the uplands, particularly near the ports of trade or capital cities, while the former trend was to be found mainly in the rice bowls (usually at mid-altitudes, but occasionally—for example, in Sumatra—at somewhat higher upland valleys) and rice-producing deltas.

This dichotomy between uplands and lowlands/midlands appears to have been accompanied by population movements, both into the highlands and into the lowlands. While the environmental consequences in the lowlands may have been at least partly beneficial, as bunded and irrigated rice fields are a sustainable form of land use, the peopling of the uplands must have been more problematic, perhaps particularly so in the areas in which Europeans organized the production of crops like coffee. European agriculture in the temperate zone is used to clean weeding between rows, which is all right under local European conditions but invites high rates of erosion in tropical upland environments.

However, some diversification took place in the lowlands and mid-altitudes. I am referring to the hinterlands of cities, which always generate a highly diverse demand for luxury crops and animals. Here, therefore, agro-diversity was stimulated by the growing population concentrations generated by increased trade and the centralization of political power.

The lowlands and uplands going their separate ways agriculturally is in keeping with (and part of) the two-track development discussed in Chapter 5, where we dealt with low and (relatively) high fertility regimes.

It seems likely that the densely settled rice bowls and deltas expanded throughout the period because of the mutual reinforcement of wet-rice cultivation and population growth. Foreign visitors often reported that the rice bowls were characterized by very fertile soils, but also that the population of those areas was quite poor and had to work hard to earn a living. In contrast, in the nonrice areas people may not have been prosperous (something quite difficult to establish), but they did not have to work very hard, either. There, agriculture could be quite diverse, with various indigenous roots and tubers (taro, yam), sago, so-called archaic cereals (millets, Job's tears), indigenous pulses, sugar cane, bananas, coconut trees, many types of fruit trees, and various types of sugar-producing palms. Here we are often dealing with mixed systems in which arboriculture and the cultivation of arable lands with annual crops were combined (agro-forestry).

A short note is in order about the production of cloves, nutmeg, and mace, the fine spices that attracted the Portuguese, Spaniards, Dutch, and English to Southeast Asia in the first place. Cloves were growing originally in the North Moluccas, and were not transplanted to the Ambon Islands, their exclusive production area from the seventeenth century, until the early sixteenth century. Nutmeg and mace both come from the nutmeg plant, which was growing on and around the Banda Islands, Central Moluccas. After having ousted their European rivals, the VOC restricted the production of these spices throughout the seventeenth and eighteenth centuries to Ambon and Banda, respectively, cutting down the trees everywhere else. As the seeds were dispersed by birds from island to island, however, the VOC had to police the area constantly for signs of young clove and nutmeg trees. This was a largely successful policy in terms of keeping prices at an elevated level, although occasionally bumper crops led to such a large supply that part of the harvest had to be burned in order to avoid a drop in prices. In contrast, therefore, to most Southeast Asian export crops, the area under cultivation shrank under European influence, although the Dutch did intensify production in the areas where it was permitted. This intensification is not very well documented, but it seems likely that a mixed system in which arboriculture played an important role—not unlike the "modern" agro-forestry systems of today—was replaced under Dutch rule by something that could be called a plantation system, including the bonded labor that historically is often linked to such systems. The shift from the original indigenous system to plantation-type exploitation was, particularly in the case of Banda, of a very violent nature. Around 1800 the British and the French got hold of cuttings, which was the beginning of the end of the VOC monopoly and of the restrictive cultivation policy.

What can we say about agricultural yields in Southeast Asia in comparison to those of other Eurasian regions? Yields of irrigated paddy fields are as a rule higher than those of rain-fed fields, while they were also more stable. It might be supposed, therefore, that yields per hectare increased during the period under consideration, at least in the irrigated areas (although some scholars think that traditional yields per hectare remained at the same level for many centuries). However, if nineteenth-century trends are anything to go by, we may also expect that the average amount of disposable arable land—the term "land ownership" might be too modern—per family or per capita decreased, which implies that rice yields per capita may not have changed much. Yields per hour worked may even have dropped, as the labor input for irrigated rice is larger than that for rain-fed rice (per unit of land).

Given the scarcity of figures—the story becomes monotonous—it would be quite difficult to quantify these impressions. Nevertheless, I have attempted to do so for Java around 1815. A number of figures for other countries at later dates are given in Table 7.1.

Table 7.1 Rice production (in the husk) per hectare, for various regions and periods.

Region	Kgs. perhectare	Year
Java	1,650	1815
Java	2,000	1900
Burma	1,430	1930
Indo-China	1,560	1910
Thailand	1,630	1930
Japan	3,000	1910

Source: Boomgaard, Peter, 2002, "Long Hours for High Yields: Agricultural Productivity in Pre-Industrial Java (Indonesia)," unpublished paper for the panel on "Agricultural Productivity in 18th and Early 19th-Century Eurasia," 13th Economic History Conference, Buenos Aires.

These data clearly suggest that Java was far ahead of most Southeast Asian countries, but that Japan was doing much better than Java in terms of yields per unit of land.

I have also compared labor productivity (based largely on the production of rice per agricultural laborer, and other products converted into rice-equivalents) in Javanese agriculture around 1815 with labor productivity in Europe around 1800. The results are presented in Table 7.2.

Java, therefore, was doing as well as Portugal in terms of labor productivity, and came fairly close to that of Finland. These were the most economically backward

Table 7.2 Productivity of labor expressed in millions of (net direct) calories per adult male agricultural laborer around 1800, Java and selected European countries.

Area	Calories
Java	3.2
Portugal	3.2
Finland	4.1
United Kingdom	13.2

Source: Boomgaard, Peter, 2002, "Long Hours for High Yields: Agricultural Productivity in Pre-Industrial Java (Indonesia)," unpublished paper for the panel on "Agricultural Productivity in 18th and Early 19th-Century Eurasia," 13th Economic History Conference, Buenos Aires.

European countries for which such data can be produced. In comparison to the United Kingdom, the most advanced country at the time, Java's labor productivity was evidently very low.

Therefore, assuming that the labor productivity of Java was not lower than that of most other Southeast Asian areas, Southeast Asia cannot claim to have been at par with the more developed European areas, a claim that is increasingly being made for parts of Japan and China, as was shown earlier in reference to the "divergence debate."

LIVESTOCK

A buffalo plowing a rice field: one would be hard pressed to find an image more typical of the lowlands of Southeast Asia (although, of course, such an image could also be found in India or China). Nevertheless, not much research has been done regarding this animal—or on any other kind of livestock—for the period under consideration. We will have to make do with a number of fairly general impressions, which, moreover, may not always be representative for the entire region and period.

As was mentioned, regarding livestock, Southeast Asia could claim elephant, buffalo, cattle (two species, of which one was indigenous), horses, pigs, dogs, goats, and sheep. Deer were captured and tamed but apparently never domesticated. Donkeys and mules were to be found only at the margins, and they did not play a role in the core areas. Chickens were probably also ubiquitous—but even less well documented than the aforementioned mammals—while ducks were certainly of local importance; in some areas (the Philippines) they appear to have

A team of water buffalo work a rice field in Southeast Asia. The water buffalo is possibly the most typical Southeast Asian animal. (Corel)

been introduced during our period. Geese may also have been introduced—at least locally—during the Early-Modern Period.

Some areas were clearly less well endowed with certain kinds of livestock than others. For instance, the regions around the equator, or the ever-wet tropical zone, did not abound in large livestock. At the beginning of the period, horses and cows—if present at all—were rare in the Philippines. That was also the case in New Guinea and much of eastern Indonesia, Borneo, Malaysia, and probably large parts of Sumatra. The reasons for this were lack of pasture, because those regions were so sparsely populated (most pastures in the tropics are man-made), and perhaps the lack of cooler areas, more appropriate for the rearing of such animals. Later on, conditions for these animals improved, because of higher population densities, and the American crops that made life in the cooler uplands possible. It would appear that horses profited most from these changes.

Foreign cattle were introduced by the Dutch and the Spaniards. In the case of the Dutch in Java, this was an attempt to crossbreed the indigenous cattle (domesticated, nonhumped *banteng*, today still in evidence as Bali cattle) with humped cattle from India, which seem to have been less sensitive to local diseases.

That reminds us of the advantages of many alien crops introduced with the Columbian exchange. The Dutch and the Spaniards also introduced foreign horses, among which were the highly prized Arabs and Persians, which were crossbred with local breeds.

The arrival of the Europeans after 1500, therefore, brought not only new people (the Europeans themselves), new diseases, and new crops (both largely part of the Columbian exchange) but also new kinds of livestock, in the meantime creating new Southeast Asian breeds. Again, it can be said that the European influence made itself felt even in the most remote areas of Southeast Asia, in ways that former historians could hardly imagine and that we are still mapping out.

Horse breeding was strongly related to the presence of courts, as the animals were needed for war and ostentation. With the introduction of firearms, horses became more important for warfare, which might explain the substitution of these animals for elephants in many Southeast Asian areas. Horses were, of course, also pack animals, but it would appear that they were not universally used as such until the nineteenth century, perhaps because they were still relatively rare and expensive (and, perhaps, locally royal monopolies).

Breeds of horses came and went. Originally, people were needed to create pasture, but once the horses were there, their grazing and browsing might have made inroads on the existing forest cover. Thus the horses had become more or less autonomous agents of environmental change. Soon, however, they would start to compete for space with people, who wanted to turn the pasture "created" by the horses into arable land. Eventually the animals would be crowded out. There are indications that, at least in Java, the number of horses grew faster than the population between 1500 and 1850. Around 1800, Java was increasingly "outsourcing" its horse rearing, importing the animals from the Lesser Sundas, to the east of Java.

In the lowlands of Java, the number of buffaloes also probably increased faster than the population (this was certainly the case between 1820 and 1850). This is arguably related to continuing sawahization; as the more than proportional increase of irrigated rice fields appears to have occurred all over Southeast Asia, the more than proportional increase of buffaloes should also have taken place across the region. Cattle numbers were probably increasing, too, but not as fast as those of buffalo. As cattle are the preferred plowing animals in the dry upland areas—insofar as any plowing was practiced there at all; in the higher areas it was often hoe agriculture—we can hypothesize that their numbers were increasing mainly in the highlands. Their meat was preferred by Europeans to that of buffalo, and so cattle were also found concentrated near the trading cities, in order to feed not only the local population but also the fleets that returned to Europe. In Java around 1850, the growth of the number of buffaloes slowed down for various reasons, to be dealt with in a later chapter, while the number of cattle started to grow. The age of the buffalo was drawing to a close, and the cattle era was about to begin. The importance of goats, today so prominent in many parts of Southeast Asia because they can be kept even on the small holdings in the densely settled areas, dates from the twentieth century; they still appear to be rare in sparsely settled areas like Cambodia.

Elephants, horses, cattle, and buffalo were used for transport of people and goods over land. As a rule, buffalo are mentioned in connection with carts; cattle and horses are usually referred to as pack animals.

If the buffalo is the most typical Southeast Asian animal, dogs, pigs, and chickens were not far behind. These animals were all eaten and have been used since time immemorial as offerings to gods and spirits. In the case of pigs, and particularly of buffalo and cattle, there are reasons to suppose that the environmental consequences of rearing these animals were considerable. Large numbers of pigs, cattle, and buffalo were slaughtered during ceremonies, particularly those relating to the death of a wealthy person, often from aristocratic descent. Particularly in the case of cattle and buffalo—which often were not used for agricultural purposes, as this type of feasting took place mainly in non–wet-rice areas—large herds were kept for the sole purpose of ostentation, the animals being used in potlatch-type ceremonies. The available documentation appears to suggest that cattle were used mainly for sacrifices in Mainland and buffalo in Island Southeast Asia. The rearing of large numbers of these animals for this type of conspicuous consumption must have influenced the natural environment negatively, as we may assume that grazing and browsing led to lasting damage to the forests, and thus a shorter rotational agricultural cycle, as is suggested by evidence from central Sulawesi. Although the animals were eaten in the end, this type of "binge eating" does not appear to have been very healthful, perhaps also because it was usually accompanied by the consumption of large quantities of alcohol, which, among other consequences, often led to violence and other forms of socially undesirable behavior.

Although, as has been argued earlier, the upland areas where animist "tribal" groups were residing became less important in terms of the proportion of people living there during the Early-Modern Period, sacrifices of large numbers of buffalo, cattle, and pigs survived up until the present. It might even be supposed that the demand from international markets gave these sacrifices a boost, as earnings from commodities (pepper, coffee, NTFPs) produced for these markets reinforced the position of the rich, made them richer yet, and therefore more inclined to organize large festivities to show off their enhanced status.

Thus a "traditional" mechanism appears to have been reinforced by modern market forces. Incidentally, the same mechanism operated as regards other traditional Southeast Asian features, such as high brideprices and slavery, which were both reinforced by international market demand. High brideprices and slavery largely vanished in the twentieth century, but sacrifices of buffalo and pigs did not.

With the coming of Islam, dogs and pigs went out of fashion in the relevant core areas, while blood sacrifices became less fashionable with the spread of Theravada Buddhism and Confucianism in the lowlands of Mainland Southeast Asia. In the uplands both the animals and the rituals persisted, reinforcing the differences between lowlands and uplands and the two-track development discussed earlier.

Livestock breeding for sacrifices was a typical cultural feature of Southeast Asia during the period. It should be briefly mentioned here that the region was also characterized by other cultural functions of livestock, such as animal fights and animal races. Examples of the latter were the Madurese bull races, for which the animals were specifically bred. Thus a Madurese cattle breed came into being typified by its large dimensions, and probably in larger numbers than if those bull races had not existed. The best-known animal fights are doubtlessly the cockfights, but elephants and water buffalo were also employed in these "spectator sports," of which the tiger-buffalo fights of Java may have been the most notorious. The fact that many tigers had to be caught for these fights had its bearing on Java's tiger population prior to European hunting.

It seems likely that all types of mammalian livestock taken together grew faster than did the population. This was, as we have seen, partly connected with the expansion of irrigated rice cultivation in the lowlands and to the feeding of the big cities and the various fleets. It was doubtlessly also linked to the growing demand for overland transport between the ports of trade and their hinterlands because of the increased importance of international trade, a typical example of a backward linkage. The relatively high rate of growth of livestock must have led to the creation of pasture on quite some scale, largely to the detriment of the forests.

A final point about livestock is that it produced manure, a point usually overlooked in the historiography. It is one of the features that make livestock useful parts of various farming complexes, some of which came into being with the introduction of new crops. The maize-livestock-tobacco complex to be found in Java perhaps as early as 1750 may serve as an example. Maize growers would feed their animals partly on maize stalks, and then use the manure of their animals for their tobacco crop.

FISHING

If livestock keeping is badly documented in the historical literature, fishing is even worse. There are reasons to assume that fishing was relatively more important in the past than it is now, both in terms of its share of GDP and the percentage of the labor force involved. The more the pity, therefore, that we know so little about it.

There are also indications that near big cities overfishing may have occurred as early as the seventeenth century. We read that fishermen of places near these towns were increasingly drawn into their commercial orbit, provisioning these establishments from increasing distances.

On the other hand we also observe that, at least in various regions of the Indonesian Archipelago, the variety of products of the sea that could be and was eaten was considerable, as was maritime biodiversity. It would appear logical that these resources could be exploited sustainably as long as people continued to have a varied diet, something that had been made possible by the high level of biodiversity in the first place. This suggests a positive feedback mechanism.

In eastern Indonesia, there is evidence for an institution of periodic closure of certain seasonally abundant natural resources called *sasi*. Many species of fish could be caught only during a certain period of the year, and the right to do so was leased out. In addition to enforcement by social control, or by actively patrolling the resources in question, supernatural means like curses were employed as well in the cases of resource management mentioned here. It is hard to say whether this institution ought to be regarded as some kind of resource management system, or whether it was aiming solely at profit maximization.

A final point is that various products of the sea were collected for the international market—pearls, mother of pearl, trepang, tortoiseshell. As we have seen, many of these products became quite sought-after almost overnight, with the "opening up" of the Chinese market in the early eighteenth century. This was an enormous boost for a number of traditional maritime and coastal communities, which could expand and prosper. There are indications that they also used slave labor, which points again at traditional mechanisms being reinforced by the international market.

In the case of pearls, a commodity that had been exploited for a millennium at the least, local depletion may have occurred. The influence of states—both foreign and indigenous—on the exploitation of such commodities could be good, but also bad. There are some examples of sustainable exploitation under the aegis of the state, but if and when it became too greedy, depletion could be the result.

A number of resources were extracted from the sea that could also be regarded as commodities pertaining to the mining sector. A modern example would be

Pearls on the half shell from Southeast Asia. Overfishing of pearls for export may have led to local depletion. (PhotoDisc)

sand, today extracted in large quantities in the region between Sumatra and the Malay Peninsula. There is not much evidence, however, that this was done on any scale during the period under consideration. We do have evidence for exploitation during this period for two other resources in this category—coral and salt.

Coral was used as an inexpensive and easily extracted building material, used as early as the seventeenth century for the building of fortifications and embankments in, for instance, the cities of Banten and Batavia, both on the island of Java. Although the quantities involved were large—vessels with coral were arriving daily in these places—one hesitates to suggest that it must have led to a considerable environmental impact; it was almost certainly dead coral that was used, excavated from islands off Java's coast.

The production of salt—in saltpans along the coast—had even less impact on the maritime environment, although locally it might have led to a shortage of firewood. It was an important item of regional trade.

Salt mounds along the coast in Thailand. Salt was an important item of regional trade in the Early-Modern Period. (Corel)

MINING

As has been shown, mining could be an important source of commodities locally, and a magnet for international trade, as was the case in Burma. Gold and silver had been mined for a long time, probably largely for the production of statues, ornaments, and jewelry, but partly also for indigenous types of coinage and other means of payment. Copper, tin, iron, zinc, and lead were also mined locally, and often at an early date. In addition, we find reports on the mining of precious and semiprecious stones, such as rubies, amber, and jade.

A number of general points can be made about these mines, but it must be kept in mind that, whereas we are relatively well informed about gold and tin mines, we do not know much about the others. Mining increased as a result of the presence first of European powers and later on of Chinese *kongsis*. Sometimes the Europeans and the Chinese had shared interests, but conflicts arose as well. Particularly tin mining was undertaken on quite some scale since the mid-eighteenth century in Thailand, Malaysia, and Indonesia. The use of Chinese and European technology must have been instrumental in the growth of the production. In many cases, however, traditional technology was employed.

An important point is that mines are often quite destructive of the natural environment, and although the scale of such ventures was restricted in comparison with what would follow after 1870, the scarred landscapes of mining operations undertaken around 1800 in, for instance, western Borneo (Kalimantan) are visible to this day. Another point is that the use of various dangerous and mordant chemicals (mercury, hydrochloric acid, saltpeter) is quite damaging to the natural environment as well. Although, generally speaking, pollution is a problem of industrialized or at least industrializing societies, here we find an early example in a region that was neither.

Two important commodities that were also mined in a sense were sulfur and saltpeter, both indispensable to the production of gunpowder. It is nevertheless a topic that is barely documented. The collection of sulfur—mined in active and extinct volcanoes—was, if not unhealthful, at least very unpleasant because of the stench.

Working in the mines appears to have been quite unhealthful in many instances. Descriptions of the most frequent complaints sound like malaria, and it is possible that the use of large quantities of water in various processes related to the production and further processing of the ores led to the presence of stagnant pools, which, in turn, attracted anopheles mosquitoes, the vectors of the malaria parasite.

CITIES

According to Anthony Reid, Southeast Asia was one of the most urbanized regions of the world around 1500. However, Jan Wisseman Christie, writing on Java prior to 1400, used the phrase "states without cities." The problem is, again, lack of reliable quantitative evidence. Reid based himself on a fairly limited set of numerical and non-numerical data, of which the reliability is often difficult to judge. Places such as Malacca, Ayudhya, Acheh, and Banten were no doubt cities with many inhabitants—of course, relatively speaking and according to the norms of the period. How big they actually were may always remain a question.

According to Reid, cities like Hanoi, Ayudhya, Malacca, Pegu, Banten, Acheh, and Macassar may all have had some 100,000 inhabitants at their peak during the period under consideration. In Reid's view, the urban population of the capital cities alone would have been at least 10 percent of the total population of Southeast Asia.

However, various historians have argued that this grossly overstates the case for the cities of Island Southeast Asia. In all likelihood places like Malacca,

Acheh, Banten, and Macassar probably never had more than between 10,000 and 20,000 inhabitants each. But even the admittedly bigger capital cities of Mainland Southeast Asia probably had a lower share of the total population than was suggested by Reid. According to Lieberman, Ayudhya and its suburbs may have had some 150,000 inhabitants around 1650. Apart from the fact that Ayudhya may have been a special case because of the large numbers of monks (around 20,000), that was about 5 percent of Siam's total population, not 10 percent.

It is unlikely that Southeast Asia could compete with either Europe or China, or even India, as its population density was so much lower than that of those regions. As a general rule, one would expect the degree of urbanization to be related to population density, in which case the degree of urbanization in Southeast Asia must have been rather low. Given the fact that the share of Europe's urban population—defined as the population living in places of 10,000 people or more—increased from 5.6 percent in 1500 to 10 percent in 1800, we would be well advised to assume that Southeast Asia's rates must have been lower than those figures. In 1815, the percentage of the population of Java living in cities of 20,000 or more was 6.7 percent, and the average figure for the whole of Southeast Asia may have been close to that.

Be that as it may, cities of considerable size were clearly part of the Southeast Asian landscape by the Early-Modern Period. Cities were engines of economic growth and therefore of environmental change. The need for wood and bamboo for construction of buildings and bridges, for timber for shipbuilding, and for fuelwood for households has been mentioned. Firewood and charcoal were also needed in large amounts for the industrial or artisanal establishments to be found in and around cities. Therefore, the surroundings of the city were deforested at an early age, and soon more remote areas had to ship their timber to the urban center. Land clearing also took place near the city at an early period, as the population had to be fed with agricultural and horticultural produce, and again faraway areas soon had to produce the bulky staple crops for which urban surroundings had become too expensive; products with a higher value per unit weight (cane sugar, vegetables, fruit) won the competition for space. Pasture for livestock was needed because the city had to be provisioned with meat, but horticultural produce with higher yields per hectare outcompeted stockbreeding, which then had to be undertaken farther afield. Overfishing near the city also led to fishermen having to come from localities farther away.

Thus the city made its presence felt and left its ecological footprint in areas located in ever widening circles around it. Cattle, buffalo, and horses came in droves from the upland areas where climate, vegetation, and the presence of large, sparsely populated areas made extensive stockbreeding a lucrative business.

This mechanism also limited the size of cities, because feeding and generally provisioning more urban people often would have been impossible at the given level of technology, particularly transport technology. As prices multiplied rapidly when bulky, low-value goods had to be transported over large distances, the moment that the average city dweller could no longer pay for rice was never far away. At the moment of sale on the urban market, transport costs usually took care of the largest share. Moreover, products like fish would not keep very long, and if cattle had to come from too far away, they would arrive as carcasses. By 1800, the sugar industry around Batavia was already in crisis because the area had been almost entirely deforested, which meant that firewood was rapidly becoming too expensive.

Cities influenced the landscape in more sense than one. They were the hubs of so much traffic over land and water that their presence and growth were conducive to the construction of good roads and the digging of canals. This, in turn, opened up new areas for all kinds of enterprise, among which was the cultivation of specialty crops for the urban market (fruit, beans, dyestuff). It also stimulated the arrival of temporary and permanent migrants from areas that had been at too great a distance for easy traveling. Thus, again, cities influenced (economic) life far beyond their immediate reach.

Finally, it must be mentioned that cities were also filthy and unhealthful places. Problems of sewage and waste disposal plagued urban concentrations from the beginning, and rivers running through cities (for example, Banten) were already seriously polluted by 1600, if not earlier. The many crocodiles to be found in the Bay of Batavia in the eighteenth century were said to have been attracted by the many human and animal carcasses that had come floating down Batavia's "Big River," the Ciliwung.

As elsewhere in the Early-Modern Period, urban mortality was as a rule higher than the death rate in rural areas, partly because people were living much closer together than elsewhere, partly because of the lack of hygiene, and in part because of the constant stream of new arrivals from outside, introducing fresh germs; cities also attracted the undernourished poor, often already in bad health, in search of work. Cities were a demographic sinkhole, siphoning off the meager rural population surpluses—if any—and depressing the already low rates of population increase even more.

In the case of Batavia there are strong indications that human actions (in this case, the construction of fish ponds near the city, which led to a long period of endemic malaria) turned the city around 1730 into a graveyard for all comers. Generally speaking, the coastal cities were the ports of entree of all epidemics that came from overseas, as was often the case with smallpox and later cholera. So again the city made its influence felt far beyond its boundaries.

ENVIRONMENTAL CONSCIOUSNESS

Is there any sense in asking whether some sort of environmental consciousness was to be encountered among the people in Southeast Asia during the Early-Modern Period? Or would that be an anachronism?

A few elements that could be considered have been mentioned earlier. We have seen that sacred groves sometimes functioned as protection forests on watersheds. There is no way of proving that such forests were preserved as a conscious effort to protect a fragile environment, but it is a possibility. Mention was also made of resources being controlled by rulers. In a number of such cases sustainable production appears to have been the result of such a situation, but it would be foolish to regard all royal monopolies as attempts at conservation. There is also documentation on the establishment of game reserves and deer parks by indigenous rulers during this period. They were largely or perhaps even entirely designed with hunting in mind, but some of these reserves survived into the twentieth century, and the perhaps unintended effect was "conservational."

Attitudes toward animals varied. On the one hand we have seen that animal fights, cruel events in the eyes of present-day observers, were nothing out of the ordinary. However, animist hunters often asked local spirits for permission to kill certain animals, and in some areas people refused to kill tigers. Again, one hesitates to call this environmental consciousness. Most of the literature suggests that this attitude was born out of fear of the spirits who resided in the area, or in the tiger.

In the late eighteenth century we also witness attempts by the Dutch in Java to organize teak production in a sustainable way, albeit unsuccessfully. Here the sources spell out in so many words that the way forest exploitation was being undertaken at the time was ruining the forests, and that a revision of those practices was needed in order to be able to benefit from them for generations to come.

In addition, it can be mentioned that from around 1840 a number of Dutch scientists and colonial civil servants warned the authorities that in Java deforestation of hill slopes was turning into a serious problem. They argued that this had led to a diminished supply of irrigation water, while at the same time the danger of flash floods had increased. One of these writers mentioned Alexander von Humboldt, a scholar whose writings had influenced "proto-conservationist" thinking in the British Empire.

Of course there were exceptions, such as the late-eighteenth-century EIC functionary William Marsden, whose aesthetic sensibilities—having enjoyed a classical education—were hurt when confronted with what he saw as wanton destruction of the magnificent Sumatran forests by slash-and-burn agriculturalists; the few other examples of possible environmental concern presented here have in common, however, that they were eminently practical. And could one really have expected more, given the lack of serious environmental problems, seen from our present-day perspective?

CONCLUSION

As we have seen, Southeast Asia was evidently much less densely populated than India and China during this period, and it still had an abundance of forests. There was no need, therefore, for cowpats as fuel (as in India) or coal (as in China). Nevertheless, considerable losses of forest cover could be observed locally, because there was a growing demand for timber (houses, palaces, religious buildings, bridges, ships), the production of which was facilitated by technological innovations like the saw and the sawmill. The collection and production of firewood, charcoal, and NTFPs increased as well. Demand was generated in all of these cases both by the subsistence sector and the international market.

Resource depletion was reported locally even prior to 1800, as was demonstrated with the example of sandalwood, where three models were presented—the sell-out model (Java), the abstinence model (Sumba), and the durable production model (Timor). Certain species of game also became locally rare or extinct around that time. On the other hand, several instances of resource management by indigenous or foreign rulers were also encountered.

Turning now to agriculture, we have seen that foreign trade led to the double cropping of rice in various areas, but also to the introduction of many new crops, the large-scale cultivation of which caused considerable changes in landscapes and agro-ecosystems. The exotic crops were doing remarkably well because they had no specialized enemies (pests, plagues) in the region. In fact, there are strong indications for two opposite effects in the development of agricultural systems—increased monocropping of rice in the irrigated wet-rice bowls, and of cloves and nutmeg in plantations, and increased agro-diversity created by the foreign crops and by the growing demand from cities.

Around 1800 agricultural productivity per hectare was lower than that of Japan, while agricultural productivity of labor was much lower than that of England, but at roughly the same level as in Portugal and Finland.

Not much livestock was to be found around the equator. Horses and cows were rare in the Philippines in the sixteenth century and in many other areas, because of a lack of pasture, because regions were sparsely populated, and perhaps owing to a lack of cooler areas. Domesticated animals usually fared better in various monsoon areas.

During the period studied in this chapter some types of livestock appear to have increased in numbers at a higher rate than did the population, partly because the Europeans introduced various new breeds. Thus hybrid livestock types gradually became distributed throughout the region, influencing behavior and the physical characteristics of farmyard animals even in remote corners of the region. The increasing numbers of livestock must have led to more permanent deforestation.

Numbers were also increasing because of the use, particularly of pigs, cattle, and buffalo, in sacrifices. It seems likely that long-distance trade stimulated the growth of this "traditional" Southeast Asian feature, as was the case with slavery and high brideprices. Finally it should be mentioned that the growth of livestock numbers must have led to increasing amounts of manure, some of which was doubtlessly used for commercial purposes, thus facilitating, for instance, the production of export-quality tobacco.

Fishing and mining are not as well documented as agriculture or forest exploitation. Nevertheless, it seems safe to say that in both cases local overexploitation occurred during the period under discussion (related to both local consumption and long-distance trade). In the case of fishing we noticed an institution that has some features of a natural resource management system. In both sectors labor was often unfree, because the work was dangerous, dirty, and unhealthful. Mining in addition destroyed the landscape and polluted the natural environment.

Finally, this chapter dealt with urbanization, which at this period had not yet reached the levels it had in Europe and China, at least partly because population densities in Southeast Asia were so much lower. Nevertheless, urban growth did occur. Cities had large "ecological footprints" because of their higher than average demand for crops, meat, and fish, as well as for timber, firewood, and charcoal. This led not only to local overexploitation but also to the expansion of natural resource exploitation in ever widening circles around the cities. When transportation costs became prohibitive because of the increased distance, this must have kept urban growth in check. Cities, therefore, were engines of economic growth and of environmental change, but they were also demographic sinkholes.

BIBLIOGRAPHICAL ESSAY

For references to the publications of the two scholars who have recently influenced our view of Southeast Asia during the Early-Modern Period most, Anthony Reid and Victor Lieberman, see the first section of the bibliography in the last chapter.

On forests and forest exploitation, see "Forest Resources of the World," *Unasylva* 2, no. 4 (1948), pp. 161–182; Ellen, Roy F., 1985, "Patterns of Indigenous Timber Extraction from Moluccan Rain Forest Fringes," *Journal of Biogeography* 12, pp. 559–587; Boomgaard, Peter, 1988, "Forests and Forestry in Colonial Java,

1677–1942." Pp. 59–88 in *Changing Tropical Forests: Historical Perspectives on Today's Challenges in Asia, Australasia and Oceania*. Edited by John Dargavel, Kay Dixon, and Noel Semple. Canberra: CRESS; Manguin, Pierre-Yves, 1993, "The Vanishing Jong: Island Southeast Asian Fleets in Trade and War (Fifteenth to Seventeenth Centuries)." Pp. 197–213 in *Southeast Asia in the Early Modern Era: Trade, Power, and Belief*. Edited by Anthony Reid. Ithaca: Cornell University Press; Boomgaard, Peter, 1995, "Sacred Trees and Haunted Forests in Indonesia, Particularly Java, Nineteenth and Twentieth Centuries." Pp. 48–62 in *Asian Perceptions of Nature: A Critical Approach*. Edited by Ole Bruun and Arne Kalland. Richmond: Curzon; Reid, Anthony, 1995, "Humans and Forests in Pre-colonial Southeast Asia," *Environment and History* 1, no. 1, pp. 93–110; Boomgaard, Peter (with the assistance of Rob de Bakker), 1996, *Forests and Forestry 1823–1941* [*Changing Economy in Indonesia: A Selection of Statistical Source Material from the Early 19th Century up to 1940*, vol. 16]. Amsterdam: Royal Tropical Institute; Boomgaard, Peter, 1998, "Environmental Impact of the European Presence in Southeast Asia, 17th–19th Centuries," *Illes i Imperis: Estudis d'Història de les Societats en el Món Colonial i Post-Colonial* 1, pp. 21–35; Boomgaard, Peter, 2000, "The Windmills of the Mind: Dutch Introductions of Pre-Modern European Technology in Asia, 1600–1800," unpublished paper for the ATMA-KITLV Colloquium, Kuala Lumpur, Malaysia; McWilliam, Andrew, 2001, "Prospects for the Sacred Grove: Valuing *Lulic* Forests on Timor," *Asian Pacific Journal of Anthropology* 2, no. 2, pp. 89–113; Boomgaard, Peter, 2003, "The High Sanctuary; Local Perceptions of Mountains in Indonesia, 1750–2000," Pp. 295–314 in *Framing Indonesian Realities: Essays in Symbolic Anthropology in Honour of Reimar Schefold*. Edited by Peter Nas, Gerard Persoon, and Rivke Jaffe. Leiden: KITLV; Bankoff, Gregg, 2004, "From Wood to Lumber: Changing Attitudes to the Forests of the Philippines 1565–1898," unpublished paper presented at the workshop Wealth of Nature, NIAS, Wassenaar, The Netherlands; Boomgaard, Peter, 2005, "The Long Goodbye? Trends in Forest Exploitation in the Indonesian Archipelago, 1600–2000." Pp. 211–234 in *Muddied Waters: Historical and Contemporary Perspectives on Management of Forests and Fisheries in Island Southeast Asia*. Edited by Peter Boomgaard, David Henley, and Manon Osseweijer. Leiden: KITLV.

On Non-Timber Forest Products, see Wheatley, Paul, 1959, "Geographical Notes on Some Commodities Involved in Sung Maritime Trade," *Journal of the Malayan Branch, Royal Asiatic Society* 32, no. 2, pp. 5–140; Villiers, John, 1985, "As Derradeiras do Mundo: The Dominican Missions and the Sandalwood Trade in the Lesser Sunda Islands in the Sixteenth and Seventeenth Centuries." Pp. 573–600 in *Il Seminário Internacional de História Indo-Portuguesa—Actas*. Lisboa: Instituto de Investigação e Cartografia Antiga; Blussé, Leonard, 1991, "In

Praise of Commodities: An Essay on the Crosscultural Trade in Edible Bird's Nests," Pp. 317–335 in *Emporia, Commodities and Entrepreneurs in Asian Maritime Trade, c. 1400–1750*. Edited by Roderich Ptak and Dietmar Rothermund. Stuttgart: Franz Steiner; Jong Boers, Bernice de, 1997, "The Exploitation of Sappan Trees in the Forests of Sumbawa, Indonesia, 1500–1875." Pp. 185–214 in *Paper Landscapes: Explorations in the Environmental History of Indonesia*. Edited by Peter Boomgaard, Freek Colombijn, and David Henley. Leiden: KITLV; Boomgaard, Peter, 1998, "The VOC Trade in Forest Products in the Seventeenth Century." Pp. 375–395 in *Nature and the Orient: The Environmental History of South and Southeast Asia*. Edited by Richard H. Grove, Vinita Damodaran, and Satpal Sangwan. Delhi: Oxford University Press; Roever, Arend de, 2002, *De Jacht op Sandelhout: De VOC en de Tweedeling van Timor in de Zeventiende Eeuw*. Zutphen: Walburg Pers; Dos Santos Alves, Jorge, M. Claude Guillot, and Roderich Ptak, eds., 2003, *Mirabilia Asiatica: Produtos Raros no Comércio Marítimo*. Wiesbaden: Harrassowitz; Lisboa: Fundação Oriente; Katz, Esther, and Marina Gouloubinoff, 2004, "Le Benjoin: Un Produit d'Exportation Ancré dans les Terroirs de Sumatra Nord," Pp. 179–196 in *Fruits du Terroir, Fruits Défendus: Identités, Mémoires er Territoires*. Edited by B. Charlery de la Masselière. Paris: IRD/Ibis; Henley, David, 2005, "Of Sago and Kings: Sustainability, Hierarchy and Collective Action in Precolonial Sulawesi." Pp. 235–258 in *Muddied Waters: Historical and Contemporary Perspectives on Management of Forests and Fisheries in Island Southeast Asia*. Edited by Peter Boomgaard, David Henley, and Manon Osseweijer. Leiden: KITLV.

On game, hunting, and trapping, see Terwiel, B. J., 1989, *Through Travellers' Eyes: An Approach to Early Nineteenth-Century Thai History*. Bangkok: Duang Kamol; Boomgaard, Peter, 1997, "Hunting and Trapping in the Indonesian Archipelago, 1500–1950," Pp. 185–214 in *Paper Landscapes: Explorations in the Environmental History of Indonesia*. Edited by Peter Boomgaard, Freek Colombijn, and David Henley. Leiden: KITLV; Kathirithamby-Wells, Jeya, 1997, "Human Impact on Large Mammal Populations in Peninsular Malaysia from the Nineteenth to the Mid-Twentieth Century." Pp 215–248 in the same volume; Boomgaard, Peter, 2001, *Frontiers of Fear: Tigers and People in the Malay World, 1600–1950*. New Haven: Yale University Press.

On the "Tragedy of the Commons" debate, see Hardin, Garret, 1968, "The Tragedy of the Commons," *Science* 162, pp. 1243–1348; Ostrom, Elinor, 1990, *Governing the Commons: The Evolution of Institutions for Collective Action*. Cambridge: Cambridge University Press; Ostrom, Elinor, et al., eds., 2002, *The Drama of the Commons*. Washington, D.C.: National Academy Press.

On the Columbian exchange (the introduction of new crops), see Crosby, Jr., Alfred W., 1972, *The Columbian Exchange: Biological and Cultural Consequences of 1492*. Westport, CT: Greenwood; Boomgaard, Peter, 1999, "Maize and Tobacco in Upland Indonesia, 1600–1940." Pp. 45–78 in *Transforming the Indonesian Uplands: Marginality, Power and Production*. Edited by Tania Murray Li. Amsterdam: Harwood; Boomgaard, Peter, 2003, "In the Shadow of Rice: Roots and Tubers in Indonesian History, 1500–1950," *Agricultural History* 77, no. 4, pp. 582–610; Hill, R. D., 2004, "Towards a Model of the History of 'Traditional' Agriculture in Southeast Asia." Pp. 19–46 in *Smallholders and Stockbreeders: Histories of Foodcrop and Livestock Farming in Southeast Asia*. Edited by Peter Boomgaard and David Henley. Leiden: KITLV.

On other agricultural matters, see Pelzer, Carl J., 1945, *Pioneer Settlements in the Asiatic Tropics: Studies in Land Utilization and Agricultural Colonization in Southeastern Asia*. New York: American Geographical Society; Tadayo, Watabe, 1978, "The Development of Rice Cultivation." Pp. 3–14 in *Thailand: A Rice-Growing Society*. Edited by Yoneo Ishii. Honolulu: University of Hawai'i Press; Ishii, Yoneo, 1978, "History of Rice-Growing." Pp. 15–39 in the same volume; Knaap, G. J., 1987, *Kruidnagelen en Christenen: De Verenigde Oost-Indische Compagnie en de Bevolking van Ambon 1656–1696*. Dordrecht: Foris; Aung-Thwin, Michael, 1990, *Irrigation in the Heartland of Burma: Foundations of the Pre-Colonial Burmese State*. DeKalb: Northern Illinois University, Center for Southeast Asian Studies; Andaya, Leonard Y., 1993, *The World of Maluku: Eastern Indonesia in the Early Modern Period*. Honolulu: University of Hawai'i Press; O'Connor, Richard A., 1995, "Agricultural Change and Ethnic Succession in Southeast Asian States: A Case for Regional Anthropology," *Journal of Asian Studies* 54, no. 4, pp. 968–996; Corpuz, O. D., 1997, *An Economic History of the Philippines*. Quezon City: University of the Philippines Press; Loth, Vincent C., 1998, "Fragrant Gold and Food Provision: Resource Management and Agriculture in Seventeenth Century Banda." Pp. 66–93 in *Old World Places, New World Problems: Exploring Resource Management Issues in Eastern Indonesia*. Edited by Sandra Pannell and Franz von Benda-Beckmann. Canberra: CRES; Boomgaard, Peter, 2002, "Long Hours for High Yields: Agricultural Productivity in Pre-Industrial Java (Indonesia)," unpublished paper for the panel on "Agricultural Productivity in 18th and Early 19th-Century Eurasia," 13th Economic History Conference, Buenos Aires; Ellen, Roy, 2003, *On the Edge of the Banda Zone: Past and Present in the Social Organization of a Moluccan Trading Network*. Honolulu: University of Hawai'i Press; Boomgaard, Peter, 2004, "From Rice to Riches? Rice Production and Trade in (Southeast) Asia, Particularly Indonesia, 1500–1950," unpublished paper written for the workshop "The Wealth of Nature: How Natural Resources Shaped Asian History, 1600–2000," NIAS, Wassenaar, The Netherlands; Henley, David, 2005, "Agrarian Change and Diversity in the Light of Brookfield, Boserup and Malthus:

Historical Illustrations from Sulawesi, Indonesia," *Asia Pacific Viewpoint* 46, no. 2, pp. 153–172.

On livestock, see Boomgaard, Peter, 1999, "Maize and Tobacco in Upland Indonesia, 1600–1940." Pp. 45–78 in *Transforming the Indonesian Uplands: Marginality, Power and Production.* Edited by Tania Murray Li. Amsterdam: Harwood; Boomgaard, Peter, and David Henley, eds., 2004, *Smallholders and Stockbreeders: Histories of Foodcrop and Livestock Farming in Southeast Asia.* Leiden: KITLV (with contributions by Martine Barwegen, Peter Boomgaard, William Gervaise Clarence-Smith, and Daniel Doeppers); Clarence-Smith, William G., 2004, "Elephants, Horses, and the Coming of Islam to Northern Sumatra," *Indonesia and the Malay World* 32, pp. 271–284.

On fishing, see Warren, James F., 1981, *The Sulu Zone 1768–1898: The Dynamics of External Trade, Slavery, and Ethnicity in the Transformation of a Southeast Asian Maritime State.* Singapore: Singapore University Press; Ptak, Roderick, 1991, "China and the Trade in Tortoise-shell (Sung to Ming Periods)." Pp. 195–229 in *Emporia, Commodities and Entrepreneurs in Asian Maritime Trade, c. 1400–1750.* Edited by Roderich Ptak and Dietmar Rothermund. Stuttgart: Franz Steiner; Sutherland, Heather, 2000, "Trepang and Wangkang: The China Trade of Eighteenth-Century Makassar c. 1720s–1840s," *Bijdragen tot de Taal-, Land- en Volkenkunde* 156, pp. 451–472; Sutherland, Heather, 2002, "The Tortoiseshell Trade of East Indonesia, 1650–1800," unpublished paper for the AAS conference, Washington, D.C.; Boomgaard, Peter, 2005, "Resources and People of the Sea in and around the Indonesian Archipelago, 900–1900." Pp. 97–120 in *Muddied Waters: Historical and Contemporary Perspectives on Management of Forests and Fisheries in Island Southeast Asia.* Edited by Peter Boomgaard, David Henley, and Manon Osseweijer. Leiden: KITLV.

On salt, see Knaap, Gerrit, and Luc Nagtegaal, 1991, "A Forgotten Trade: Salt in Southeast Asia, 1670–1813." Pp. 127–158 in *Emporia, Commodities and Entrepreneurs in Asian Maritime Trade, c. 1400–1750.* Edited by Roderich Ptak and Dietmar Rothermund. Stuttgart: Franz Steiner; Le Roux, Pierre, and Jacques Ivanoff, eds., 1993, *Le Sel de la Vie en Asie du Sud-Est.* Bangkok: Prince of Songkla University.

For studies on mining, see Wong Lin Ken, 1965, *The Malayan Tin Industry to 1914, with Special Reference to the States of Perak, Selangor, Negri Sembilan and Pahang.* Tucson: University of Arizona Press; Cushman, Jennifer, 1991, *Family and State: The Formation of a Sino-Thai Tin-Mining Dynasty 1797–1932.* Edited by C. Reynolds. Singapore: Oxford University Press; Rueb, Patricia, 1991, "Une Mine d'Or à Sumatra: Technologie Saxonne et Méthodes Indigènes au XVIIe siècle," *Archipel* 41, pp. 13–32; Heidhues, Mary F. Somers, 1992, *Bangka Tin and Muntok Pepper: Chinese Settlement on an Indonesian Island.* Singapore: ISEAS;

Henley, David, 1997, "Goudkoorts: Mijnbouw, Gezondheid en Milieu op Sulawesi (1670–1995)," *Spiegel Historiael* 32, pp. 424–430; Heidhues, Mary Somers, 2003, *Golddiggers, Farmers, and Traders in the "Chinese Districts" of West Kalimantan, Indonesia.* Ithaca, NY: Cornell Southeast Asia Program.

On cities and urbanization, see De Vries, Jan, 1984, *European Urbanization 1500–1800.* London: Methuen; Boomgaard, Peter, 1989, "The Javanese Rice Economy, 800–1800." Pp. 317–344 in *Economic and Demographic Development in Rice Producing Societies: Some Aspects of East Asian Economic History.* Edited by Akira Hayami and Yoshihiro Tsubouchi. Tokyo: Keio University; Boomgaard, Peter, 1989, *Children of the Colonial State: Population Growth and Economic Development in Java, 1795–1880.* Amsterdam: Free University Press; Boomgaard, Peter, 1993, "Economic Growth in Indonesia, 500–1990." Pp. 195–216 in *Explaining Economic Growth: Essays in Honour of Angus Maddison.* Edited by Adam Szirmai, Bart van Ark, and Dirk Pilat. Amsterdam: North-Holland; Nagtegaal, Luc, 1993, "The Pre-Modern City in Indonesia and Its Fall from Grace with the Gods," *Economic and Social History in the Netherlands* 5, pp. 39–59; Brug, P. H. van der, 1994, *Malaria en Malaise: De VOC in Batavia in de Achttiende Eeuw.* Amsterdam: De Bataafsche Leeuw; Nagtegaal, Luc, 1995, "Urban Pollution in Java, 1600–1850." Pp. 9–30 in *Issues in Urban Development: Case Studies from Indonesia.* Edited by Peter J. M. Nas. Leiden: Research School CNWS; Talens, Johan, 1999, *Een Feodale Samenleving in Koloniaal Vaarwater: Staatsvorming, Koloniale Expansie en Economische Onderontwikkeling in Banten, West-Java (1600–1750).* Hilversum: Verloren.

On the early stirrings of environmental consciousness in the region, see Grove, Richard H., 1995, *Green Imperialism: Colonial Expansion, Tropical Island Edens and the Origins of Environmentalism, 1600–1860.* Cambridge: Cambridge University Press; Boomgaard, Peter, 1999, "Oriental Nature, Its Friends and Its Enemies: Conservation of Nature in Late-Colonial Indonesia, 1889–1949," *Environment and History* 5, no. 3, pp. 257–292.

PART FIVE

THE LATE-COLONIAL PERIOD AND THE EARLY YEARS OF INDEPENDENCE

8

POLITICAL, DEMOGRAPHIC, AND AGRICULTURAL DEVELOPMENT

During the nineteenth and early twentieth centuries most Southeast Asian regions came under European overlordship, particularly between the 1870s and World War I (1914–1918), the period of Modern Imperialism. Burma and the Malay Peninsula came under the British sphere of influence, as did, under various titles, northern Borneo. The present-day countries of Laos, Cambodia, and Vietnam came to constitute French Indochina. The Dutch expanded their rule in what is now Indonesia, as did the Spanish in the Philippine Archipelago. After around 1900 the United States took over the position of the Spaniards. East Timor remained in Portuguese hands. Only Thailand kept the Westerners out and remained independent, although the Thais had to cede some territory to the British and the French and lost their right to unilaterally establish import and export duties through the conclusion of the Bowring Treaty with the British (1855). World War II in the Pacific (1941–1945) spelled the end of colonial rule as "business as usual" in the region, and between 1946 and 1957 almost all Southeast Asian countries became independent.

This chapter deals mainly with the years from the 1870s to the 1940s, often called the Late-Colonial Period, and with the period of war, revolutions, and early independence, lasting until the early 1970s. In most studies of Southeast Asia and its various countries, the Pacific War is taken as a watershed between colonial and postcolonial periods, and it is used as a convenient end or start of chapters on these periods. However, in a study dealing with environmental history, another periodization appears to make more sense. The years between about 1870 and 1930 constitute a period of almost uninterrupted growth, globalization, and the ever-increasing exploitation of natural resources. The years between around 1930 and 1970, in contrast, were a period of (relative) stagnation, caused by the Depression of the 1930s, the War in the Pacific, the Japanese occupation of much of Southeast Asia, and the various wars of independence and other uprisings and conflicts that characterized so much of the years between 1945 and 1970. Even during the 1970s war and violent conflict predominated in some Southeast Asian countries (Cambodia, Vietnam). Only around

Philippine soldiers killed by American troops during the Philippine-American War (1896–1898). Spain's position as colonial power was taken over by the United States. (National Archives)

the 1960s or 1970s would a new period of high growth rates get underway in most countries in the region.

This chapter, therefore, deals with a period of growth between about 1870 and 1930, and a period of relative stagnation from the 1930s to the 1960s.

POPULATION GROWTH

It must be stressed that the stagnation between the 1930s and the 1960s was relative. It would be more accurate, perhaps, to say that economic growth, in terms of increasing income or production per capita, slowed down considerably and in some instances came to a standstill. However, not everything stagnated: the average annual population growth rate between 1930 and 1960 was higher than between 1900 and 1930, as can be seen in Table 8.1.

It will be recalled that prior to 1800 average growth rates of the population must have been below or around 0.2 percent per year. Therefore, the average annual rate that was obtained during the nineteenth century (0.9 percent) was obviously

Table 8.1 Population numbers and average annual growth rates, Southeast Asia, 1880–2000.

Year	Population (millions)	Annual growth rate from last year (%)
1800	35	—
1900	86	0.9
1930	129	1.4
1960	221	1.8
1990	423	2.2
2000	522	2.1

Source: Elson, Robert E., 1997, The End of the Peasantry in Southeast Asia: A Social and Economic History of Peasant Livelihood, 1800–1990s. *Houndmills: Macmillan; New York: St. Martin's;* The Future of Population in Asia. *Honolulu: East-West Center (2002).*

much higher. However, this does not mean that the year 1800 constituted a sharp and sudden dividing line between very low and considerably higher population growth rates. In all likelihood, the acceleration of the rate of growth was gradual, as witness population data from the Philippines and Java—shaky though they are—for the eighteenth and nineteenth centuries.

The Philippines and Java are the best-documented areas as regards population data, no doubt because they were under tighter colonial control—colonial regimes loved statistics—at an earlier stage than other Southeast Asian regions. Admittedly, compared with real census data, available for the Philippines and Java in the early twentieth century, the population documentation prior to 1900 for these two areas leaves much to be desired. Used with discrimination, however, these data are nevertheless informative.

The Philippines and Java were also the regions with the highest growth rates for population in Southeast Asia—at least, that is the current assumption. The average annual growth rate in the Philippines during the second half of the eighteenth century may not have been much less than 1.0 percent per annum on average, while Java's growth rate during that period was probably somewhat lower, perhaps closer to 0.5 percent. Between 1820 and 1870 the average annual increase of the population in the Philippines may have been as high as 1.6 percent, while owing to various disasters, including war, it was around 1.2 percent between 1870 and 1900. In Java it appears to have been the other way around—1.4 percent before 1850, and perhaps as much as 1.75 percent (but possibly less) thereafter.

High rates were also obtained in some other areas. For the period 1840 to 1900 an average annual population growth rate can be calculated for what is now Vietnam, to the tune of 1.3 or 1.4 percent—figures, however, that are deemed too high by some demographers. The population of the Minahasa, northern Sulawesi, was probably growing at a rate of 1.5 percent on average per year between 1860 and 1900.

Clearly, the figures for the Philippines, Java, Minahasa, and, if confirmed, Vietnam in the nineteenth century are much higher than the averages for the entire Southeast Asian region. This implies, of course, that figures for some other areas must have been much lower. Alas, data for other areas prior to the twentieth century are rare, and if they are not, their reliability has been called into question. The problem is compounded by the fact that the best data we have are for bureaucratically administered regions, while there are reasons to believe that in such regions population growth rates were much higher than in areas without a bureaucracy interested in population figures.

Nevertheless, we do have some data that suggest lower average annual population growth figures in various areas. There is an estimate for Southeast Borneo (now Kalimantan) of an average annual population growth rate of 0.8 or 0.9 percent between ca. 1840 and ca. 1870, and therefore close to the average nineteenth-century rate for Southeast Asia as a whole. In the area of Bolaang Mongondow—adjacent to the Minahasa—the growth rate appears to have been close to zero during the period 1850 to 1900. In Bolaang Mongondow, Dutch colonial rule had been virtually absent until 1901, while Minahasa had had a strong Dutch presence ever since the seventeenth century. However, the Dutch had been close enough to the area of Bolaang Mongondow to monitor population data.

It may be assumed that similar very low growth rates were obtained in many other Southeast Asian areas during the nineteenth century. In fact, the situation appears to be a continuation of the two-track Southeast Asian demographic pattern, postulated in Chapter 5 for the Early-Modern Period. There, a pattern was suggested of fast-growing and densely populated wet-rice bowls in the lowlands and mid-altitudes, versus slow-growing and sparsely populated upland areas, where rice and other crops were cultivated under slash-and-burn conditions. This pattern appears to have persisted during the colonial period. The wet-rice bowls were taken over, administrated directly, and controlled tightly by European colonizers, while the upland areas, if formally colonized at all, were often administered indirectly and controlled more loosely. Such lightly populated areas were often of only marginal interest to the colonizers, as they were difficult to penetrate militarily, peopled by "savage tribes" (sometimes real or imagined headhunters and cannibals) and equally savage animals, and not easily exploited owing to the lack of a resident labor force and to transportation problems. Of

A group of tribal warriors in the Philippines with spears, blowpipes, and shields are seen ca. 1900. Colonial powers tended to ignore sparsely peopled upland areas in Southeast Asia, often inhabited by "savage peoples," if they were not easily exploited, continuing the two-track demographic pattern of the region. (Ridpath, John Clark, Ridpath's History of the World, 1901)

course, if the area contained valuable resources (minerals), or if high-value, low-bulk crops (like opium) could be grown there, that situation would change.

The average figures presented in Table 8.1 therefore mask the underlying lower figures for the "marginal" upland areas and the higher growth rates of the densely populated rice bowls. It is possible that the disparity between the two tracks increased for some time, but the figures are not sufficiently robust for proof of that supposition. However, there are also indications that at the end of the nineteenth century a combination of impoverishment and increased morbidity in densely settled areas like Java, northern Vietnam, and parts of the Philippines led to higher mortality, a drop in the number of marriages, lower fertility rates, and increased out-migration (a "Malthusian" response), and therefore to lower rates of population growth. This might be seen as the beginning of a trend toward the narrowing of the gap between the low growth rates in the uplands and the high rates in the lowlands.

Between 1900 and 1930, and again between 1930 and 1970, average annual population growth rates would continue to rise.

MORTALITY

What are the factors behind these ever-increasing (average) population growth rates? The ones most often mentioned are the "Pax Imperica," the introduction of Western medicine and hygiene, and improved communications and transportation.

The term *Pax Imperica* ("imperial peace") reflects the notion that the presence of the colonial powers imposed peace on a region where armed conflict had been ubiquitous, a topic dealt with in earlier chapters. After the colonial wars were over, mortality owing to wars and other conflicts dropped to much lower levels, which, in combination with more or less constant fertility levels, led to higher rates of natural increase. This would explain the higher population growth rates in Java and the Philippines in the late eighteenth century, when the other factors mentioned cannot yet have had much influence. It must be remembered, however, that colonial wars did occasionally interrupt such peaceful periods, which then led to a temporary drop in the population growth rates. The Java War, between 1825 and 1830, may serve as an example.

The introduction of Western medicine and hygiene in Southeast Asia was probably not of much importance prior to the discovery of smallpox vaccination in Europe in 1798. Vaccination spread surprisingly rapidly to Southeast Asia, having been introduced in various areas by the first decade of the nineteenth century. However, it was a long time before effective vaccination campaigns, covering areas large enough to have any impact on the incidence of smallpox (and thus on mortality levels), were in place. In Java this was probably the case around 1850, and not much later the effects of vaccination made themselves felt in Minahasa as well. In most other areas not much progress was made prior to 1900, and during the War in the Pacific and its aftermath smallpox returned to many areas that had been almost free of it by the 1920s. It was not until the early 1970s that the disease was finally eradicated in the region.

The role of hygiene in Southeast Asia in the nineteenth and early twentieth centuries is not much researched. European doctors attempted to introduce notions of hygiene in the region at a fairly early stage, but the effects of these actions are difficult to gauge and should not be overestimated.

Another important introduction was quinine, for use against malaria. Although introduced in Java as early as 1764, it did not become available in large quantities until the 1850s, when the Dutch started cinchona plantations in the area. But even then an effective program of quinine distribution, including attempts to make people take the drug regularly, was not in place until much later. Therefore, the influence on the death rate prior to 1900 must have been rather limited.

Vaccination, hygiene, and quinine were, in a sense, all chance discoveries. The real revolution in Western medical theory, the "microbial revolution," did not come about until late in the nineteenth century with scientists such as Koch and Pasteur. With them, the "germ theory" triumphed over the "miasma theory" (bacilli versus vapors). Knowledge of the causative organisms of the major endemic and epidemic diseases provided medical science with the tools needed for preventive or curative measures. During the first half of the twentieth century there were some successes with injections against the plague, syphilis, and yaws, but vaccinations aimed at cholera were probably useless. Hygiene measures and malaria prevention—in the meantime extended to species sanitation (killing the larvae of anopheles mosquitoes, the carriers of the malaria parasite)—were now in all likelihood more effective.

The revolution in medicine, however, came after World War II, with the so-called miracle drugs (antibiotics), of which penicillin is the best known. In addition, spraying of DDT against anopheles mosquitoes was introduced. This was a very effective measure, but it had to be stopped (but locally often was not) because it was also harmful to humans. Now the death rate dropped rapidly and continuously, which explains the high population growth rates after 1960, as presented in Table 8.1. While in the 1950s the death rate in Southeast Asia was usually between 20 and 30 per thousand, by the year 2000 it had dropped to between 5 and 15.

The third factor usually mentioned in connection to increased population growth rates in Southeast Asia is better communications and transportation. All colonial states undertook road-construction programs, which made hinterland areas more accessible from the coast and the capital. As time went by more all-weather roads were constructed, which made traveling possible even during the rainy season. From the second half of the nineteenth century, railway construction was undertaken as well. Steamships began connecting the various parts of Maritime Southeast Asia, and they also speeded up travel along the big rivers of Mainland Southeast Asia. From around the 1920s, trucks arrived on the scene as well. All of this made it easier for the states involved to send food—usually rice—to areas threatened by famine owing to harvest failures or other disasters, while the time it took the relief goods to reach the disaster-stricken areas was much reduced. News of disasters also traveled faster from the peripheries to the centers because of the construction of telegraph and, later, telephone networks.

FERTILITY

Up to now we have discussed dropping mortality rates, but what about fertility? Higher growth rates of the population can be caused by dropping mortality but also by increased fertility rates, or, of course, a combination of both (and by a positive migration balance, a topic dealt with shortly).

Data on fertility are imprecise and hard to come by. However, there are indications that the following developments have taken place. Generally speaking, fertility rates were, as we have seen earlier, in all probability low in the upland and inland areas where foragers and slash-and-burn agriculturists predominated, and high in the wet-rice–producing lowlands and mid-altitudes. Now, ever since the high population growth rates of the 1960s, some scholars have been talking about "unchecked" fertility rates, but that is clearly a myth, as is their supposition that such high rates were relics of older times. Unchecked fertility would mean some twenty children per woman, whereas seven (of which, prior to the drop in mortality, perhaps three survived) would be closer to the real historical value.

It seems likely that during the period under consideration a pincer movement occurred—birth rates were going up in parts of the low-fertility areas, while in some of the densely settled, high-fertility areas, birth rates were going down. The higher birth rates may have occurred because the transition from foragers and swidden agriculturists to settled wet-rice cultivators continued, often in tandem with conversion to Islam or Christianity, all of them factors that are supposed to drive the birth rate up. In addition, increased economic opportunities or increased labor burdens may have led to a drop in the age of marriage, and to shorter periods of breast-feeding, and thus to higher fertility. This is called the "demand for labor" hypothesis.

In contrast, in densely settled areas, where land was becoming scarce and landlessness was increasing, later (and perhaps fewer) marriages, fewer births, and out-migration appear already to have occurred during the last decades of the nineteenth century. In addition, the incipient drop in the death rate was strongly reflected in the lowered rate of infant mortality. With fewer deaths of breast-fed children (breast-feeding of up to two years was quite usual in Southeast Asia), the average period of lactation increased, thus influencing fecundity.

It is not clear how, on balance, this pincer movement influenced the average fertility rate. What we do know is that in the 1950s, birth rates in Southeast Asia varied from forty to almost fifty per thousand, which is probably not far from the figures for 1870, or, for that matter, for 1800. Although modern research suggests that Southeast Asian women had been using "traditional" methods of birth control as far back as the records allow us to go (and supposedly further), it was not until the 1960s that modern family planning methods, which were probably safer and more dependable, became available in the region. It would take much longer before these methods were being used by so many women that the fertility rates started to drop considerably. As long as this was not the case—and countries like Laos and Cambodia still have very high birth rates—the constantly dropping death rate in combination with a still high birth rate led to higher rates of natural increase in the region than ever before, as is shown in Table 8.1.

MIGRATION

Finally, a few words on migration, which has always played a role in Southeast Asia, even if it would be very difficult to quantify. There was always some migration in the region, both within and between areas that we now regard as countries. For instance, Minangkabau young men, from western central Sumatra, came to the Malay Peninsula, as did Buginese people, from southern Sulawesi. However, if we are interested in migration to and from Southeast Asia as a whole, in order to establish whether migration led to higher or lower population growth rates, such internal migratory movements are irrelevant.

Migration from Southeast Asia to other regions was rare prior to about 1870, and between the 1870s and the 1960s it was not numerically significant. Immigration was of some importance, however, both before and after the 1870s. In both periods, Chinese migrants (or rather migrants from what is now China) constituted doubtlessly the largest category of immigrants. Broadly speaking, one can distinguish two types of migration—overseas migration from southern China to the Philippines, Indonesia, Malaysia, and Thailand, and overland migration of "tribal" groups from China's southwestern provinces such as Yunnan to Burma, Thailand, and Vietnam. Merchants were an important category among the former type of migrants, but poor "coolies" (unskilled laborers) from the overpopulated southern Chinese provinces dominated numerically. They came to work as miners, plantation workers, and stevedores, and were also employed in the construction of harbors, buildings, canals, roads, and railroads. The more skilled Chinese migrants found work as artisans in the cities. The tribal groups from southwestern China were mainly shifting cultivators looking for "empty" spaces where they could continue their way of life.

People from the Indian subcontinent arrived in large numbers in Southeast Asia as well, particularly in Burma and Malaya. Indian capital and Indian labor gangs were active in the rice fields of Lower Burma, while Indian coolies were employed by Western rubber plantations on the Malay Peninsula.

Europeans also came in notable numbers to the region, as sailors, soldiers, and as civil servants in the employ of the colonial state, as well as employees of private companies. Many of these people were temporary migrants, who returned to

Europe after their tour of duty, but a sizable minority stayed on, often marrying or living in concubinage with local women. Thus a group of Eurasians came into existence, of which a part would remain in Southeast Asia after decolonization.

Finally, mention should be made of Arabs who migrated to the Indonesian Archipelago and the Malay Peninsula. Among them, traders-cum-moneylenders were an important category.

Census data from the 1920s suggest that there were around 4 million Chinese in Southeast Asia, in addition to 1.5 million Indians and perhaps 0.25 million Europeans, or 5.7 million migrants from the largest categories altogether—between 4 and 5 percent of the total population.

Why were so many Asians emigrating from their areas of origin and immigrating to Southeast Asia? The short answer is that many areas in India and China were so densely populated that, at the current level of technology, many people could no longer make a living. In contrast, Southeast Asia was sparsely populated. As will be shown presently, the Western world was generating a huge and increasing demand for raw materials, which could be supplied by Southeast Asia, provided that sufficient labor be made available.

On balance, then, immigration must have made a positive but moderate contribution to the rate of growth of the population of Southeast Asia between the 1870s and the 1960s. This occurred mainly between the 1870s and 1930, as after 1930 many Chinese and Europeans left the region, because the Depression of the 1930s, the War in the Pacific, and the troubled period of decolonization made the area much less attractive to them.

High and increasing rates of population growth were typical for the region between the 1870s and the 1960s. This was, as we have seen, largely the result of dropping death rates. Immigration must have contributed as well, but we are uncertain about the role of the birth rate prior to, say, 1960.

THE DEVELOPMENT OF AGRICULTURE

Given the almost continuously increasing population growth in the region between the 1870s and the 1960s, it was only to be expected that agriculture would expand considerably. This is demonstrated by the phenomenal expansion of cultivated land, and the equally impressive drop in land under forest cover, illustrated in Table 8.2.

It is, of course, marvelous that we have these figures, taken from Richards and Flint (1994), but, as is so often the case with historical statistics from Southeast Asia, we must be aware that the further back in time we go, the less reliable are the statistics. Nevertheless, they are sufficiently robust to draw a number of conclusions.

Table 8.2 Land use categories in Mainland Southeast Asia (MSA) and Island Southeast Asia (ISA), 1880–1980, for benchmark years, in millions of hectares.

Category	Region	1880	1920	1950	1980
Temporary crops	MSA	7.7	15.6	19.5	32.9
	ISA	6.4	13.0	17.9	26.9
	Entire region	14.1	28.6	37.4	59.8
Permanent crops	MSA	0.4	0.9	1.5	2.9
	ISA	1.7	4.5	6.9	15.0
	Entire region	2.1	5.4	8.4	17.9
Cultivated area	MSA	8.0	16.5	21.0	36.0
	ISA	8.1	17.5	24.7	42.0
	Entire region	16.1	34.0	45.7	78.0
Wooded area	MSA	150.3	134.7	123.3	107.8
	ISA	215.4	202.8	190.0	166.4
	Entire region	365.7	337.5	313.3	274.2
Interrupted	MSA	49.2	45.0	44.1	44.8
wooded area	ISA	21.7	20.8	27.5	32.5
	Entire region	70.9	65.8	71.6	77.3
Total area	MSA	193.9	193.9	193.9	193.9
	ISA	256.4	256.4	256.4	256.4
	Entire region	450.3	450.3	450.3	450.3

Source: Richards, John F., and Elizabeth P. Flint, 1994, "A Century of Land-Use Change in South and Southeast Asia." Pp. 15–66 in Effects of Land-Use Change on Atmospheric CO_2 Concentrations: South and Southeast Asia as a Case Study. Edited by Virginia H. Dale. New York: Springer.

Note: Figures do not always add up because of rounding.

While the total cultivated area expanded from 16 to 78 million hectares (an increase of 60 million), forest cover decreased from 365 to 274 million hectares (a drop of 90 million). The difference between the two figures (60 and 90) is accounted for by the increase in grass and shrub lands of the built-up area.

In the meantime the population of the region had grown from some 60 to 355 million people, which implies that the cultivated area grew more slowly (about five times) than did the population (about six times). However, as both population and arable figures become less reliable the further we go back in time, we must not put too much faith in the 1880 data. Nevertheless, if we may assume that they are off by roughly the same percentage, the differences in the growth rates between 1880 and 1980 would remain of the same order of magnitude.

No matter how unreliable these figures are, however, it is clear that in comparison with the post-1950 cultivated area and population data, around 1880 people had had a modest (though, as we have seen, not negligible) impact on their natural environment. That becomes even more apparent if we look at the changes in the percentages of the total surface area under forest cover or cultivated, as presented in Table 8.3.

Table 8.3 Proportions of total surface area wooded and cultivated in Mainland Southeast Asia (MSA) and Island Southeast Asia (ISA), 1880–1980, for benchmark years (in percent).

Category	Area	1880	1920	1950	1980
Cultivated area	MSA	4.1	8.5	10.8	18.6
	ISA	3.2	6.8	9.6	16.4
	Entire region	3.6	7.6	10.1	17.3
Wooded area	MSA	77.5	69.5	63.9	55.6
	ISA	84.0	79.0	74.1	64.9
	Entire region	81.2	75.0	69.6	60.9

Source: Richards, John F., and Elizabeth P. Flint, 1994, "A Century of Land-Use Change in South and Southeast Asia." Pp. 15–66 in Effects of Land-Use Change on Atmospheric CO$_2$ Concentrations: South and Southeast Asia as a Case Study. *Edited by Virginia H. Dale. New York: Springer.*

The figures in Table 8.2 also enable us to calculate differences between periods in the growth rates of the cultivated areas. The results of these calculations are given in Table 8.4.

Evidently, and in accordance with what was said at the beginning of this chapter, average annual growth rates for the cultivated area in Southeast Asia between 1880 and 1980 were high between 1880 and 1920 and again during the period from 1950 to 1980, while they were low between 1920 and 1950.

The average annual growth rates of the periods 1880 to 1920 and 1950 to 1980 are quite similar, which is remarkable, to say the least, given the fact that the population was growing so much faster between 1950 and 1980 (around 2 percent) than it did between 1880 and 1920 (between 1 and 1.5 percent). This has been interpreted generally as a sign that the frontier of the (lateral) expansion of lowland cultivation was reached in the 1960s and 1970s (depending on the area). Since then, expansion of production had to come from intensification, a notion to be explored presently.

Table 8.4 Average annual rates of growth of cultivated area (temporary and permanent crops) in Mainland Southeast Asia (MSA) and Island Southeast Asia (ISA), 1880–1980, for four periods (in percent).

Category	Area	1880–1920	1920–1950	1950–1980	1880–1980
Temp. crops	MSA	1.78	0.75	1.76	1.46
	ISA	1.79	1.07	1.37	1.45
	Entire region	1.78	0.90	1.58	1.46
Perm. Crops	MSA	2.05	1.72	2.22	2.00
	ISA	2.46	1.44	2.62	2.20
	Entire region	2.39	1.48	2.55	2.17
Cult. area	MSA	1.83	0.81	1.81	1.52
	ISA	1.94	1.16	1.79	1.66
	Entire region	1.89	0.99	1.80	1.59

Source: Richards, John F., and Elizabeth P. Flint, 1994, "A Century of Land-Use Change in South and Southeast Asia." Pp. 15–66 in Effects of Land-Use Change on Atmospheric CO_2 Concentrations: South and Southeast Asia as a Case Study. Edited by Virginia H. Dale. New York: Springer.

It is also clear from Table 8.4 that the expansion of permanent crops (perennials) was faster than that of temporary crops (annuals). This holds true for all periods dealt with here, and for both parts of Southeast Asia. This may be interpreted as faster expansion of crops for the world market than of those grown mainly for local demand. The most important permanent crops were pepper, coconut, abaca, coffee, tea, rubber, oil palm, and cacao. Production from these trees is destined almost entirely for the world market, and a big share of the production is taken care of by large-scale plantations, often laid out with Western capital. Nevertheless, smallholders play an important and probably increasing role as well, particularly regarding coconuts and rubber.

Among the annual crops, the most important, such as rice and maize (corn), are grown for local consumption, partly even for the consumption of the producers themselves. Therefore, the expansion of the area under temporary crops can be used as a proxy for local food production, which may be supposed to correlate highly with population growth. I hasten to add that rice and maize are also produced for the market, even for the international market, which is also the case with other annuals such as sugar and tobacco, crops grown by smallholders but also by "Western" plantations. However, rice and maize are today, and have been for the last hundred years or so, without doubt the dominant annual crops. In 1993, of the 61.3 million hectares cultivated with temporary crops in Southeast Asia, 38.5 million were planted with rice and 8.2 million with maize, together covering 75 percent of all arable lands. The next group of crops, pulses, were grown on 2.8 million hectares, and therefore not in the same league as rice or maize.

Although, therefore, the area under perennials (the commercial sector) was growing faster than that under annuals (the subsistence sector), the latter was still three times as large as the former in 1980 (see Table 8.2). However, in 1880, the difference had been almost a factor of seven.

To phrase it differently—although the share of the agricultural production geared toward international commerce, and therefore the influence of the latter on Southeast Asian land use, evidently was growing during the period under consideration, by 1980 the role of population growth as represented by the share of subsistence crops was still much more important.

The distinction between "commercial" and "subsistence" crops—both terms used rather loosely—is a useful one for a first impression of long-term developments in the agrarian sector, but it does not tell the whole story. As we are interested mainly in the question of what all of these developments did for the natural environment, it will be necessary to look in more detail at various types of land use.

FORAGERS

The oldest means of land use, and the least burdensome to the environment, is foraging (hunting and gathering), practiced by people at densities of below one person per square kilometer. Although we tend to think of foragers as Stone Age people, we really do not know all that much about them before nineteenth-century missionaries, anthropologists, and civil servants started to document their lives in some detail. Nevertheless, we do know that many real Stone Age foragers were living in places where they can no longer be found—in the coastal lowlands and at higher elevations in some of the fertile river valleys of Mainland Southeast Asia. Today, almost all foragers—or rather, what is left of them since the last time they were reported on—live in the much less fertile inland areas of Island Southeast Asia. It is probably not a coincidence that they are to be found almost exclusively in the ever-wet tropical forest zone around the equator, as the more fertile areas have all been claimed by agriculturists.

The foraging groups that have survived (although sometimes barely) until the end of the period under consideration include the Semang of the Malay Peninsula, the Kubu of Sumatra, the Punan or Penan of Borneo, the Agta (or Ata) and the Batak of the Philippines, and the Togutil or Tobelo of the Moluccas. Closer study of some of these groups has led to a number of insights that are of some relevance to our theme.

One of the most interesting points made over the last decades is that not all of these peoples appear to be "pristine" hunter-gatherers. The term *pristine* is applied particularly to the so-called Negritos (Semang, Agta), a collective name for those groups—closely related to the Papuans of New Guinea—that were present in the area prior to the arrival of the Austronesian speakers several millennia ago. Pristine hunter-gatherers are supposed to have been foragers since they came to the region. In contrast, peoples like the Kubu or the Punan are regarded as "devolved" hunter-gatherers, who arrived in the area as agriculturists, but for some reason subsequently became foragers. They live in fairly poor environments, of which it is doubtful whether they ever produced sufficient carbohydrates that could be digested by humans. Therefore—or so the argument goes—"devolved" hunter-gatherers have always depended on exchange with neighboring settled agriculturalists and coastal traders, with whom they bartered tradable, nonedible forest products (for the international market) for food. In fact, an increased demand for forest products generated by international trade could have been the very reason that such groups started to specialize in foraging, thereby saying farewell to existence as agriculturalists.

Another point is that boundaries between hunter-gatherers and swidden agriculturalists are sometimes blurred. Although, contrary to popular stereotypes, most Papuans are and have been agriculturists for thousands of years, some groups living in New Guinea are also (pristine) hunters. The Nuaulu (Seram, Moluccas) have been classed as relatively sedentary agriculturalists in some ethnographic surveys, while most of their calories come from mostly uncultivated sago palm and two-thirds of their protein from wild resources. However, they started planting cassava, sweet potato, and maize perhaps as long as a century ago. It will be noted that these are all American crops.

The latter case, and the points made about the "devolved" foragers, are illustrations of how the arrival of foreign crops and foreign trade have influenced, and perhaps even called into existence, "traditional" groups, of whom it is usually assumed that they have not changed their way of life since the Stone Age.

Be that as it may, the numbers involved here were always very small during the period under consideration, and they appear to have declined throughout the entire period. This happened partly because Western medicine hardly reached them, but also because their habitat was destroyed; many of them turned agriculturist.

SHIFTING CULTIVATORS

The number of people who could be called shifting cultivators (swidden cultivation, slash-and-burn agriculture) was much larger. As with foragers, many people will probably associate shifting cultivators with the remote past. Although it is true that the first shifting cultivators arrived on the scene many thousands of years ago, our information on Southeast Asian swiddeners dates overwhelmingly from the post-1870 period. They were living at higher population densities than foragers, of up to twenty-five or even fifty people per square kilometer. In fact, during part of the period being discussed here, slash-and-burn agriculture was the dominant mode of land use in large areas, particularly in Island Southeast Asia. It is, however, far from easy to find good statistics—or, usually, any statistics at all—on the amounts of land under shifting cultivation or on the numbers of swiddeners.

Shifting cultivators clear a plot of most of its growth—usually primary or secondary forest—set fire to the cut (and dried) vegetation of the plot, and then use a dibble stick to make holes into which they drop a number of seeds. During one, two, and occasionally more years the same plot is planted with a variety of crops, of which "dry" rice is and was usually the most important. Often, more than one crop is grown simultaneously on the same plot, crops that mature at different moments in time. After a few harvests, the plot is abandoned, although people occasionally return to it for the fruit trees and other perennials they had planted during the time the field was used for the cultivation of annuals. The plot is then essentially left alone until the fertility of the soil is sufficiently restored—swiddeners use no manure and undertake no tillage—for another series of harvests. This may take as long as thirty years, but much shorter cycles, of often less than ten years, have usually been observed.

Ever since Harold Conklin introduced these terms in the 1950s, a distinction has been made between "integral" shifting cultivation systems and "partial" systems, though other ways of categorizing swidden systems have been developed since. In the former case, shifting cultivation is a way of life, and those who practice it rely on it almost entirely for their subsistence. This does not mean that they are (or were) planting only subsistence crops; it seems highly likely that the proportion of commercial crops has increased greatly between the 1870s and the 1960s. Those who practice partial slash-and-burn cultivation are originally settled permanent field agriculturalists for whom swiddening is a sideline (supplementary swiddening), or landless lowlanders who want to become shifting cultivators (incipient swiddeners). This is a phenomenon that appears to be fairly recent, at least on a scale that makes it noticeable to outsiders. Since lateral expansion in the lowlands is no longer possible in most areas of Southeast Asia, a state of affairs that should be dated, depending on the area, between the 1950s and the 1970s, land-poor lowlanders have increasingly sought additional cultivable land in the more sparsely settled uplands, thus becoming partial swiddeners. The expression "reluctant swiddeners" has been used for them as well.

Usually, scholars distinguish two types of land use among the integral swiddeners. There are "pioneer" swiddeners, who prefer to clear primary forests for their plots of crop land, and there are "established" swiddeners, who use a regular system of rotation, preferring to clear secondary forest for cultivation purposes, or having to make do with it because primary forest is no longer available.

There appear to be various reasons for a preference for clearing either primary or secondary forest. One of these factors is that in the areas of the ever-wet rain forest depending on burning the remains of a cleared old-growth forest is risky, because one cannot be certain of a sufficiently long dry spell. The cut vegetation of a secondary forest takes much less time to dry, and therefore the cultivator is more certain of a good burn. The cutting of old-growth forests is also more labor intensive, a factor that was even more important in the pre-chainsaw era. Another factor is the distance one is prepared to walk to a cultivated plot from one's main settlement. When all climax forests in the proximity of one's village had been cut, old-growth forests could be found only at a greater distance. This is a factor that in the past was often related to feelings of insecurity because of the presence of inimical tribes in the neighborhood. When strong central—often colonial—rule created safer conditions, the cutting of remoter primary forests became less of a problem. The preference for cutting old-growth forests is also reported to have been based on considerations of soil fertility and the expectation of fewer problems with weed control. Over the last decades, of course, the main factor has become in many instances whether there is any old-growth forest left.

Generally speaking, "established" swiddeners are fairly sedentary, as their forest-fallow rotation cycle does not take them far from home. The tendency to remain in one location was almost certainly stimulated by the increased interest of swiddeners in the growing of perennials, the products of which were destined for the international market. This had begun long before the 1870s with pepper and picked up speed with coffee, and even more so around 1900 with rubber and coconuts. This phenomenon, today called agro-forestry, in principle tied the shifting cultivator much more strongly to a circumscribed area—at least as long as the trees that had been planted survived. Thereafter the soil where these perennials had been growing must have been exhausted, as the use of manure was far from general.

In addition to all of this, it should be mentioned that during the period dealt with here, swiddeners also started growing more annual commercial crops, such as maize, cassava, and poppies.

Having discussed various types of swiddening, it should be pointed out that even a period of thirty years is not enough to restore the original (climax) vegetation of an old-growth forest. If that were the purpose of the cycle—which it is not—the soil would have to be left alone for, say, at least a hundred years. Therefore, any plot of land taken up in the swidden cycle cannot be expected to revert to climax forest ever again, unless a group of shifting cultivators leaves the area altogether because of conflicts with neighboring groups. That happened regularly in the past, but as the "forest frontier" is closing, most shifting cultivators have no other place to go—at least not as shifting cultivators.

Nevertheless, as a rule, a cycle of thirty years is much better in environmental terms than is a cycle of, say, five years. During a thirty-year cycle on an abandoned plot a secondary forest springs up that, in terms of biodiversity, protection of the soil, habitat for wildlife, the presence of non-timber forest products and timber, and as a CO_2 "sink," is superior to most other types of vegetation. What is at least equally important is that with a fallow cycle of between, say, eight and thirty years, shifting cultivation appears to be a sustainable form of land use, although under certain circumstances weed problems even then assume serious proportions when the land is cultivated for too long. If a shorter cycle is practiced, there is a much greater risk that the abandoned fields will not revert to forest cover. Eventually, such soils will often become covered in tall grasses (*Imperata cylindrica* [cogon grass], *Saccharum spontaneum*), and weed control will be so burdensome that such plots can no longer be part of a forest-fallow cycle. Recently—that is, after the 1950s—this weed control problem, at least from the point of view of the cultivator (but not from an environmental point of view), seems to have been solved in some areas by the presence of an invasive species from the New World, *Chromolaena odorata* (formerly *Eupatorium odoratum*), a plant that suppresses *Imperata cylindrica*, thus acting as a (nonchemical) weed control. Nevertheless, such plots are no longer part of a forest-fallow cycle.

Generally speaking, therefore, the recently observed trend toward a much shorter rotation cycle means that fewer soils revert to secondary forest, with all that this implies regarding erosion, forest resources, wildlife habitat, and CO_2 uptake. The shorter rotation cycle is caused primarily by the lack of space, which in turn is caused partly by the natural increase of the resident population, but probably to a larger extent by the arrival of migrants and by the loss of forest cover owing to logging and the laying out of plantations. One assumes that increased involvement with the market has been another factor leading to shorter cycles, as growing demand led to the wish to produce more on the same amount of land—although, as will be shown presently, stronger market forces have also been known to lead to sustainable production in vulnerable uplands. It is not easy to date this trend toward a shorter rotational cycle, but it seems safe to assume that it was not very important prior to the 1960s or 1970s, and therefore falls largely outside the period under discussion.

Finally, it should be mentioned that in some areas swiddeners, for various reasons, have turned to permanent-field agriculture. Higher population densities may have played a role, but also market forces (including better connections to harbors) and attempts by the state to settle shifting cultivators, turning them into wet-rice producers. Both indigenous and colonial states had a preference for settled wet-rice–cultivating peasants over shifting cultivators, for a variety of reasons, of which the most important may have been that they were easier to locate and therefore to tax, and that yields per unit of land were higher. Both carrot and stick were used by the state to that end.

It is virtually impossible to quantify the influence of the various trends discussed here, and very few authors attempt to do so. An exception to this rule is Robert Elson, who assumes a "vast but unmeasurable increase in the rate of swiddening across Southeast Asia over the last two centuries." This assessment is, alas, not backed up by any quantitative evidence.

As there is no direct way to measure the changes over time in the activities of shifting cultivators between the 1870s and the 1960s, one should look for a proxy variable. Perhaps the data on "interrupted wooded areas" presented in Table 8.2 could be used for that purpose. The amount of interrupted wooded area should correlate highly with slash-and-burn activities, as this is precisely what one gets when a swiddener abandons a plot that has been used for a couple of years. The fit will be far from perfect, as one assumes that mature and very young secondary forest will not show up as "interrupted wooded area," which implies that the figures in the table underrate the amount of swiddening. However, this would not invalidate the trend suggested by the data in the table.

The figures on "interrupted wooded areas" in Table 8.2 certainly do not suggest a vast increase of swiddening. In fact, the amount of interrupted woodland appears to have shrunk in Mainland Southeast Asia. However, in Island Southeast Asia we do witness an increase, particularly between 1920 and 1980. For Southeast Asia as a whole, the growth between 1880 and 1950 is nil (with even a dip between those two years), and with only a modest increase between 1950 and 1980. Therefore, during the period under consideration (the 1870s to the 1960s), these figures do not suggest an increase. Looking back at the various trends discussed above, such a development—no strong increase—would certainly appear to be a possibility.

"WET" CULTIVATION

The largest increases during this period were recorded in agriculture on permanent fields. Settled agriculturalists also formed the largest group of cultivators in the region as a whole, certainly in the 1960s, but probably also around the 1870s.

However, their numbers locally would have been smaller than those of the shifting cultivators, particularly at the beginning of the period.

Regarding permanent field agriculture, Table 8.2 distinguishes between temporary and permanent crops. Temporary crops are discussed here first, making a distinction between "dry" and "wet" types of cultivation. "Wet" cultivation is the best-known type of agriculture in Southeast Asia. It refers to bunded fields, often planted with rice (hence the term *rice paddies*), that can be inundated, usually during the period that the land has to be plowed (or churned by letting buffaloes loose on it) and during a considerable part of the growing season. There are various sources of water—rain, the rising waters of lakes and rivers, swamp water, and irrigation water from tanks (reservoirs) or from gravitation flows from hills and mountains, with or without a weir as go-between.

The latter subcategory of "wet" agriculture—artificial irrigation—has always received the most attention from scholars. Gravity-flow irrigation systems are often spectacular both regarding their aesthetic aspects—the beautiful landscapes of hills and mountain slopes covered with terraced rice paddies arranged in the form of an amphitheater—and as regards the imposing technology of large weirs, dams, and sluices. The latter technology was introduced largely by the Western powers, from the beginning of the period discussed here, but the rice-paddy-covered mountain slopes locally date back at least some ten centuries.

Irrigated wet-rice cultivation, therefore, appears to be a sustainable system par excellence. It is a system that in principle hardly required bought inputs—at least, as long as livestock could be fed from the village commons, and local rice strains were used (although one should not underestimate the nineteenth-century trade in strains between areas). The water that was led onto the fields and remained on them until the crop had ripened and had to be harvested not only guaranteed sufficient moisture for the growth of the crop but also suppressed the growth of competing weeds. In addition, it carried many nutrients, at least in areas like Sumatra, Java, and Bali, where the soil is volcanic and hence fertile. It lent cohesion to the terraced soils, thus making collapse of the terraces less likely, and it was instrumental in making the nutrients found in the soil available to the plants. In many cases, inundated soils form a so-called plow pan or tillage pan, a compacted layer of soil just below the maximum depth of tillage, which was instrumental in keeping water and nutrients from seeping into the subsoil. Furthermore, nitrogen-fixing blue-green algae were growing naturally in these inundated fields, which, in combination with the dung of the buffaloes that were used for plowing and that were pastured on these fields after the harvest, together with the nutrients supplied by the water itself (from the volcanic upper slope areas), and with the rice straw that was plowed under, made the use of fertilizer

Planting rice almost a century ago as seen in this photograph is still done the same way today in much of the Philippines. This kind of "wet" cultivation is the most common agricultural technique in Southeast Asia. (Library of Congress)

from outside redundant. Moreover, from the early nineteenth century, many observers mentioned fish being kept in inundated rice fields, the feces of which were additional nutrients, while more recent reports suggest that occasionally small outhouses were built over the inundated rice fields as well, thus supplying human manure. It was a sustainable system with very little erosion, capable of feeding people well from a small acreage and needing few outside inputs. In fact, in environmental terms, the only problem with wet-rice cultivation was (and of course, still is) its production of methane (CH_4), a greenhouse gas.

However, during the period under consideration, external inputs became more important. In the first place, during the latter part of the nineteenth century, large-scale irrigation works, designed and carried out by the various colonial states, were needed as population growth was increasing, and because of the requirements of Western agricultural enterprise in Southeast Asia. Even then, the growth of arable lands under artificial irrigation could often not keep up with population growth, particularly after 1900. Nevertheless, the period from 1870 to 1930 can be regarded as the heyday of colonial irrigation, while relative and sometimes absolute stagnation or even decline set in between 1930 and the 1960s.

At the end of the period we are dealing with, many things started to change more or less at the same time. Owing to a lack of pasture, among other things, it became increasingly difficult to keep buffaloes, while fodder for stall-fed cattle now often had to be bought. Thus mechanization became more attractive, but it also led to a shortage of manure. At the same time the so-called Green Revolution took place, when high-yielding rice varieties (HYVs) became available that, however, had to be accompanied by liberal doses of fertilizer and pesticides. After a period of stagnation—sometimes even decline, as irrigation works fell into disrepair—governments started to invest in the agricultural sector again, with the proceeds of growing exports of oil and forest products among other commodities. Much of this investment was spent on technical irrigation, which shows up in the statistics of arable land under irrigation from around the 1960s. Thus, irrigated wet-rice cultivation was no longer a more or less self-contained, environmentally "clean" activity.

Another development that occurred during the period between the 1870s and the 1960s, and that appears to have accelerated after the 1960s, was the increase of double cropping. In almost all areas it became customary to grow a second crop after the first rice crop. This might also be rice, but it was often something else (diversification). A second rice crop would also be grown on an inundated field, but most second crops were not. They were grown as "dry" crops, a topic dealt with presently. Therefore, as the frontier of lateral expansion in the lowlands had been reached around this time, intensification of land use was one of the answers, an option that had been made possible by the expansion of technical irrigation. Insofar as these second crops were "dry," the possibility of erosion, which is slight under "wet" conditions, increased considerably.

Wet-rice cultivation, other than that carried out with artificial irrigation, has many of the same advantages—and the disadvantage—of methane emission. Generally speaking, though, its yields are less predictable, and, given the fact that the water used for this purpose is usually not carrying particles of volcanic soil, lower. However, as a rule it also involves less labor, and therefore, on balance, the productivity of labor need not be much lower than in the case of artificial irrigation.

"DRY" CULTIVATION

Finally, a few words are in order about "dry" cultivation (of temporary crops) on permanent fields—or, as it is sometimes formulated, "dry" crops. This type of agriculture is well known in many areas of the world, and it is the predominant, almost exclusive, form of agriculture in Europe and America. In Southeast Asia it is much less usual, and its history is not well documented. Dry cultivation (on permanent fields) refers to the cultivation of annuals (as first and usually only crop) on soils that cannot be inundated, although, of course, they need (rain) water for their growth. Second crops on bunded fields are also usually grown as dry (that is, not inundated) crops. Rice may be grown as a dry crop on permanent fields. Other dry staple crops are maize, cassava, and sweet potato. Dry rice grown on permanent fields was probably never more than a small share of all cultivated rice in Southeast Asia, but there are local exceptions (for example, Sumatra).

This type of dry-rice cultivation is not to be confused with swidden cultivation of rice, often called dry upland rice. The main difference between the two types used to be that swiddening was undertaken by cultivators whose entire way of life could be called "tribal," while the cultivation of dry rice on permanent fields was usually undertaken by the same villages where wet-rice was being grown. In such a village many cultivators possessed both wet and dry fields, although the poorer segment of the villagers had to make do with dry fields only. Today, the difference between the two types is less absolute, as poor lowlanders move into the uplands, while "original" swiddeners apply shorter fallow cycles.

Permanent fields under temporary dry crops produce, in principle, less sustainably than do wet fields and fields under shifting cultivation with long fallow periods. The possibility of erosion is greater, and more fertilizer must be applied. Harvest failures—at least as regards rice—may be expected to be more frequent.

A man gathers latex from a rubber tree in Java, Indonesia, in the early 20th century. Perennial crops, like rubber, were largely grown on plantations for exportation. (Library of Congress)

PERENNIAL CROPS

Having dealt with permanent fields under temporary (annual) crops, we now turn to permanent (perennial) crops on permanent fields. The most important ones were cloves, nutmeg, pepper, coffee, tea, abaca, coconut, cinchona, rubber, oil palm, and cacao. As was mentioned earlier, these crops were grown for the most part on large-scale plantations, financed by Western capital and producing for the international market. However, peasant smallholders had an important share in the production of some of them, notably coconut, rubber, and cacao.

The export crop sector was growing at a very high rate between about 1850 or 1870 (depending on the area) and 1930. Although high growth rates could be observed during earlier periods, they were usually confined to selected areas and to shorter periods of time. The period between 1870 and 1930 should be regarded as the third round of globalization, as the share of international trade in world income and production increased considerably during that epoch, reaching unprecedented levels. World demand increasingly made itself felt in the most remote corners of Southeast Asia. However, this was more a matter of degree than of kind, as many remote places had already been confronted with world market demand in the Early-Modern Period—what I have called the second wave of globalization. Between 1930 and the 1960s, the growth of trade stagnated; the forth round of globalization did not take off until the 1970s.

The origins of the high growth rates between 1850–1870 and 1930 are to be found largely in Europe and North America, where self-sustained economic growth, urbanization, and industrialization went into higher gear during this period. European and U.S. economic growth led to a continuously and rapidly increasing demand for raw materials from the region, supported by improvements in transport and communications (iron steamships, telegraph, the opening of the Suez Canal in 1869) between those regions and Southeast Asia; by the construction of canals, all-weather roads, and railways; by free trade; and by the introduction of the internal combustion engine in Southeast Asia itself. The term often used for the foreign-demand-stimulated, export-led agricultural developments during this period is *vent for surplus*.

Tables 8.2 and 8.4 show clearly that the area under perennials grew faster between 1880 and 1980 than did the area under annual crops. However, it is also clear that, even in 1980, the area under permanent crops was still much smaller than the one under temporary crops. No matter how high the rate of growth of agricultural export commodities produced (largely) by plantations may have been, the share of small-scale peasant agriculture in the area under cultivation, and therefore the influence of population growth, was still the more important factor if we look at the expansion of the cultivated area and the decrease of forest cover.

By around 1870, cloves and nutmeg had lost much of their importance, while the production of pepper (originally coming from India), already an important Southeast Asian export product prior to 1500, had been increasing during most of the nineteenth century. Coffee was introduced in the eighteenth century and tea in the early nineteenth century. Abaca and coconut were indigenous products, but their production expanded considerably after around 1870 because of increased international demand. From around 1870 onward, cinchona, rubber, oil palm, and cacao were introduced from outside and became quite successful.

Most of the products mentioned here were more or less recent introductions, and it is no exaggeration to state that the expansion of export crops produced by Western enterprise in Southeast Asia was based largely on commodities originally alien to the area. This is a phenomenon that we have come across earlier—for instance, when discussing the role of the American crops maize, sweet potato, cassava, peanuts, chilies, and tobacco in Southeast Asia after 1500. It is no coincidence, as was also explained earlier, that the alien crops were so successful, given the fact that they were usually introduced in areas where they had no natural (specialized) enemies, while the enemies they had had at home were left behind.

I think it fair to say that the introduction and expansion of "exotic" permanent crops, particularly (but not exclusively) during the period under consideration, together with the introduction of alien capital and management, constituted not only a major economic and social transformation in Southeast Asia but also one of the major environmental revolutions in the region. Large areas of old-growth forest were cut down and replaced by monocultures of rubber or oil palm. Such areas still looked like forest—and in some respects could still be regarded as forest—but they had lost much of their original biodiversity, which influenced features like the presence of wild animals and NTFPs.

Finally, mention should be made of small-scale agro-forestry systems other than smallholder rubber, cacao, or clove plantations. We should differentiate between home gardens and what for Java has been called the *talun-kebun* system. Home gardens are, not surprisingly, plots of land with a house where both annuals and perennials, including fruit trees, are cultivated. They may contain fish ponds and chicken pens and are characterized by a high plant diversity and density. Home gardens are well documented from the nineteenth century but are without any doubt much older. In many densely settled areas they provided the peasantry with perhaps 20 percent of their income and 20 percent of their food. The talun-kebun was also a mixed annual-perennial plot, at the fringe of the village, geared largely toward the market. They often yielded, among other products, bamboo, timber, and fuelwood, in addition to annuals such as beans, tobacco, and onions. Both systems appear to score very well as regards sustainability and agro-diversity.

BIOINVASIONS

The role of exotic plants was certainly not restricted to the export sector and to foodcrop agriculture. Of the 5,011 species of noncultivated plants recorded in Java around 1960, 413 were introduced. In addition, there were 1,523 species of cultivated plants, many of which were, as we have seen, introduced as well.

Introduced species are a mixed blessing. On the one hand, subsistence and export agriculture have been evidently enriched by the many exotic species that are now, if not dominant, at least of great importance in these sectors. Many of these crops, however, are dry crops, and are not cultivated in a sustainable way. Another negative point is that some exotics have run wild, a phenomenon for which the term *bioinvasions* is particularly apt.

A good example of such a bioinvasion is *Lantana camara*, an American species that was present in India by 1810, in the Philippines by 1845, and was introduced in Java from Singapore around 1860. It rapidly colonized many deforested areas. If everything else failed, it was sometimes used by European foresters for reforestation purposes, as it was capable of destroying alang-alang (*Imperata cylindrica*). However, fields invaded by (toxic) *Lantana* could no longer be used for grazing. Other examples are several species of *Chromolaena* (formerly called *Eupatorium*), also originating from America. *C. odorata* was spreading from Thailand to Malaysia by 1930 and somewhat later to Indonesia, where it soon became a pest, making whole areas unpalatable to livestock (which, in environmental terms, might be rather good than bad). As we have seen earlier, it also destroys *Imperata cylindrica*, but it is, according to some authors, unsuited for reforestation; others argue that reforestation is unproblematic. Somewhat earlier, *C. pallescens* had made its way to Java, where it became a pest on plantations, but it was also used as green manure and as a suppressant of alang-alang (cogon grass). Discussions about the pros and cons of particularly *Lantana camara* and *Chromolaena odorata* have been raging ever since they were introduced in Southeast Asia.

Another infamous example of a bioinvasion in Southeast Asia is the water hyacinth (*Eichhornia crassipes*), also an American species. It was introduced to the Botanical Gardens of Buitenzorg (now Bogor) in Java in 1894 through the agency of the wife of the Brazilian ambassador. It is an attractive blue-flowered

The water hyacinth is an attractive, blue-flowered plant, but also an invasive, tropical aquatic weed. Introduced in Southeast Asia in the late 1800s as a decorative plant, it quickly spread and began clogging canals, slow-moving rivers, and even rice fields. (Corel)

plant that was popular as a pool ornament. A surplus of these plants was dumped in the Ciliwung River, and it quickly became a pest, fully deserving its reputation of being possibly the worst tropical aquatic weed. It was even worse in Mainland Southeast Asia, where it was probably introduced in 1902, in what was then called Tonkin (Vietnam). It forms dense, matted masses that cover tanks, slow-flowing rivers, irrigation canals, and even rice fields. It has its uses, though, being employed as pig fodder and, after having been burned, as manure. Not even a decade ago it was discovered in Thailand to be excellent raw material for wickerwork.

Whereas the spread of the plants just mentioned was at least partly unintended, the introduction and expansion of various species of eucalyptus, an Australian tree genus, took place deliberately. In Java, *Eucalyptus globulus*, the blue gum tree, was introduced in the 1870s by the Colonial Forest Service, based on positive experiences in India. The main purpose of its introduction was reforestation, as it was a fast growing tree, but its wood was also deemed fairly good timber. Nowadays, eucalyptus can be found all over the world, and for the same reasons that the trees were introduced in India and Java. The tree is a bit of a Trojan horse, however, as it consumes inordinate amounts of water (apart from conspicuously littering the landscape with branches and bark), which allegedly led locally to problems with wet-rice production. In many places, therefore, the tree has become a political issue, and people are trying to get rid of it or at least to keep the government or private enterprise from planting more of it. Recently, plans for more eucalyptus plantations in Thailand sparked off riots and led to the burning of eucalyptuses.

Although foreign introductions of plants (and animals), as we have seen, have been going on since time immemorial, it is also clear that considerable acceleration took place during the period under consideration. During the last decades this process has accelerated even more, particularly owing to the enormous expansion of air traffic and of the use of large ships with ballast tanks filled with water. Unless rigorous measures are taken to prevent them, bioinvasions are here to stay, and they may be expected to produce many more unpleasant surprises. Nevertheless, it would be wrong to regard the presence of exotics as purely negative. It would appear that areas like Java, where bioinvasions have always been one of the environmental facts of life, have often put exotics to good use. It is probably in "pristine," remote areas that bioinvasions on balance may have had a negative impact.

LIVESTOCK

The introduction and expansion of livestock in Southeast Asia could be regarded as a bioinvasion writ large, although perhaps not on a scale like that which took place in the Americas, where livestock as we know it was virtually absent prior to 1492. As during the eighteenth and earlier parts of the nineteenth centuries, foreign livestock breeds were imported after around 1870, but now on a much larger scale. They brought diseases like rinderpest with them, which may have caused larger epidemics than in the past. At least, rinderpest epidemics were raging in many parts of Southeast Asia in the 1880s and 1890s. There are also indications that such epidemics were transmitted to wild ungulates (hoofed mammals), as more and more livestock invaded the shrinking forests, while it is even possible that some epizootics were transmitted to nonungulate mammals as well. This would have contributed to the disappearance of wildlife in the more densely settled areas. In Table 8.5, quantitative data on livestock are presented.

As can be calculated from the data in the table, the growth of the number of animals was somewhat higher in Mainland Southeast Asia (some 1.55 percent per year, on average) than it was in Island Southeast Asia (1.35 percent). As a result, while there was only a slight difference in the average number of animals per

Table 8.5 Number of livestock in Mainland Southeast Asia (MSA) and Island Southeast Asia (ISA), 1880–1980, for benchmark years, in millions.

Region	1880	1920	1950	1980
MSA	10.3	19.0	28.9	48.5
ISA	11.2	16.8	27.7	43.3
Entire region	21.5	35.8	56.6	91.8

Source: Richards, John F., and Elizabeth P. Flint, 1994, "A Century of Land-Use Change in South and Southeast Asia." Pp. 15–66 in Effects of Land-Use Change on Atmospheric CO$_2$ Concentrations: South and Southeast Asia as a Case Study. *Edited by Virginia H. Dale. New York: Springer.*

hectare between the two areas ca. 1880 (0.6 versus 0.5), the livestock density is now much higher in Mainland Southeast Asia (0.33 versus 0.21). It is probably the case that the initial difference and the higher growth rates in Mainland Southeast Asia have something to do with ever-wet tropical areas being less favorable to livestock keeping.

We also observe the by now familiar pattern of high growth rates between 1880 and 1920 and from 1950 to 1980, and lower rates between 1920 and 1950. Between 1880 and 1950 the population in the region was growing somewhat faster than livestock numbers. This was probably caused by the arrival of alternative means of transportation. After 1950 that difference increased considerably, no doubt because of urbanization, agricultural mechanization, the ongoing expansion of train and truck transportation, and possibly also a lack of space in the most densely settled areas.

As regards type of livestock, it is clear that horses were doing rather badly, as they became increasingly redundant for military purposes and as a means of transportation. It is also likely that the growth of the number of buffalo slowed down, while cattle and even more goats were doing very well. All-weather roads are not suitable to buffalo (being too hard on their hooves), and stall-feeding of buffalo is problematic, two factors that may explain why cattle are doing better than buffalo. The proportion of cattle is also growing because their meat is preferred to buffalo meat by those with a rising income. Goats thrive in densely settled areas, where there is no room for cattle or buffalo.

As has been said, the role of livestock in the (environmental) history of Southeast Asia has hardly been studied. Therefore, much of what can be said about it is conjecture. Nevertheless, it may be safely stated that the presence and expansion of livestock greatly influenced land use and landscape change. In Southeast Asia livestock seldom if ever grazed on formal meadows. The animals were pastured on "waste" land, which is often a mixture of forested and grassy areas. This was to the detriment of the forest cover found there. In addition, areas under tall grasses like *Imperata cylindrica* or *Saccharum spontaneum*, which are palatable to livestock only when the grasses are young and green, were often set afire at the beginning of the wet monsoon. This also appears to have happened to swidden areas after the harvest. In all of these cases, possible forest regrowth was impeded. No wonder, then, that grass and shrub complexes increased in size during the period under consideration, although not at the same rate as did the number of livestock.

As waste land was increasingly cleared for agriculture, stall-feeding livestock became more and more usual. Fodder was increasingly cultivated or bought on the market, but during the period dealt with here there were continuous complaints that people were cutting grass and leaves for their animals in the forests, particularly in reforestation projects.

Sacrifices of large numbers of livestock, a topic dealt with in the previous chapter, were still in evidence during the period from 1870 to the 1960s—and still today. It is even possible that locally the numbers involved have increased, as market involvement made people richer and therefore capable of more conspicuous consumption. This occurred particularly in upland areas, where livestock was (and is) not needed for agricultural purposes, and the animals were therefore reared explicitly for sacrifices, a practice regarded as (environmentally) wasteful by many observers, now and in the past. What can be said in favor of such practices, however, is that the presence of large numbers of livestock, apart from being money on the hoof, may have acted as a shock absorber in times of dearth.

CONCLUSION

Summing up this chapter in a few lines, it can be said that environmental change took place at a higher rate than ever before, particularly between approximately 1870 and 1930. This was the result of both internal and external causes. The internal cause was land clearing for subsistence agriculture as a result of higher population growth rates than prior to 1870. Insofar as the rates of population growth have been influenced by Western medicine, this internal cause has partly external roots. The external cause was the growth of the European and North American markets, which led to a continuous and increased demand for primary products from the region, and thus to land clearing for perennial (plantation) crops.

Among the types of agriculture, wet cultivation became more important. As such, wet-rice cultivation is a sustainable activity. But its expansion is to the detriment of forested areas, and it leads to the increased production of methane, a greenhouse gas. It is unclear what happened to the amount of land under swidden cultivation, but it is generally assumed that fallow periods became shorter, which makes it an unsustainable system.

Livestock, expanding in numbers at a slightly lower rate than that of the population, contributed considerably to deforestation and landscape change. Various kinds of livestock expanded at different rates, depending upon market forces and environmental circumstances.

The importance of exotics increased, but also the threat of detrimental bioinvasions. In addition, foreign capital, foreign technology, foreign medicine, and foreign management played a greater role than ever before, driving many of the changes we have dealt with here. It seems fitting, therefore, to dub this period the third globalization phase.

BIBLIOGRAPHICAL ESSAY

A very short introduction to the political history of the region during this period is to be found in Dixon, Chris, 1991, *Southeast Asia in the World-Economy: A Regional Geography*. Cambridge: Cambridge University Press, pp. 69–84 and 134–148. For a more detailed treatment, see Tarling, Nicholas, 1999, "The Establishment of the Colonial Régimes." Pp. 1–74 in *The Cambridge History of Southeast Asia*. Vol. 2, Part 1: *From c. 1800 to the 1930s*. Edited by Nicholas Tarling. Cambridge: Cambridge University Press; and Stockwell, A. J., 1999, "Southeast Asia in War and Peace: The End of European Colonial Empires." Pp. 1–58 in *The Cambridge History of Southeast Asia*. Vol. 2, Part 2: *From World War II to the Present*. Edited by Nicholas Tarling. Cambridge: Cambridge University Press.

A very short, recent introduction to the demographic history of Southeast Asia during the period under consideration is to be found in Owen, Norman G., ed., 2005, *The Emergence of Modern Southeast Asia: A New History*, Honolulu: University of Hawaii Press, pp. 224–228. An earlier attempt to make sense of population development after 1830 in the region as a whole is Fisher, Charles A., 1964, "Some Comments on Population Growth in South-East Asia with Special Reference to the Period since 1830." Pp. 48–71 in *The Economic Development of Southeast Asia: Studies in Economic History and Political Economy*. Edited by C. D. Cowan. London: Allen and Unwin.

More recent studies, covering much of the region, are Owen, Norman G., ed., 1987, *Death and Disease in Southeast Asia: Explorations in Social, Medical and Demographic History*. Singapore: Oxford University Press; Hirschman, Charles, 1994, "Population and Society in Twentieth-Century Southeast Asia," *Journal of Southeast Asian Studies* 25, no. 2, pp. 381–416; Boomgaard, Peter, 2004, "Demographic Transition in Southeast Asia." Pp. 414–418 in *Southeast Asia: A*

Historical Encyclopedia, from Angkor Wat to East Timor. Vol. 1. Edited by Ooi Keat Gin. Santa Barbara, CA: ABC-CLIO.

Demographic historical studies on specific countries and smaller regions are Boomgaard, Peter, 1989, *Children of the Colonial State: Population Growth and Economic Development in Java, 1795–1880.* Amsterdam: Free University Press; Doeppers, Daniel F., and Peter Xenos, eds., 1998, *Population and History: The Demographic Origins of the Modern Philippines.* Madison: Center for Southeast Asian Studies, University of Wisconsin; Gooszen, Hans, 1999, *A Demographic History of the Indonesian Archipelago 1880–1942.* Leiden: KITLV; Knapen, Han, 2001, *Forests of Fortune? The Environmental History of Southeast Borneo, 1600–1880.* Leiden: KITLV; Banens, Maks, 2002, "Vietnam: A Demographic Reconstitution," unpublished paper; Boomgaard, Peter, 2003, "Bridewealth and Birth Control in the Indonesian Archipelago, 1500–1900," *Population and Development Review* 19, no. 2, pp. 197–214; Henley, David, 2005, *Fertility, Food, and Fever: Population, Economy and Environment in North and Central Sulawesi c. 1600–1930.* Leiden: KITLV.

On diseases, see particularly Owen, Norman G., ed., 1987, *Death and Disease in Southeast Asia: Explorations in Social, Medical and Demographic History.* Singapore: Oxford University Press. Other publications on disease and medicine are Boomgaard, Peter, 1989, "Smallpox and Vaccination on Java, 1780–1860: Medical Data as a Source for Demographic History." Pp. 119–132 in *Dutch Medicine in the Malay Archipelago 1816–1942.* Edited by A. M. Luyendijk-Elshout et al. Amsterdam: Rodopi; Boomgaard, Peter, 1993, "The Development of Colonial Health Care in Java: An Exploratory Introduction," *Bijdragen tot de Taal-, Land- en Volkenkunde* 149, no. 1, pp. 77–93; De Bevoise, Ken, 1995, *Agents of Apocalypse: Epidemic Disease in the Colonial Philippines.* Princeton: Princeton University Press; Boomgaard, Peter, 1996, "Dutch Medicine in Asia, 1600–1900." Pp. 42–64 in *Warm Climates and Western Medicine: The Emergence of Tropical Medicine, 1500–1900.* Edited by David Arnold. Amsterdam: Rodopi; Manderson, Lenore, 1996, *Sickness and the State: Health and Illness in Colonial Malaya, 1870–1940.* Cambridge: Cambridge University Press; Monnais-Rousselot, Laurence, 1999, *Médecine et Colonisation: L'Aventure Indochinoise 1860–1939.* Paris: CNRS; Boomgaard, Peter, 2003, "Smallpox, Vaccination, and the Pax Neerlandica: Indonesia, 1550–1930," *Bijdragen tot de Taal-, Land- en Volkenkunde* 159, no. 4, pp. 590–617; Richell, Judith, 2006, *Disease and Demography in Colonial Burma.* Copenhagen: NIAS.

Fairly recent general introductions on economic (mainly agricultural) development in the region between the 1870s and the 1960s are Richards, John F., and Elizabeth P. Flint, 1994, "A Century of Land-Use Change in South and Southeast Asia." Pp. 15–66 in *Effects of Land-Use Change on Atmospheric CO_2 Concentrations: South* *and Southeast Asia as a Case Study*. Edited by Virginia H. Dale. New York: Springer; Brown, Ian, 1997, *Economic Change in South-East Asia, c. 1830–1980*. Kuala Lumpur: Oxford University Press; Elson, Robert E., 1997, *The End of the Peasantry in Southeast Asia: A Social and Economic History of Peasant Livelihood, 1800–1990s*. Houndmills: Macmillan; New York: St. Martin's; Hill, R. D., 1998, "Stasis and Change in Forty Years of Southeast Asian Agriculture," *Singapore Journal of Tropical Geography* 19, no. 1, pp. 1–25.

Monographs on economic development, including agricultural development, in individual countries are Cheng, Siok-Hwa, 1968, *The Rice Industry of Burma 1852–1940*. Kuala Lumpur: University of Malaya Press; Ingram, James C., 1971, *Economic Change in Thailand 1850–1970*. Stanford: Stanford University Press; Adas, Michael, 1974, *The Burma Delta: Economic Development and Social Change on an Asian Rice Frontier, 1852–1941*. Madison: University of Wisconsin Press; Murray, Martin J., 1980, *The Development of Capitalism in Colonial Indochina (1870–1940)*. Berkeley: University of California Press; Feeny, David, 1982, *The Political Economy of Productivity: Thai Agricultural Development, 1880–1975*. Vancouver: University of British Columbia Press; Booth, Anne, 1988, *Agricultural Development in Indonesia*. Sidney: Allen and Unwin; Manarungsan, Sompop, 1989, *Economic Development of Thailand, 1850–1950: Response to the Challenge of the World Economy*. Bangkok: Chulalonkorn University; Eng, Pierre van der, 1993, "Agricultural Growth in Indonesia since 1880," Ph.D. dissertation, University of Groningen; Brocheux, Pierre, 1995, *The Mekong Delta: Ecology, Economy, and Revolution, 1860–1960*. Madison: Center for Southeast Asian Studies, University of Wisconsin; Corpuz, O. D., 1997, *An Economic History of the Philippines*. Quezon City: University of the Philippines Press; Booth, Anne, 1998, *The Indonesian Economy in the Nineteenth and Twentieth Centuries: A History of Missed Opportunities*. Houndmills: Macmillan; New York: St. Martin's; Dick, Howard, et al., 2002, *The Emergence of a National Economy: An Economic History of Indonesia, 1800–2000*. St. Leonards: Allen and Unwin; Leiden: KITLV.

A recent introduction on hunter-gatherers in Southeast Asia is Junker, Laura L., 2002, "Introduction." Pp. 131–165 in *Forager-Traders in South and Southeast Asia: Long-Term Histories*. Edited by Kathleen D. Morrison and Laura L. Junker. Cambridge: Cambridge University Press.

Other recent studies on hunter-gatherers (and hunting agriculturalists) are Sellato, Bernard, 1994, *Nomads of the Borneo Rainforest: The Economics, Politics, and Ideology of Settling Down*. Honolulu: University of Hawaii Press; Early, John D., and Thomas N. Headland, 1998, *Population Dynamics of a Philippine Rain Forest People: The San Ildefonso Agta*. Gainesville: University

Press of Florida; Chazée, Laurent, 2001, *The Mrabri in Laos: A World under the Canopy*. Bangkok: White Lotus; Sillitoe, Paul, 2003, *Managing Animals in New Guinea: Preying the Game in the Highlands*. London: Routledge; Pasveer, Juliette, 2004, *The Djief Hunters: 26,000 Years of Rainforest Exploitation on the Bird's Head of Papua, Indonesia*. Leiden: Balkema.

Classic studies on shifting cultivators are Pelzer, Karl J., 1945, *Pioneer Settlement in the Asiatic Tropics: Studies in Land Utilization and Agricultural Colonization in Southeastern Asia*. New York: Institute of Pacific Relations; Izikowitz, Karl Gustav, 1951, *Lamet Hill Peasants in French Indochina*. Göteborg: Etnografiska Museet; Geddes, W. R., 1954, *The Land Dayaks of Sarawak*. London: Her Majesty's Stationery Office; Leach, E. R., 1954, *Political Systems of Highland Burma: A Study of Kachin Social Structure*. London: Bell; Freeman, J. D., 1955, *Iban Agriculture: A Report on the Shifting Cultivation of Hill Rice by the Iban of Sarawak*. London: Her Majesty's Stationery Office; Condominas, Georges, 1957, *Nous Avons Mangé la Forêt de la Pierre-Genie Goo*. Paris: Mercure de France; Conklin, Harold C., 1957, *Hanunóo Agriculture: A Report on an Integral System of Shifting Cultivation in the Philippines*. Rome: FAO; Spencer, J. S., 1967, *Shifting Cultivation in Southeastern Asia*. Berkeley: University of California Press.

Later studies on swiddeners include Ellen, Roy F., 1978, *Nuaulu Settlement and Ecology: An Approach to the Environmental Relations of an Eastern Indonesian Community*. The Hague: Nijhoff; Kazutaka, Nakano, 1980, "An Ecological View of a Subsistence Economy Based Mainly on the Production of Rice in Swiddens and in Irrigated Fields in a Hilly Region of Northern Thailand," *Southeast Asian Studies* 18, no. 1, pp. 40–67; Scholz, Ulrich, 1982, "Decrease and Revival of Shifting Cultivation in the Tropics of South-East Asia—The Examples of Sumatra and Thailand," *Applied Geography and Development* 20, pp. 32–45; *Swidden Cultivation in Asia, Volume Two*. Bangkok: UNESCO Regional Office (1983); Dove, Michael Roger, 1985, *Swidden Agriculture in Indonesia: The Subsistence Strategies of the Kalimantan Kantu'*. Berlin: Mouton; Uhlig, James, Charles A. S. Hall, and Tun Nyo, 1994, "Changing Patterns of Shifting Culti-vation in Selected Countries in Southeast Asia and Their Effect on the Global Carbon Cycle." Pp. 145–199 in *Effects of Land-Use Change on Atmospheric CO_2 Concentrations: South and Southeast Asia as a Case Study*. Edited by Virginia H. Dale. New York: Springer; Martin, Marie Alexandrine, 1997, *Les Khmers Daeum, "Khmers de l'Origine": Société Montagnarde et Exploitation de la Forêt: De l'Écologie à l'Histoire*. Paris: EFEO; Brookfield, Harold, 2001, *Exploring Agrodiversity*. New York: Columbia University Press; Momose, Kuniyasu, 2002, "Ecological Factors of the Recently Expanding Style of Shifting Cultivation in

Southeast Asian Subtropical Areas: Why Could Fallow Periods Be Shortened?" *Southeast Asian Studies* 40, no. 2, pp. 190–199.

Most studies mentioned above regarding general economic and agricultural development between the 1870s and the 1960s also contain relevant information on permanent field agriculture. In addition the reader is referred to Uhlig, Harald, 1983, "Reisbausysteme und -ökotope in Südostasien; Geowissenschaftliche Methoden in der Reisbauforschung und die Ökosysteme des Überschwemmungsreisbaues," *Erdkunde: Archiv für Wissenschaftliche Geographie* 37, pp. 269–282; Soemarwoto, Otto, and Idjah Soemarwoto, 1984, "The Javanese Rural Ecosystem." Pp. 254–287 in *An Introduction to Human Ecology Research on Agricultural Systems in Southeast Asia*. Edited by A. T. Rambo and P. E. Sajise. Los Baños: University of the Philippines; Bray, Francesca, 1986, *The Rice Economies: Technology and Development in Asian Societies*. Oxford: Basil Blackwell; Uhlig, Harald, 1990, "Der Trockenland Reisbau—Ein wenig bekanntes Potential der tropischen Entwicklungsländer." Pp. 375–398 in *Festschrift für Wendelin Klaer zum 65. Geburtstag*. Edited by M. Domrös et al. Mainz: Geographisches Institut der Johannes Gutenberg-Universität; Tran Duc Vien, 2003, "Culture, Environment, and Farming Systems in Vietnam's Northern Mountain Region," *Southeast Asian Studies* 41, no. 2, pp. 180–205; Boomgaard, Peter, 2004, "From Rice to Riches? Rice Production and Trade in (Southeast) Asia, particularly Indonesia, 1500–1950." Unpublished paper written for the workshop "The Wealth of Nature; How Natural Resources Have Shaped Asian History, 1600–2000," NIAS, Wassenaar.

A general text on bioinvasions is Bright, Chris, 1998, *Life out of Bounds: Bioinvasions in a Borderless World*. New York: Norton; a much briefer account, concentrating on the twentieth century, is McNeill, J. R., 2000, *Something New under the Sun: An Environmental History of the Twentieth-Century World*. New York: Norton, pp. 252–264. More specific texts on Indonesia, particularly Java, are Steenis, C. G. G. J. van, and A. F. Schippers-Lammertse, 1965, "Concise Plant-Geography of Java." Pp. 1–72 in *Flora of Java*, Vol. 2. Edited by C. A. Backer and R. C. Bakhuizen van den Brink. Groningen: Noordhoff; Dove, Michael R., 1986, "The Practical Reason of Weeds in Indonesia: Peasant vs. State Views of *Imperata* and *Chromolaena*," *Human Ecology* 14, no. 2, pp. 163–190; Whitten, Tony, Roehayat Emon Soeriaatmadja, and Suraya A. Afiff, 1996, *The Ecology of Java and Bali*. Singapore: Periplus, pp. 182–190.

Detailed historical data on introductions, by species, in Indonesia are found in Heyne, K., 1927, *De Nuttige Planten van Nederlandsch Indië*. 3 vols. Buitenzorg: Departement van Landbouw, Nijverheid en Landbouw; and for Malaysia (and occasionally other Southeast Asian countries) in Burkill, I. H.,

1935, *A Dictionary of the Economic Products of the Malay Peninsula*. 2 vols. London: Crown Agents for the Colonies.

On livestock, see Boomgaard, Peter, and David Henley, eds., 2004, *Smallholders and Stockbreeders: Histories of Foodcrop and Livestock Farming in Southeast Asia*. Leiden: KITLV.

9

EXPLOITATION, MANAGEMENT, AND CONSERVATION OF NATURAL RESOURCES

FORESTS AND FORESTRY

In Chapters 7 and 8 it has been shown that the reduction of forest cover during the period between the 1870s and the 1960s happened faster than during earlier periods of comparable length. The rate of deforestation picked up speed between the 1870s and the 1920s, but slowed down somewhat between 1930 and the 1960s. After around 1960 the rate of deforestation would be even higher than between the 1870s and 1930. While around 1870 "national" averages of forest cover of between 70 and 75 percent of the surface area had been the rule, by the 1960s that had dropped to figures of usually between 60 and 65 percent. Locally, however, such figures could be much lower, as was the case in Java and Bali, and in the Red River delta of northern Vietnam.

It was also shown that during this period the rate of deforestation could be explained largely by the expansion of agriculture and livestock. Most of this was caused by the needs of subsistence agriculture, driven by more or less continuously increasing rates of population growth. However, a relatively small but increasing proportion of cleared forest is to be explained by the activities of plantations and smallholders producing for the international market.

These activities, however, do not explain the entire loss of forest cover. A considerable proportion of the loss was caused by the fact that forests were cut for timber, firewood, and wood for the production of charcoal.

Part of this production was undertaken by local people, for their own use. As we have seen in Chapter 7, the indigenous population used almost exclusively wood and bamboo for the building of houses (even for the rich) and other buildings, such as mosques or houses for the village council, male bachelors, and guests. Firewood was still the main fuel for the local population, although from 1900 onward the use of kerosene, derived from the oil reserves that had been recently discovered and put into production, was on the increase. However, after the world market crisis of 1929, this process was reversed for a time.

Four men stand on a large stump of a recently cut tree, Mindoro, Philippines.
Deforestation in Southeast Asia picked up speed between 1870 and 1920.
(Library of Congress)

Generally speaking, the increase in the consumption of timber and firewood by the local population grew at the same pace as did the population itself. Given the fact that population growth was continuously on the increase, the same observation is bound to have applied to their consumption of lumber and firewood as well.

In addition to the growth of wood consumption for subsistence purposes, industrial and commercial uses increased as well. The period from 1870 to 1930 was a time of rapid economic growth, driven largely by foreign investment and foreign trade, often under the aegis of the colonial states in the region. The laying out of plantations, the growth of the number of small indigenous, medium Chinese, and large Western industrial establishments (often connected to plantations), urbanization, the expansion of mining, the building of small vessels and buffalo carts, and the construction of bridges and railways were all part of this process. It is important to note that the introduction of advanced, "modern" machinery and equipment, often made of metal, thus also stimulated, as backward and forward linkages, the proliferation of "traditional" wooden technology. It is another example of the spread of traditional features as a result of international market forces.

Increasing quantities of timber, firewood, and charcoal were needed for the enterprises mentioned here, and they all had to come from the local woods. It will be recalled that, prior to 1850, coal was virtually absent in the area. Although some coal mines had been established in the region since then, the proportion of steam mills operating on the basis of firewood and charcoal as fuel was and would remain much higher than it was in Europe or the United States during the same period. Gas (produced from coal), oil, and electricity would all start playing a role during the period under consideration, and this must have slowed down the consumption of firewood and charcoal for industrial purposes to some extent. But the proportion of the latter remained relatively high.

Another factor that would become much more important between 1870 and 1930 was the production of timber for export. Some timber, particularly hardwoods, had always been exported to India or China, regions with much higher population densities and therefore far lower percentages of forest cover. However, the scale of the exports increased considerably, partly because export to Europe became more attractive with the coming of regular steamship connections between Southeast Asia and Europe, and with the opening of the Suez Canal. Countries like England and The Netherlands had hardly any hardwood forests left, and the costs of importing timber species like teak (*Tectona grandis*) from Southeast Asia were now no longer prohibitive.

It is unlikely that sufficient data will ever come to light for the reconstruction of all timber exports from Southeast Asia for this entire period, but data

Table 9.1 Export of timber from the Netherlands Indies, 1874–1938, selected years, in Netherlands Indies Guilders (in thousands).

Year	Amount
1874	20
1884	39
1894	547
1904	2,143
1914	2,585
1926	4,777
1929	10,666
1938	3,491

Source: Boomgaard, Peter (with R. de Bakker), 1996, Forests and Forestry 1823–1941. (Changing Economy in Indonesia, *Vol. 16). Amsterdam: Royal Tropical Institute.*

Table 9.2 Export of firewood and charcoal from the Netherlands Indies, 1900–1938, selected years, in Netherlands Indies Guilders (in thousands).

Year	Firewood	Charcoal
1900	—	—
1910	—	14
1920	—	121
1926	818	789
1930	1,162	1,839
1938	417	779

Source: Boomgaard, Peter (with R. de Bakker), 1996, Forests and Forestry 1823–1941. (Changing Economy in Indonesia, Vol. 16). *Amsterdam: Royal Tropical Institute.*

from the Netherlands Indies (the Indonesian Archipelago), presented in Table 9.1, give an impression of the high growth rates that were typical for the region.

Table 9.1 demonstrates very high growth rates, particularly in the 1880s and 1890s, and also in the 1920s. Even during the Depression of the 1930s, though, exports remained quite high.

A similar story can be told regarding the export of firewood and charcoal, as is demonstrated in Table 9.2.

A word of caution is in order here. Although in absolute terms the exploitation of forests and the value of the export of forest products were rising between about 1870 and 1930, that was not necessarily the case with the share of forest exports in total exports. In the Netherlands Indies this share fluctuated strongly between around 1870 and around 1900, but with an upward tendency, from between 3 and 4 percent around 1875, reaching percentages of some 10 percent by about 1900. Thereafter the share dropped fairly rapidly to a level of between 2 and 3 percent. This drop after 1910 was largely caused by the rapidly increasing importance of oil and sugar exports.

Looking at timber exports alone, we find a share of less than 1 percent in the Netherlands Indies throughout the period 1870 to 1940. The Philippines were doing somewhat better (in economic terms) than the Indonesian Archipelago, and in 1940 timber there constituted 3 percent of all exports in terms of value. Compared with the figures after the start of the timber boom of the 1950s and 1960s, such figures are clearly very modest. In 1973, Indonesian timber exports constituted 18 percent of all exports. Between 1960 and 1975, Philippine timber export revenue fluctuated between 20 and 30 percent of all export revenue.

Therefore, during the period dealt with here, the export of forest products was increasing in absolute terms up until 1930, but its share in the economy was rather modest (and more fluctuating than rising) than the next period, from the 1960s onward. Nevertheless, the environmental impact of the exploitation of the forests during the period between 1870 and 1960 was considerable.

CHANGES ON THE SUPPLY SIDE

Such increasing levels of forest exploitation can be explained from the demand side, as was done above, but there were also notable transformations on the supply side. The four most important were technological change; the further expansion of colonial states; high rates of investment of foreign capital by private, usually Western (and in the 1930s also Japanese) lumber companies, but occasionally also furnished by government; and the establishment of (mainly) colonial forest services, which introduced "scientific" forestry.

Starting with the technological factors, it can be said that up until the middle of the nineteenth century, the European presence had led to only three major changes: the introduction of the saw, the large-wheeled timber wagon, and the wind- or water-driven sawmill. During the second half of the century steam-driven sawmills were introduced in addition to narrow-gauge railway tracks on the forest range. Both were important but expensive inventions, which greatly facilitated the processing and transportation of logs. Particularly transport had been a big problem, as the mortality of buffaloes (in areas where elephants or timber wagons were not available), owing to the dragging of heavy logs, had always been rather high. Transport of logs from the forest range to cities and harbors, for a long time mainly either shipped or rafted, improved considerably with the arrival of "normal" railways, steamships, and, later, trucks.

After about 1900 steam winches and tractors were used for the dragging of logs in certain areas, and electrical sawmills increasingly took the place of steam mills. During the second half of the twentieth century, chainsaws and bulldozers led to spectacular growth in productivity, but that occurred largely after the period we are concerned with here. Thus, the technological barriers to cutting, transporting, and processing trees in once isolated forest areas had been progressively removed, lowering costs and improving access to even the remotest parts of the region.

This often coincided with the expansion of the effective authority of the (colonial) states into these remote areas—remote from the point of view of the central state, of course. A good example is Burma, a country with rich teak resources that, partly or perhaps even largely because of these resources, was conquered in three stages by British forces (in the three Anglo-Burmese wars of 1824–1826, 1852, and 1885–1886). British India needed a regular supply of teak, particularly for shipbuilding, and the erratic performance of independent Burma in that respect prompted drastic action. Burma is also a good example of the presence of powerful Western timber companies, backed by the colonial state. The fact that the Burmese king Thebaw had imposed an "unfair" fine on the British-owned Bombay Burmah Trading Corporation, which had obtained lumber concessions in that part of Burma that was still independent at the time, was a handy pretext for the Third Anglo-Burmese War.

Although Thailand was not a colony, the story of teak in Thailand is not dissimilar to that in Burma. Traditionally, the teak forests here had been under the control of Lao chiefs, who issued permits to Burman, Shan, and Chinese traders. In order to be better able to control (the income generated by) such resources, the central Thai ruler had to consolidate his power in these remote areas. Here, too, British capital played a major role in the exploitation of these resources, no doubt partly because the Thai king had brought in a British forester from Burma to regulate the cutting of teak in 1896.

A different pattern was found in the Indonesian Archipelago. Originally, the exploitation of a number of teak forests in Java had been in the hands of the VOC after the 1670s. That expanded considerably during the next century, and under the Dutch colonial state (from 1816 onward) government exploitation remained the rule. From around 1855, the colonial government concluded contracts with Dutch capitalists for the exploitation of teak concessions in Java. However, after prolonged discussions in government circles, the decision was taken in 1897 to return to government exploitation, mainly because sustainable forest production did not appear to be possible if private enterprise exploited the forests. Government exploitation of the Java teak forests would be continued until the end of the colonial period, and even for a long time under independent Indonesian rule. In the case of Java, therefore, investment came from the government, for which the forest establishment itself footed the bill.

THE FOREST SERVICES

Finally, the role of the various forest services should be discussed. In the three countries just dealt with, there was a close relationship between teak and the forest services. Teak forests in Southeast Asia are encountered mainly in Burma, Thailand, and Java; it is a monsoon forest tree, and therefore not to be found in the evergreen forests around the equator. The presence of teak, as an easily exploitable and valuable hardwood, to a certain extent shaped the forest services in those countries.

The earliest European-inspired forest service in Southeast Asia (and perhaps in the whole of Asia) was established in Java in 1808, but it was largely dismantled in 1811 when the British temporarily took over Java from the Dutch. The service was resurrected with the return of the Dutch in 1816, and again abolished in 1826, this time owing to budget cuts. It was finally revived in 1869, and then it was there to stay, surviving even the end of colonial rule.

On Burmese soil, a European-inspired forest department had been called into being in 1856, under the direction of a German forester with experience in British India (where extensive teak forests were to be found as well). In the Philippines a similar institution, the Inspección General de Montes, was established in 1863. Later in the nineteenth century, and in the first two decades of the twentieth century, forest services or forest departments were established across the remainder of Southeast Asia. Foresters from the Burma Forest Service would play a role in setting up forest departments in Thailand and Malaysia.

In all Southeast Asian cases, (mainly colonial) governments had established forest departments or forest services because they felt that the forests were under threat of overexploitation. As teak had already been cut for quite some time in these countries on a considerable scale, it was the teak forests that prompted the earliest measures against overcutting. In the early nineteenth century, most non-teak species were regarded as more or less worthless—in British India they were called "junglewood"; in Java the term used was *wildwood*. But as exploitation and transportation became less expensive for reasons we have seen earlier, other types of forests or trees would soon follow suit, such as the dipterocarp forests of Malaysia, Sumatra, Borneo, and the Philippines. Early forestry measures, therefore, were often species specific.

Destruction of teak forests on a considerable scale had caught the attention of the colonial authorities (or quasi-colonial authorities, as in the case of the VOC) quite early. Up until the 1770s, the VOC had filled its teak requirements (in Java) by means of a system of compulsory deliveries. They told the local power-holders how much teak they had to cut annually, and the company would then send ships to pick up the beams, planks, and other types of timber. In the

1770s the VOC became concerned over what they called the wanton destruction of the teak forests, which, in their view, had been caused by wasteful indigenous forest exploitation. In order to curb this destruction they started to lay down regulations for sustainable production. However, during the later decades of the eighteenth century, things got only worse when private (Dutch and Chinese) capital became interested in the teak forests as well, largely in order to acquire timber for shipbuilding. In order to put a stop to all nongovernmental teak cutting, a forest board was called into being in 1808, as was mentioned earlier.

In most Southeast Asian countries, governments were not only concerned about the damage done by private companies; they also worried, perhaps even more, about the activities of slash-and-burn agriculturalists, either traditional ("integral") swiddeners, or recent ("partial") ones, such as the Chinese cassava growers in Malaysia. With their European backgrounds, colonial governments regarded all types of temporary cultivation as wasteful, and insofar as no use was made of the large trees that were burned before cultivation could get started, they did have a point—although in the early days there was not much of a market for most types of timber. Their main concern, however, was that forest land was being exploited in such a way that it did not revert to secondary forest, something that happened if fallow periods were too short, cultivation periods too long, or fields lying fallow were set afire repeatedly, in order to attract game or create pastures for livestock. However, as was argued in the last chapter, most integral swiddeners practiced genuine forest-fallow cultivation, which was a sustainable system, a fact that governments and foresters in those days do not appear to have been fully aware of.

"SCIENTIFIC" FORESTRY, SUSTAINABLE PRODUCTION

Forest services in Southeast Asia, therefore, were established to save the forests from destruction, but paradoxically, they themselves were the institutions that facilitated the unsustainable exploitation of these forests.

According to the official rhetoric, the forest services were called into being in order to manage the forests on a scientific basis, thus arriving at sustainable yield and sustainable exploitation. The need for sustainable exploitation of the teak forests was recognized early in Java, under the VOC in the eighteenth century, and various measures were taken—on paper—to arrive at such a system. An important feature of these measures was that instead of selective cutting—taking only those trees that had reached specified dimensions—a system of clear felling should be introduced, which meant that all trees in a certain block of forest should be cut. Thereafter, that territory should be reforested, which would make it possible to harvest it again in, say, fifty years. Some replanting of teak was indeed carried out after 1800, but not in a systematic way and not on a scale sufficiently large to make sustainable production possible.

In fact, the Dutch (and this applied to the British in India and Burma as well), coming from a country that had been largely deforested for quite some time, lacked the expertise for such an undertaking. It was not until the 1850s, when the Dutch colonial state started to ship German foresters to Java, that foresters with experience with scientific forestry, as it was called, could make a start by introducing those measures that were needed for sustainable forest exploitation. Germany had a tradition of this type of forest exploitation, and there were forestry colleges in which scientific forestry was being taught.

The problem, of course, was that the knowledge and expertise of the German foresters (and this applied to French foresters in French Indochina as well) was based on forests in the temperate zone; the foresters had no experience with monsoon or wet tropical evergreen forests. That showed up, for instance, in their opinion about the use of fire in the forests. Asian teak forest people, and in their wake the European overseers in teak forest ranges, used fire to clear the forests of debris caused by cutting and to get rid of undesirable undergrowth. Fire also stimulated natural regeneration, because teak seeds germinate better after having been in a fire. However, to a German forester, used as he was to the coniferous forests of Germany, where fire led to terrible conflagrations, partly because of the highly resinous character of the trees, the use of fire in the forests was anathema. The use of fire in the Southeast Asian forests led to heated debates that were even continuing in the era of independence. Of course, the German foresters were equally appalled at the practices of slash-and-burn agriculturalists. Such practices, therefore, were often forbidden in or near state-owned forests—usually to no avail. However, the effect was that in some areas of the Malay Peninsula and the Philippines, swidden cultivation had become a criminal offense.

Be that as it may, the scientific exploitation of forests as propounded by the German foresters and foresters trained in the German tradition, who were sent not only to Java but also to Burma, Malaysia, and Thailand, entailed in the first place the measurement and demarcation of all forest areas. Then, those forests that were found to be worth preserving, either for production, for watershed protection, or, rarely, as communal forest or for scientific research, were officially given protected status and were thus gazetted.

The exact status of forest lands varied from one area to another, but as a rule all forests were either by definition state property, because all soils not used for agriculture were regarded as eminent domain/regalian domain (land that belonged to the indigenous ruler and had been taken over by the colonizer in most cases), or they were declared off bounds for everyone except the state and its licensees once they had been gazetted as forest reserve. In that case the reserved forests were de facto state property. In fact, therefore, in all Southeast Asian countries a large proportion of the forests were made either de facto or de iure state-owned lands, a state of affairs that differed markedly from the legal status of forests in the colonizing countries themselves, which were more often than not private property. The current opinion among many Southeast Asia specialists appears to be that the colonizers introduced the land rights they had back home, but, as we have just seen, in the case of the forests this is obviously an erroneous notion.

The process of the creation of state forests in Southeast Asia has been called the territorialization of forests, or the creation of "political forests." Wooded areas in which the state had an interest, for whatever reason, were now defined and gazetted as "forests," while other wooded areas were not. Such arrangements, however, were not hewn in stone, and forests could be "degazetted," while other, nongazetted wooded areas could be gazetted at a later stage.

Between 1900 and 1940 the notion that all forests were eminent domain was challenged in the Netherlands Indies and in the Philippines by new legal opinions originating from their mother countries. In the Philippines this seems to have been largely ignored, but in the Outer Provinces of the Netherlands Indies it led to problems regarding forest reserves, mainly in Sumatra. The discussion centered around the question of what the role and status of the forest service were going to be if forests were not declared eminent domain and would remain part of the commons of the local communities—and whether local communities were able to manage such community forests in a sustainable way. These problems had not been solved when the colonial era drew to a close. Later on the independent Indonesian government would continue the creation of "political forests."

PRIVATE COMPANIES AND DEFORESTATION

Forests that had been gazetted as production forests were slated for sustainable timber production. In most countries in the region, concessions were given out to private companies, of which the larger ones were usually in the hands of European entrepreneurs. But interest in such permits was increasingly shown by U.S., Chinese, Japanese, and indigenous firms as well. In some areas, however, the forest service itself took care of (part of) the timber production.

Usually, private lumber companies were under the obligation to replant the area they had cut over, in order to arrive at a rotational system that would allow sustained yields. The problem was that the forest service proved to be unable to make the private companies adhere to their contractual obligations regarding replanting. Only if the forest service took the exploitation of a forest range upon itself, and consequently the replanting of it as well—as was the case in the teak forests of Java—was reforestation as a rule successful. Generally speaking, however, the attempts to reforest the cut-over areas in Southeast Asia, one of the main tenets of scientific forestry, were a dismal failure.

This was, of course, also clear at the time, but given the fact that the forest departments of the various countries generated considerable income for the state, largely based on payments from concessions given out to private companies, criticism was rather muted. The notion that unsustainable exploitation of forests would ruin them was in a region where, after all, the proportion of forest cover was still much higher than in most areas in Europe and China—and anyway was not all that common a perspective beyond a small group of concerned foresters. For instance, in many of these countries, those colonial civil servants who were in charge of agricultural extension services were sometimes inclined to see the creation of forest reserves as a threat to agricultural expansion, or "development," as it would be called from the 1950s onward.

Actually, the forest services can be said to have paved the way to growing deforestation as a result of increased timber cutting by private firms. The former had measured, mapped, and delimited the forests. Often, they also made a rough inventory of the types of wood to be found there, and they gave equally rough estimates of the quality of the wood. All of this served the timber companies well, as they could base their bids for concessions on much better data than they could have had without the activities of the forest service.

However, the forest service was also responsible for environmentally valuable activities. They mapped, delimited, and proposed forest reservations for the hydrological purpose of watershed protection. Watershed protection was and is essential for the proper functioning of gravity-flow irrigation in wet-rice–growing societies. It is also vital in preventing erosion and flooding. Part of the income generated by giving out cutting licenses and timber concessions, and by selling timber and other forest products, was spent on watershed protection.

Watershed protection included attempts at reforestation of denuded mountain slopes, in some areas as early as 1875 (Java), but elsewhere not earlier than 1940 (the Philippines). In many cases, forest services experimented with fast-growing foreign tree species for such projects. As an example of this tendency, eucalyptuses have been mentioned in the last chapter.

A major task of all forest services was the sustainable management of the production forests. This implies that they should have seen to the reforestation of logged-over areas. In the case of teak (and other single-species woods), teak plantations theoretically should have taken the place of "natural" teak forests (but they may not have been as "natural" as most observers assumed). The impression conveyed by most of the literature is that in those areas where private enterprise was involved in the exploitation of the teak forests, such plantations were not a success. Where it was undertaken by the forest service itself, as was the case in Java, replanting appears to have functioned much better. By the beginning of the War in the Pacific, Java had a well-functioning system of rotational harvesting in place—clear-cut areas were replanted on a routine basis, and could be cut again after some fifty years, with thinning going on during that period.

On balance, I am inclined to say that forest services have done much to preserve and restore the so-called protection forests, which might have been damaged more without their activities. But in the case of production forests, their role appears to have been negative, as a rule.

FOREST OFFENSES, PUNISHMENT, PROTEST, AND THE "RESOURCE CURSE"

This brings us to a point that has received much attention in the recent literature—that colonial forest policy, and particularly the creation of "political forests," has taken the forests from the people who were living there. People living close to or surrounded by forests obtained part of their livelihood from them—sometimes a considerable part. This was obviously true for foragers and shifting cultivators, but it also applied to villages of sedentary peasants with permanent fields, located at the forest margins. Originally, when the forests had just been gazetted, most forest regulations permitted the local population free access to the woods, including the cutting of timber and the collection of firewood for their own use. In time, however, access became more and more restricted, and people had to acquire a license for timber cutting. Grazing of animals was also increasingly forbidden in the forest reserves.

All of this could be enforced only if the gazetted forests were patrolled on a regular basis, and that was often not the case. However, as the forest department grew, so, as a rule, did the number of forest guards, which made for a stricter application of the rules. Thus, activities that had been part and parcel of the daily lives of forest people had become offenses, for which punishment (fines, prison) was meted out by the organs of the (colonial) state. It is a process that had occurred much earlier in more densely settled areas such as Western Europe, and which is not typical for a colonial society, but takes place whenever population density increases and forest cover retreats.

There are indications that the number of forest offenses increased during periods of depressed agricultural prices or dropping agricultural exports. During such periods the exports of forest products increased as well, because peasant-cultivators who could no longer survive on their farm income would increasingly turn to these "free gifts of nature," which had become forbidden fruits.

Occasionally, the closing of the forests by the colonial state led to local or even supralocal unrest, as was the case with the so-called Samin movement in Java around 1900. The people who participated in this movement were active mainly in and around Rembang, an area where the forest service had concentrated its activities. The movement, which started around 1890, did not come to the attention of the authorities until around 1905; its activities, which peaked at the eve of World War I, demanded that the forests be open to all, and it preached disobedience as regards the cutting of wood in these forests. Again, such actions are very similar to the way in which peasants had reacted earlier in Western Europe (Germany 1525, England ca. 1800).

We are used to looking at natural resources as "free gifts of nature." The presence of such resources provides a living for many people, and if the income generated by resources is spent wisely—both in a micro- and in a macro-economic sense—they can be a boost to private and national income. However, governments often do not spend such income wisely, and effects on the economy at large may be negative (the so-called Dutch disease). Stories about gold miners who gamble and drink away all of their income are all too familiar. Recently, it has been suggested that there are good reasons not to regard such natural resources as a blessing any longer, but, in contrast, as a curse. The term *resource curse* has been introduced to emphasize the negative effects of the presence of valuable natural resources on many people who get in the path of those who want to exploit those resources. The term appears to apply to many people who happened to live in or near a Southeast Asian forest at the moment that the state moved in, in order to facilitate its exploitation.

Most people, however, appear to have adapted their way of life to the altered circumstances. The closing of the forests (if well patrolled), and the large-scale exploitation of the same (production) forests, must have led to structural changes in the forest communities nearby. Locally, it may have compelled shifting cultivators to make the shift to permanent field agriculture. Villagers with very small holdings ("dwarfholdings") and landless people may have tried to get work with the forest service, or else tried to become tenants (sharecroppers), or sought wage labor, on a temporary or permanent basis, in the growing nonagricultural sectors. In the latter case it is likely that they moved to the growing cities and towns. These people, therefore, participated in the processes for which we have designed terms like "migration," "industrialization," and "urbanization," when they had to give up their forest village way of life.

NON-TIMBER FOREST PRODUCTS

During the period under discussion, the share of Non-Timber Forest Products (NTFPs) in exports was usually more important than that of timber. The importance of NTFPs varied widely between areas and countries. While in the Netherlands Indies the share in exports of NTFPs was never more than 10 percent during this period, and usually less, around 1900 the share in total exports from lower Laos was around 75 percent.

As was mentioned in Chapter 7, NTFPs are often collected by "tribal" forest people, constituting a link with the world economy. But when demand was strong, sedentary peasants and foreign migrants (Chinese) could also join the collectors. Demand for individual products often fluctuated strongly, partly owing to fashions or to new applications resulting from technological development, but also because better or cheaper natural or synthetic substitutes eventually became available.

A good example was the boom in the export of gutta percha (various latex-producing plants, species of *Palaquium* and *Payena*), between ca. 1860 and 1900. Demand was generated around 1850 because the latex could be used for coating underwater telegraph cables, and soon the word was spreading from Singapore and the Malay Peninsula to the Indonesian Archipelago that gutta percha was fetching a good price. Its collection was undertaken unsustainably, as the tree was cut before the latex was extracted, and soon colonial officials in those areas where colonial rule was more or less effective were trying to stop the onslaught. In fact, attempts to protect gutta percha–producing trees in the Malay Peninsula played a role in the establishment of a central Forest Administration for the Federated Malay States and the Straits Settlements in 1902. When prices increased in 1870, many people started looking for it, creating rice shortages in Sarawak (northern Borneo), where it was the leading forest product for a while. During years of harvest failures of food crops, as occurred during the double ENSO years of 1877–1879, gutta percha collection increased even further. In various places attempts were made to cultivate some of the species that produced it. By 1900 the gutta percha craze was over, when prices started to drop.

CONSERVATION AND WILDLIFE PROTECTION

Even though around 1870 some 70 to 75 percent of most Southeast Asian countries was still covered with woodlands—a luxury commented upon by all visitors from countries where the forests had almost disappeared entirely—it was during the closing decades of the nineteenth century that colonial civil servants and scientists with an interest in botany, zoology, or hunting started to get worried about the disappearance of certain plant and animal species. For instance, in 1896, two Dutch amateur naturalists wrote alarming accounts about a number of locally threatened species of plants and, particularly, animals to be found in the Indonesian Archipelago. They mentioned orchids, birds of paradise, the Javan peacock, the argus pheasant, rhinoceros, banteng (wild cattle), and orangutan. Such species-directed concerns were originally voiced by a tiny European minority, but often by people who were placed in high positions, while they sometimes influenced indigenous rulers who shared their hunting passion.

Owing to accelerated economic and population growth, topics dealt with in the last chapter, forest destruction was locally threatening the survival of specific species of plants and animals. As regards plants, this was reinforced by the increased "commodification" of certain wild plants, such as orchids and the gutta percha–bearing trees just mentioned. Regarding wild animals, increased unregulated indigenous and Western hunting added to the loss of habitat.

Hunting and trapping had always been part of indigenous life, and locally some animals had become rare prior to ca. 1850. However, during the second half of the nineteenth century two things happened that were potential threats to wildlife numbers. In the first place the number of Europeans increased considerably, and among them those people who may be regarded as avid hunters—the military, who were equipped with firearms and knew how to use them, and the planters, who often had to defend their crops or even their personnel against marauding animals. Second, the quality of firearms improved considerably, as did the numbers of improved firearms in the hands of indigenous hunters.

It is a law of nature that large animals are rare, and hunting pressure on large animals, therefore, leads quickly to the endangerment of those species. Examples are elephant, rhinoceros, wild buffalo, wild cattle, tapir, orangutan, and tiger. However, unregulated shooting of various kinds of birds was also considered a threat to their survival, of which the various species of birds of paradise, mainly from New Guinea, are good examples. Gradually, it began to dawn on a small number of people—often the same hunters who had been responsible for this in the first place—that the numbers of these animals were dropping. The first conservationists, therefore, were often also hunters (one author has called them, somewhat unkindly, "repentant butchers").

This particular strain of environmental awareness—if it is permitted to use this term for such early stirrings of conscience—was then grafted onto an older stem of "conservational" thinking—the notion, already mentioned in Chapter 7, that deforestation was influencing local climates. At the beginning of the nineteenth century this idea had been propagated by Alexander von Humboldt, based on his experience in tropical and subtropical America. It had struck a chord with colonial civil servants in Asia, who reported on the dangers of watershed denudation.

They too had seen signs that the deforestation of mountain slopes and peaks led to the drying up of water springs and watercourses, with all of the negative effects that such developments entailed for irrigated agriculture. It was even argued that the disappearance of these wooded areas made for less rain, while the rains that did fall were no longer retained by the trees, thus causing flash floods.

By the late nineteenth century, then, concerns about disappearing forests and the fear that a number of (the larger) game species were under threat of becoming extinct were voiced by increasing numbers of people. Articles appeared in local newspapers and in those of the colonial mother countries, influencing public opinion in the colonies and in Europe.

Generally speaking, species protection was the first measure taken in regions like the Malay Peninsula and the Indonesian Archipelago. In Malaysia, local regulations for the protection of specific birds and specific big mammals were issued in the 1880s. All other conservation measures in Southeast Asia postdated 1900, with one exception. In the Netherlands Indies the Cibodas (Java) forest reserve was created in 1889. This was neither a production forest, nor a (watershed) protection forest, but a wooded area with relatively unspoiled vegetation, which was reserved for purely scientific reasons. This was an exceptional measure in more than one sense—it was much earlier than other reserves of a similar nature, and it was not a species-protection measure.

After 1900 most Southeast Asian countries produced game laws that can be considered to be one step beyond species-protection measures. Nature reserves, wildlife reserves, and national parks were usually established at a later date, from around 1920 onward. In addition, regulations were established in some countries for the trade in wildlife products, forbidding the export of threatened species and the products derived from them. In a number of cases the reverse occurred, and Western countries forbade the import of products from animals perceived to be endangered, as was done by the United States and Britain regarding birds of paradise. During this period trade regulations may have been the most effective means of nature protection.

By 1941, at the eve of the Pacific War, Thailand appears to have been the only major Southeast Asian country without nature reserves (but they had had an elephant protection act from 1921). Its first national park came in 1962.

In fact, during the period between 1940 and the 1960s or 1970s, not much went on in any Southeast Asian country in the field of conservation legislation. Both the colonial states and the newly independent nations had other things on their minds during this period of war and revolution.

The conservation "movements"—often more small groups of like-minded individuals than anything else—and the conservation practice in the various Southeast Asian countries had a number of features in common. The people who put pressure on the colonial administration were overwhelmingly Europeans, with the support of small numbers of indigenous rulers and aristocrats, who often shared the hunting passion of the Europeans. To my knowledge, the only country with a real organization (an NGO, as we would call it today) in this field was the Netherlands Indies, where in 1912 The Netherlands Indies Society for the Protection of Nature was established.

Originally, governments were often unsympathetic to conservationist pleas. Realization of their plans, so the colonial civil servants argued, would always cost money (for instance, for game wardens). Protection of big game was also deemed a threat to agriculture. Elephants—already in need of protection in some areas in the late nineteenth century—did damage to the crops of smallholders and plantations alike, as was the case with various types of deer and wild boar. Particularly plantation owners, therefore, were vociferous adversaries of the conservationists. As all these regions were ruled in a rather authoritarian manner, much depended on the willingness of the highest government officials to side at least partly with the conservationists.

The debates between the various groups were given space in the press, which brought the controversies into the open and stimulated public discussion (insofar as it was not reigned in by laws that restricted the freedom of the press). As the colonial press was also read in the various mother countries, debates spilled over into Europe and the United States. Here, public opinion was alerted to the conservation issues at stake in the Southeast Asian colonies, and local conservationist organizations often started lobbying for better colonial legislation. Thus, in a number of cases press coverage in the mother country, occasionally resulting in questions in parliament, led to pressure being brought to bear by governments in Europe on their representatives in the colonies for more legislation favoring the environmental cause.

By the 1920s, by which time many countries in Southeast Asia had issued game laws that limited hunting and had passed legislation for the creation of nature reserves, a spirit of conservationist competition could be felt between the various "empires." Europeans were well aware of the fact that the United States had been quite early with its Yellowstone National Park (1872), while colonial functionaries in Southeast Asia were impressed by the national parks in Congo (1925) and South Africa (1926), at a moment that legislation on this point was lagging behind in Southeast Asia.

This was also the period in which international organizations and international conferences started to discuss nature protection issues on a global scale, including those regarding tropical countries. In 1923 there was an international conference on the protection of nature in Paris, and in 1926 an international organization for the protection of nature was established in Brussels. In 1929, the

Fourth Pacific Science Congress, held in Java, discussed the theme of conservation of nature in the region, including the establishment of nature reserves. All this influenced legislation in the mother countries and in the colonies.

Aside from the one exception—the Cibodas forest reserve—mentioned above, attempts at conservation evolved from a species orientation to the awareness that conservation should take place at the ecosystem level. Given the limited knowledge available prior to the Pacific War on this score, this must be regarded as a fairly rapid and surprisingly "modern" development. Another modern notion that was discussed and locally acted upon in the 1930s was that nature reserves, up until then conceived of mainly as relatively "pristine" areas without people (who, if needs be, had to be removed first), might or even should include the "wild" people residing there.

There were also fairly rapid changes in the perceptions of the colonial civil servants regarding the dangers represented by local wildlife. While, as was suggested earlier, many large mammals had been regarded as "vermin" at the beginning of the period being discussed, awareness of the importance of their survival was spreading after the turn of the century. In some cases it took somewhat longer—while during the first decade of the twentieth century many governments still paid out bounties for the killing of tigers and leopards, in the 1930s some naturalists were arguing that the big cats should be protected. However, that was too wild a notion at that moment, and protection would not be granted to leopards and tigers until the 1970s.

How successful were the conservation measures discussed here? On the eve of the War in the Pacific, many nature and wildlife reserves had been established, some of them measuring hundreds of thousands of hectares. However, very few of them were patrolled effectively, and particularly the larger ones in fact existed mainly on paper. It is, therefore, questionable whether most of the measures taken had much of an impact. Perhaps the best that can be said for them is that they may have raised environmental awareness, that they established a framework that is still in use today, and that there are small local successes, where it can be demonstrated that legislation did matter. A good example of the latter is the Ujung Kulon nature reserve in western Java, established in 1921. It was well patrolled, and the herds of large mammals (rhinos, wild cattle) to be found there were constantly being monitored. It is generally accepted that the survival of the rhinoceros in Java is the result of the creation of this nature reserve.

Finally, it should be mentioned that there was often a link between forestry and conservation. The management of nature reserves was often in the hands of foresters being employed by the forest service. The same foresters were engaged in botanical research and in monitoring wildlife in the reserves, thus contributing to the development of scientific botany and zoology, and to the understanding of the working of ecosystems.

MINING

There is a remarkable contrast between the literature on forest exploitation in Southeast Asia between the 1870s and the 1960s on the one hand, and the writings on mining during that same period on the other. In the first place, there is a huge difference in the numbers of recent titles, with a clear preponderance of books and articles on forests. In addition to that, while almost all forest exploitation studies are concerned with environmental questions, those dealing with mining seldom are. Somehow, the not unimportant environmental effects of various types of mining have not really captured the imagination of scholars writing about its history, perhaps because the effects are rather localized. Another reason may be that, in contrast to trees, minerals are nonrenewable resources, so there is no point in studying the absence or presence of sustainable production and management, which is almost a matter of course among forestry historians.

What can be said in general about mining in Southeast Asia during the period under consideration is that the scale of operations increased considerably. The by now familiar story of growing European and U.S. demand, which was met by an increased supply because of the influx of Western capital and Western technology, supported by growing numbers of (often Chinese) workers, applies to this sector as well.

What almost all mines had in common is that they were established in rather inaccessible areas. This meant that hitherto more or less "empty" areas—that is, areas with very low population densities—had to be provided with an infrastructure that would make transportation of people and commodities possible—in other words, with roads, often all-weather roads, and narrow-gauge railways. Areas close to the mines had to be cleared for housing and for small-scale agriculture. In a number of cases space was also needed for processing plants, although these could be located at some distance, servicing more than one production area, as was the case with oil and occasionally also with tin. Therefore, the environmental impact of mining operations was not restricted to the effects of the mine itself.

Some minerals were "new" to the region, in the sense that they had never been exploited on any scale to speak of prior to the 1870s, as was the case with coal, oil, mica, and nickel. Particularly oil would turn out to be of great economic importance for various countries in the region.

Burmese women hunt for rubies that may escape from a mining company's washing pans into the mud outside the wire line. Technological improvements led to more mining in Southeast Asia from 1870 onward. (Library of Congress)

However, the industry that arguably had the largest environmental impact in the region was probably not oil, but tin. Tin, as was shown earlier, had been mined in Thailand, the Malay Peninsula, and the Indonesian Archipelago prior to the 1870s. During the period under discussion, production expanded considerably. During the nineteenth century, it had been mainly open-cast mining, done by hand, partly by indigenous and partly by Chinese labor and with Chinese technology. However, slowly but surely, Dutch and British capital moved in, and Western technology was introduced increasingly from around 1870 onward. Thus, the steam engine and the centrifugal pump were introduced, which made possible operations on a much greater depth. Also adopted were hydraulic sluicing and gravel pumping, and, finally, dredging. The latter technique was available in the region from around 1910. However, traditional mining methods—open-cast mining, lode mining, and *dulang* washing—were still being employed alongside the modern ones, not only until the end of the period but also up to the present.

Dredging was a dramatic change of technology for various reasons. Dredges were monstrously large machines that could be regarded as the physical representation of the quantum leap in scale of capital invested and of operations. Nevertheless, it is questionable whether the use of dredges also led to visual environmental damage on a much larger scale. Open-cast mining with manual labor had always been an environmentally destructive undertaking, leaving denuded landscapes that had lost not only their forest cover but much of their soils as well. Dredges operated usually in bays and river estuaries, which meant that the damage they did—which, of course, may have been considerable—was hardly visible to anyone who did not look for it underwater. However, from the 1930s, dredges were also floated on pontoons in flooded open-cast mines, which must have added to the moonlike landscape, with artificial lakes, which an area that had been mined over would acquire.

In addition to all of this, the tin ore was often converted to tin in local smelters, for which large amounts of charcoal were needed. This was locally produced, which led to further forest destruction. In areas in which lode mining was undertaken, timber was needed to shore up the shafts and galleries. Small wonder, therefore, that tin mining was a nightmare to the forest services.

The combination of open-cast mining on a considerable scale and the use of wood for timber and charcoal have locally led to a thorough transformation of the landscape. Particularly on relatively small islands like Indonesian Bangka, tin mining has thus led to an almost entire makeover of the natural environment, which was even more pronounced if we include the land cleared for agriculture by the families of the miners and former miners, mainly Chinese immigrants.

Finally, tin mining involved the use of massive quantities of water. The waste products (tailings) and the sand and gravel that were washed away by the water led to large-scale pollution and siltation of streams and estuaries.

Coal mining was more or less new to the area—a few very small-scale local operations apart—but its environmental impact was less dramatic in scope than was the case with tin mining. Locally, however, coal mines could certainly transform the natural environment for many kilometers around. In the case of open-cast mining the effects were very similar to the "moonscapes" mentioned above, while underground mines needed lots of timber, again much like what occurred near tin mines.

Although it is not directly connected to the mining activities, it should be mentioned in passing that the increased use of coal also had its impact on the environment. Air pollution caused by increased sulfur dioxide emissions in cities and near industrial establishments must have been the result, although apparently not on a large enough scale to lead to complaints, something that would change drastically during the next period.

Oil mining was also new to the region. After a modest start with a few small-scale operations from the 1870s onward, the importance of the industry increased rapidly when gasoline, originally a worthless by-product, turned out to be the most appropriate fuel for the internal combustion engine of cars. Today we read a lot about giant oil spills from tankers, problems with pipelines, oil rigs that have to be disposed of, burning oil wells, and other examples of pollution, but the historical sources regarding the oil industry in Southeast Asia are silent on this point. Nevertheless, when we see photographs of the refineries of those days, air pollution must have been considerable, although, again, complaints appear to be lacking.

Finally, gold and silver mines should be mentioned. As we have seen earlier, they had been operating for centuries. There were some attempts to increase the scale of operations with Western capital and machinery, but indigenous and Chinese gold panning activities, based on traditional methods, continued unabated. What also continued, therefore, was the use of various dangerous and mordant chemicals (mainly mercury), to the detriment of the natural environment. It led—and still leads—to water pollution, and renders fish in the downstream areas unfit for human consumption.

FISHERIES

If in comparison with the abundant information on forest exploitation, we have little data on the environmental impact of mining in Southeast Asia during this period, there is even less information on fishing and the exploitation of other resources of the sea. Fish and other marine products are renewable resources, just like forests, so this cannot explain the difference in scholarly attention between the two. Perhaps it has played a role that overfishing and the overexploitation of other marine resources is much less visible to the layperson than is deforestation, while many people do not even appear to have believed prior to the 1950s or 1960s that overfishing was possible at all. Even fishermen, who are confronted with diminishing catches, appear to find it hard to believe that fish, if not collected sustainably, are a finite resource, although in principle a renewable one.

Be that as it may, we do know that, as in almost all other branches of the economy, production was rising between the 1870s and 1930, while it stagnated during the 1940s. Expansion was driven partly by new technology and investment (some of it from Japan), but also by the growth of the number of traditional fishermen. New technology and new methods were applied to previously untapped ecosystems. Marine catches picked up again in the late 1950s, and higher growth rates would be obtained into the 1970s. This expansion was partly the result of conscious efforts at "development" by the newly independent states, partly the result of the introduction and spread of trawlers.

There are signs that some local overfishing may have occurred during the period we are dealing with. This could have been as early as the 1920s, but as we

A Myanmarese fisherman rows with his leg as he casts his nets in Inle Lake. Overfishing was partly driven by new technology in the early 1900s, but also by the growth of the number of traditional fishermen. (iStockPhoto.com)

have no estimates of the potential catch for those years, it is unlikely that we will ever know for sure. However, it is possible that marine reproduction had been influenced by the removal of the coastal mangrove forests (used for stakes, firewood, and the production of charcoal), areas in which many maritime species spawn; by siltation caused by mining, logging, and erosion resulting from unsustainable agriculture; and by pollution caused by human and industrial waste. As an example a report on the problems of shad should be mentioned, dated ca. 1920, suggesting that the numbers of these fish along the west coast of the Malay Peninsula were declining because of the pollution of the rivers in which they spawned by the tailings of the tin mines. It is possible, therefore, that at least locally, the expansion of agriculture, forestry, mining, and industry had some dampening effects on marine reproduction and catches even prior to the War in the Pacific.

It is also likely that the ocean ecosystem underwent changes induced by human behavior, even if they may not have been immediately visible at the time. Fishermen tend to target specific species, and when the numbers of the latter drop, we may expect others to take their places, because a predator was removed from the system, or more food became available as competing species were eliminated. Human fishing activities and the effects of pollution might reinforce the damage done.

There are early, local reports, dating from the beginning of the period dealt with here, about overfishing or overhunting of certain species. In two cases we are dealing with mammals—sperm whale and dugong. However, as a rule, human fishing activities probably did not have much impact at the time. Fishing was restricted mainly to a small proportion of the sea—inshore waters near heavily populated areas. Other factors may also have kept overfishing at bay, such as the cost of salt and technological limitations.

After around 1900, a number of factors led to concerns that things were not all well. These included the enormous demand for pearls from Europe (particularly in the Aru Islands), and the role of Japanese fish consumption and Japanese fishermen using large trawlers in Southeast Asia. Fuseliers (fish of the *Caesionidae* family) were being caught in large numbers from among the reefs by Japanese fishermen as well, and this appears to have led to local overfishing. The proliferation of fishing stakes with fine mesh was harmful to young fry and breeding stock.

Finally it should be pointed out that coral mining—which, as the term suggests, is both mining and the exploitation of a product of the sea—destroyed so many reefs in the Thousands Islands near Batavia that around 1940 one observer compared it to deforestation.

POLLUTION

A few words are in order about pollution. Although globally not much thought was given to pollution until the 1960s, it had been around for a long time. Locally, it had been a problem from the nineteenth century in many developed countries. The "first industrial nation," Great Britain, had to undergo "the penalty of the pioneer" by having the dirtiest capital city in the world, in terms of air and water.

Small-scale pollution in Southeast Asian cities prior to the 1870s was described in Chapter 7, and as urbanization progressed during the period under discussion, so did pollution. As has already been made clear, the increased exploitation of land, forests, and mines led to growing water pollution, as did the effluvia of an incipient industry. The next chapters deal with industry in more detail, as industrialization did not really take off until after the 1960s; even the limited number of industrial establishments predating the period of independence, however, made for increased pollution of air, land, and water.

CONCLUSION

During this—what I would call—third globalization phase, considerably increased production of timber, charcoal, and firewood took place, both for subsistence purposes and for commercial reasons, as well as increasing exports. NTFPs, mainly for export, were also collected in growing quantities. All of these developments led to a notable reduction in the region's forest cover, although compared with China and Europe it was still strongly forested. Increased production had been made possible by the investment of more foreign capital and by the introduction and spread of various Western innovations. One could regard the various (colonial) forest services as one such innovation that facilitated the exploitation of the forests.

Forest services or departments were called into existence in order to supervise forest exploitation, which ideally should be carried out in a sustainable way, particularly regarding valuable hardwoods like teak. Another function of the forest services and forest departments was the preservation of watershed protection forests. While the latter function was in many instances carried out satisfactorily, sustainable production was a goal seldom attained, mainly because these government services were unable to hold private enterprise to its legal obligations of reforestation.

The forest services turned woodlands into "political forests," thereby limiting local people in their rights of access, which ultimately contributed to fundamental changes in the way of life of forest people. The term *resource curse* has been used to express the often negative impact of resource exploitation on the people living nearby.

This period witnessed the beginnings of European-inspired attempts at conservation and wildlife protection on a modest scale, at a moment that a number of animals and plants, and occasionally entire local ecosystems, were becoming endangered and local extinctions occurred. Such attempts usually started with species protection programs, followed by game laws. From around 1920, nature reserves, wildlife reserves, and national parks were established. It was also attempted to curb depletion of certain species by prohibiting exports of endangered animals and plants. Importing countries occasionally installed import bans, as in the case of birds of paradise feathers.

What applies to forest exploitation also holds true for mining and fishing—increased scale of operations as a result of population growth and investment of foreign capital, leading to more landscape destruction, pollution, and local depletion.

Between around 1930 and 1970, the growth of most of these activities slowed down, and it was not until the fourth and present globalization phase that followed from the 1970s that the rate of growth of production, depletion, and pollution would become even higher.

BIBLIOGRAPHICAL ESSAY

Regarding forests and forestry between the 1870s and the 1960s, a fairly early example of a historical monograph (partly) written with an environmental focus in mind is Keeton, Charles Lee, 1974, *King Thebaw and the Ecological Rape of Burma: The Political and Commercial Struggle between British India and French Indo-China in Burma 1878–1886*. Delhi: Manohar.

More recent studies in this field include Dargavel, John, Kay Dixon, and Noel Semple, eds., 1988, *Changing Tropical Forests: Historical Perspectives on Today's Challenges in Asia, Australasia and Oceania*. Canberra: CRESS; Poffenberger, Mark, ed., 1990, *Keepers of the Forest: Land Management Alternatives in Southeast Asia*. West Hartford, CT: Kumarian; Aiken, S. Robert, and Colin H. Leigh, 1992, *Vanishing Rain Forests: The Ecological Transition in Malaysia*. Oxford: Oxford University Press; Dargavel, John, and Richard Tucker, eds., 1992, *Changing Pacific Forests: Historical Perspectives on the Forest Economy of the Pacific Basin*. Durham: Forest History Society; De 'Ath, Colin, 1992, "A History of Timber Exports from Thailand with Emphasis on the 1870–1937 Period," *Natural History Bulletin of the Siam Society* 40, pp. 49–66; Peluso, Nancy Lee, 1992, *Rich Forests, Poor People: Resource Control and Resistance in Java*. Berkeley: University of California Press; Pouchepadass, Jacques, ed., 1993, *Colonisations et Environnement*. Paris: Société Française d'Histoire d'Outre-Mer; Boomgaard, Peter, 1994, "Colonial Forest Policy in Java in Transition, 1865–1916." Pp. 117–138 in *The Late Colonial State: Political and Economic Foundations of the Netherlands Indies 1880–1942*. Edited by Robert Cribb. Leiden: KITLV; Boomgaard, Peter (with R. de Bakker), 1996, *Forests and Forestry 1823–1941*. [Changing Economy in Indonesia, Vol. 16]. Amsterdam: Royal Tropical Institute; Bryant, Raymond L., 1997, *The Political Ecology of Forestry in Burma 1824–1994*. London: Hurst; Boomgaard, Peter, Freek Colombijn, and David Henley, eds., 1997, *Paper Landscapes: Explorations in the Environmental History of Indonesia*. Leiden: KITLV; Kaur, Amarjit, 1998, "A History of Forestry in Sarawak," *Modern Asian Studies* 32, no. 1, pp. 117–147; Peluso, Nancy Lee, 2001, "Genealogies of the Political Forest and Customary Rights in Indonesia, Malaysia, and Thailand," *Journal of Asian Studies* 60, no. 3, pp. 761–812; Tuck-Po, Lye, Wil de Jong, and Abe Ken-ichi, eds., 2003, *The Political Ecology of Tropical Forests in Southeast Asia: Historical Perspectives*. Kyoto: Kyoto University Press/Trans Pacific Press; Bankoff, Greg, 2004, "'The Tree as the Enemy of Man'; Changing Attitudes to the Forests of the Philippines, 1565–1898," *Philippine Studies* 52, no. 3, pp. 320–344; Boomgaard, Peter, 2005, "The Long Goodbye? Trends in Forest Exploitation in the Indonesian

Archipelago, 1600–2000." Pp. 211–234 in *Muddied Waters: Historical and Contemporary Perspectives on Management of Forests and Fisheries in Island Southeast Asia*. Edited by Peter Boomgaard, David Henley, and Manon Osseweijer. Leiden: KITLV; Cleary, Mark, 2005, "Managing the Forest in Colonial Indochina c. 1900–1940," *Modern Asian Studies* 39, pp. 257–284; Kathirithamby-Wells, Jeyamalar, 2006, *Nature and Nation: Forests and Development in Peninsular Malaysia*. Copenhagen: NIAS.

On the notion of "resource curse," see Auty, R. M., 1993, *Sustaining Development in Mineral Economies: The Resource Curse Thesis*. London: Routledge.

A classic on conservation history is Hoogerwerf, A., 1970, *Udjung Kulon: The Land of the Last Javan Rhinoceros*. Leiden: Brill. More recent studies on hunting, wildlife protection, and other forms of conservation are Aiken, S. Robert, 1991/1992, "The Writing on the Wall: Declining Fauna and the Report of the Wild Life Commission of Malaya (1932)," *Wallaceana* 66/67, pp. 1–6; Aiken, Robert S., 1994, "Peninsular Malaysia's Protected Areas' Coverage, 1903–92: Creation, Rescission, Excision, and Intrusion," *Environmental Conservation* 21, no. 1, pp. 49–56; Cubitt, Gerald, and Belinda Stewart-Cox, 1995, *Wild Thailand*. London: New Holland; Boomgaard, Peter, Freek Colombijn, and David Henley, eds., 1997, *Paper Landscapes: Explorations in the Environmental History of Indonesia*. Leiden: KITLV; Boomgaard, Peter, 1999, "Oriental Nature, Its Friends and Its Enemies: Conservation of Nature in Late-Colonial Indonesia, 1889–1949," *Environment and History* 5, no. 3, pp. 257–292; Boomgaard, Peter, 2001, *Frontiers of Fear: Tigers and People in the Malay World, 1600–1950*. New Haven: Yale University Press; Kathirithamby-Wells, Jeyamalar, 2006, *Nature and Nation: Forests and Development in Peninsular Malaysia*. Copenhagen: NIAS.

An older title with details on mining is Allen, G. C., and Audrey G. Donnithorne, 1954, *Western Enterprise in Indonesia and Malaya*. London: Allen and Unwin. More recent titles include Loh, Francis Kok Wah, 1988, *Beyond the Tin Mines: Coolies, Squatters and New Villagers in the Kinta Valley, Malaysia, c. 1880–1980*. Singapore: Oxford University Press; Somers Heidhues, Mary F., 1992, *Bangka Tin and Muntok Pepper: Chinese Settlement on an Indonesian Island*. Singapore: ISEAS; Kaur, Amarjit, and Frits Diehl, 1996, "Tin Miners and Tin Mining in Indonesia, 1850–1950," *Asian Studies Review* 20, no. 2, pp. 95–120; Erwiza, 1999, "Miners, Managers and the State: A Socio-Political History of the Ombilin Coal-Mines, West Sumatra, 1892–1996," Ph.D. dissertation, University of Amsterdam; Aragon, Lorraine V., 2002, "In Pursuit of Mica: The Japanese and Highland Minorities." Pp. 81–96 in *Southeast Asian Minorities in the Wartime Japanese Empire*. Edited by Paul H. Kratoska. London:

RoutledgeCurzon; Somers Heidhues, Mary, 2003, *Golddiggers, Farmers, and Traders in the "Chinese Districts" of West Kalimantan, Indonesia*. Ithaca, NY: Southeast Asia Program Publications.

The main text on fishing for this period is Butcher, John G., 2004, *The Closing of the Frontier: A History of the Marine Fisheries of Southeast Asia c. 1850—2000*. Singapore: ISEAS; Leiden: KITLV. Other recent titles on fishing include Masyhuri, 1995, "Pasang Surut Usaha Perikanan Laut: Tinjauan Sosial-Ekonomi Kenelayanan di Jawa dan Madura, 1850–1940," Ph.D. dissertation, Free University, Amsterdam; Masyhuri, 1997, "Fishing Industry and Environment off the North Coast of Java." Pp. 249–260 in *Paper Landscapes: Explorations in the Environmental History of Indonesia*. Edited by Peter Boomgaard, Freek Colombijn, and David Henley. Leiden: KITLV; Semedi, Pujo, 2001, "Close to the Stone, Far from the Throne: The Story of a Javanese Fishing Community, 1820s—1990s," Ph.D. dissertation, University of Amsterdam; Butcher, John, 2005, "The Marine Animals of Southeast Asia: Towards a Demographic History, 1850–2000." Pp. 63–96 in *Muddied Waters: Historical and Contemporary Perspectives on Management of Forests and Fisheries in Island Southeast Asia*. Edited by Peter Boomgaard, David Henley, and Manon Osseweijer. Leiden: KITLV.

PART SIX

THE YEARS OF THE ECONOMIC MIRACLE AND ENVIRONMENTAL DEBACLE

10

DEMOGRAPHIC, ECONOMIC, AND AGRICULTURAL GROWTH

Southeast Asia entered the Modern Era around the 1970s. The first big city in Southeast Asia that started to look like New York, Tokyo, or Hong Kong was Singapore, followed by Jakarta, Bangkok, and Manila. Modern high-rise buildings now dominate their skylines. Large numbers of people live in these cities in apartment buildings, often working in air-conditioned offices. It could be—and has been—argued that nothing has stimulated the increase in productivity of (white-collar) labor in Southeast Asia as much as did the introduction and spread of the air-conditioner.

All-weather roads and highways being plied by regular transportation services connect cities, towns, and many villages. The larger cities are also connected by airlines. In many houses electricity and running water are available, as are showers, washing machines, and flush toilets. Increasing numbers (and percentages) of people have some sort of formal education, and can read and write. They own (Japanese or Korean) cars, motorcycles, radios, televisions, videos, satellite dishes, refrigerators, freezers, and telephones (nowadays often mobile telephones), as well as, lately, CD and DVD players, personal computers and laptops connected to the Internet, and digital cameras. Prior to the 1970s only a small minority of the population was used to those amenities, which were available at the time in the developed world. After the 1970s, life in Southeast Asia more and more started to look like life in the more advanced countries; globalization had done its work.

However, there are notable differences within Southeast Asia, between and also within countries. There are various upland and inland areas, sometimes not too far from the teeming capital cities, where people are still living primarily as hunter-gatherers or slash-and-burn agriculturalists. There are also a great many villages with permanent-field cultivators that are not well connected to the outside world. This is not to say that they have not been influenced at all by external forces of modernization, because that is rarely the case. It is simply that for various reasons—perhaps mainly geographical and economic reasons—they are still living more "traditional" lives than many other Southeast Asians.

The skyline of Manila bristles with high-rise buildings. Many cities in Southeast Asia are now starting to resemble those in more advanced countries. (PhotoDisc)

Returning now to the "modern" aspects of life in Southeast Asia, from the 1970s, production took place increasingly in large factories equipped with state-of-the-art machinery, although there are still large numbers of smaller factories with obsolete, often polluting technology. Tractors—particularly the so-called walking tractors, or two-wheeled tractors—have often replaced livestock. Logging operations are carried out largely with chainsaws and bulldozers.

However, this increased productivity and prosperity have a downside, which is also part of modern life. I am referring to hugely increased deforestation and erosion. I am also referring to the state of more or less permanent traffic conges-tion in cities like Jakarta and Bangkok, and to the increasing amount of air pol-lution that covers those cities as a semipermanent blanket of smog most of the time. The conditions of inland waterways are equally appalling, as are those of the estuaries of the big rivers. Water is increasingly polluted not only by waste from the continuously growing cities but also by the still-increasing application of artificial fertilizers, pesticides, and herbicides. Pollution of water, land, and air, familiar features of the industrial landscapes of the West for more than a century—the classical London smog, for instance—has finally reached Southeast Asia as well.

Most of this is of recent date, and was the product of unusually high rates of sustained economic growth and development. The last three decades of the twentieth century were, generally speaking—at least until the financial crisis of 1997—characterized by high rates of economic growth in many countries of the region, varying from 5 to 10 percent growth of GDP annually. And although Southeast Asian population growth rates were high as well (on average above 2 percent per year), real income per capita increased greatly. Thus, the percentage of people below the poverty line dropped considerably as well. While in the 1960s, 50 percent or more of the population in many Southeast Asian countries were reckoned to be below the poverty line, by the 1990s that percentage had dropped to between 10 and 20 percent in most areas. Therefore, the number of people increased at an unprecedented pace, consumption per capita was higher than ever before, and production levels, needed for the maintenance of those higher levels of prosperity, were also higher than ever, based largely on equally unprecedented levels of exploitation of natural resources.

POPULATION GROWTH

As was shown in Table 8.1, population growth in Southeast Asia as a whole was 2.2 percent between 1960 and 1990, and 2.1 percent between 1990 and 2000. Those rates were higher than ever before, but it is also clear that the years after 1990 show a slight drop in the growth rate, which spells the end of two centuries of increasing population growth rates. Both this drop, although still very small, and the peak that preceded it are signs that the so-called demographic transition, also called epidemiological transition, has now reached Southeast Asia.

The demographic transition is a process whereby high levels of mortality and fertility—in other words, high death and birth rates—slowly but irreversibly drop to much lower levels. Mortality levels drop first, which leads to higher rates of natural increase of the population. At a later stage, fertility levels start to decline, eventually leading to lower rates of natural increase. In Europe, the first part of the world to undergo this process, the transition, which started around 1800, was a long and drawn-out affair. At a much later point in time, East Asia went through the same transitional phase, albeit at a higher pace; now it is Southeast Asia's turn.

Conventional wisdom has it that the fertility decline in Southeast Asia did not begin before the 1960s. Mortality levels had already started to drop slightly at the beginning of the twentieth century, but an acceleration of the downward trend occurred in the 1950s. In the short run, that led to very high rates of natural increase. However, around the year 2005, the total fertility rate (TFR: average number of children per woman) had dropped in some cases below replacement level, which is usually pegged at 2.1 or 2.15. At this level, the rate of natural increase of the population is close to zero. That was the case in Singapore, Thailand, Vietnam, and Burma, while Brunei and Indonesia were getting close. In Southeast Asia as a whole the TFR was around 2.5 in 2000, while it had been 4.8 in 1975 and 6.0 in 1950.

However, in Laos and Cambodia the TFR was still at a very elevated 4.8 and 3.4, respectively, in the year 2005, which means that in those countries the demographic transition still has a long way to go. This state of affairs once again confirms the existence of large differences in way of life and livelihood levels within Southeast Asia. Generally speaking, the poorer, less developed countries have high rates of fertility, while the rapidly developing economies have dropping birth rates.

In fact, it could be argued that we still witness a two-track demographic development, but now the roles of the "modern" and the "traditional" sectors have been reversed, so to speak. While I argued in previous chapters that there was a "traditional," isolated (tribal, upland) sector with low fertility rates, and a market-oriented (lowland, settled peasant, and urban) sector with relatively high birth rates, the present "modern" sector is now characterized by low birth and death rates, while the more backward areas still have high rates of mortality and fertility, even if the death rate has been dropping for some time even there.

It is not difficult to explain why the death rate has dropped more or less continuously across the board in Southeast Asia, particularly over the last fifty years or so. The two main factors are the growing availability (and affordability) of the so-called miracle drugs (antibiotics), and the improved calorie intake per capita. Both factors are closely related to higher income per capita, particularly from the 1970s onward. The figures presented in Table 10.1 illustrate these points well.

It is particularly revealing that while life expectancy (at birth) increased by only three years between 1950 and 1975, it made a jump of twenty-three years between 1975 and 2000. It would be difficult to find a better way of illustrating the acceleration of "modernization" from the 1970s onward.

While this was going on, birth rates were dropping as well, but fertility levels started to go down later than mortality rates; the effect of the lower birth rate on the rate of natural increase was for a long time barely visible. Nevertheless, the Southeast Asian (crude) birth rate did drop, from 44 per thousand in 1950, to 35 in 1975, and 21 in 2000. Here, again, the reader should be reminded that such averages mask large differences. Around 2000, the birth rate in Singapore was close to 10, while in Cambodia it was around 35.

Table 10.1 Crude death and birth rates (per thousand) and life expectancy (in years) in Southeast Asia, 1950–2000, benchmark years.

	1950	1975	2000
Crude death rate	24.7	13.1	7.0
Crude birth rate	44.3	35.3	21.4
Life expectancy	40.5	43.6	67.0

Source: Based on data taken from The Future of Population in Asia. *Honolulu: East-West Center (2002).*

What are the factors behind this amazingly rapid fertility transition? The "proximate factors" affecting the process are clear. Women marry later, and their marital fertility is lower than it used to be because they are using methods of family limitation, which they can do because reliable methods of birth control are readily available. For the search for the "ultimate" or "underlying" factors, we will have to look at things like economic development, modernization, and mass communications.

Not so long ago, demographic orthodoxy insisted that in order to achieve lower rates of fertility, all that was needed was economic development. Even though now it is recognized that other factors have contributed to this process, it is still rather obvious that economic growth is the real motor behind the demographic transition. But the first step is to look at the proximate factors.

One of the most obvious factors, of course, is the use of modern methods of contraception, including the pill, condoms, sterilization, and IUDs. Singapore, Thailand, and Indonesia are good examples of countries in which the acceptance of modern birth control methods has been increasing steadily. In Indonesia and Singapore this has been strongly stimulated by the state. In the Philippines, where the influence of the Roman Catholic Church is strong, antinatalist policies are far less popular; that is reflected in the fairly high total fertility rate (TFR of 3.16 in 2005). Increased use of modern methods of birth control is reflected in lower rates of marital fertility. Early termination of pregnancies (induced abortions) may play a role as well, as is shown in the case of Singapore, where the law, originating in Victorian Britain, was changed in 1970. This brief discussion illustrates the potential importance of the state in these matters. It also suggests that variation between regions is to be expected, based on differences in culture, religion, and ethnic reproductive patterns.

The effects of family planning are reinforced by the increasing age at first marriage of women. In 1960, the proportion of women in Thailand between the ages of twenty-five and twenty-nine who were married was 87, which had dropped to 75 by 1990. For Indonesia, those figures are 96 and 89. If experiences in East Asia are anything to go by (and they seem to be), this proportion might drop as low as 60 (Japan in 1990). Generally speaking, past experience has shown that a higher age at first marriage leads to lower numbers of children per woman.

Later age at marriage is generally assumed to be related to a drop in arranged marriages and to higher proportions of women being educated beyond primary school. Education for women also had other implications for the birth rate. Educated women were, at least in an early stage of the fertility transition, quicker to accept modern family planning methods. More education for girls and young women also implies that they are no longer available to their parents as cheap labor. On the contrary, they are now costing money. It is assumed by many scholars that these considerations have played, and still are playing, an important role in the declining birth rate, as children are turning from being an economic asset into a liability, at least in the short run. However, the role of education for women varies from country to country. Where Singapore, the leader in fertility transition, shows the expected combination of low TFR and high proportion of women in secondary education, the runner-up, Thailand, combines a low TFR with a low rate of female participation in education.

But how do we explain the sudden interest of the state in birth control, the success in the adoption of family planning methods, the growing interest in education, and the postponement of marriages? The keywords here are modernization, economic development, and mass communications.

Economic growth, in the sense of an almost continuous increase in real income per capita, is probably the most important driving force behind the success story of the demographic transition in (parts of) Southeast Asia. Economic development was largely responsible for the lower rate of mortality (better diet and better medical care), and it is also one of the main forces behind the fertility transition. Economic development implies, among other things, urbanization and a shift from agriculture to industry and the service sector, with young women migrating temporarily to urban areas. This may have influenced the age of marriage and the arrival of a first child, as children could be combined with agricultural activities, but far less with working in the factory. Economic development also implies schooling.

"Modernization," admittedly a rather vague notion, follows in the wake of economic growth. Notions of individual choices and destinies, of better education, of higher aspirations, particularly for women, are all part of the "Western" ideology; it is more or less identical with the modern way of life, which has been "globalized" during the last few decades.

The enormous impact of modernization and the rapid adoption of methods of birth control would have been unthinkable without the spread of radio and television. It would appear that this goes a long way toward explaining why the fertility transition could have taken place so quickly in a number of countries.

In Europe and East Asia, the demographic transition, once it had been set in motion, was unstoppable. And while mortality can hardly be expected to drop any further—in fact, because of an aging population crude death rates may be expected to rise somewhat—fertility levels are still dropping. We cannot be sure that this will also be the case in Southeast Asia—there is anecdotal evidence that in some rural areas in Indonesia people are having more children than the two they are supposed to get, perhaps a combination of devout Muslim influence and the end of the Old Order under President Suharto—but chances are that the process in the region will be quite similar. It is to be expected, therefore, that the natural increase of the population will continue to slow down. From an environmental point of view, that may be regarded as a positive development.

However, a number of developments in the public health sphere are cause for concern. It is too early to judge their effects on the death rate in the long run, but the possibility that they could lead to somewhat higher levels of mortality is not to be excluded. Most of these factors are not restricted to Southeast Asia, but it would appear that, along with other developing countries, the region is particularly vulnerable to their effects. Three factors will be mentioned.

The first is the advent of multiple-drug-resistant (MDR) bacteria strains. With the application of antibiotics, strains resistant to specific drugs soon emerged. As long as they could be killed by other drugs, this was not much of a problem. But careless prescription, lack of discipline in taking the drugs, and insufficient supervision of the use of antibiotics, in addition to the sale of antibiotics as nonprescription drugs, led to the evolution of bacteria resistant to all drugs in existence. An example from the region is that the Indonesian Ministry of Religion in 1981 provided 100,000 pilgrims to Mecca with tetracycline against cholera, thus hastening the evolution of tetracycline-resistant bacteria.

Prime examples of MDR strains are tuberculosis and malaria, while some species of the anopheles mosquito, the vector of malaria, became resistant to the insecticide DDT. It stands to reason that these developments may reverse the downward trend of the mortality rates reported from Southeast Asia.

A second threat is the rapid spread of old and new diseases across the globe owing to an increase in the volume and velocity of transport, particularly through the air and over sea. This can be considered as another example of bioinvasion, a topic dealt with in the last chapter. Examples are SARS (Sudden Acute Respiratory Syndrome), a disease transmitted through droplet infection that at one point was spreading rapidly by way of air traffic but was fortunately nipped in the bud, and HIV/AIDS, a disease transmitted through sexual intercourse and through the use of (hypodermic) needles. In the case of AIDS, unfortunately not as easily

Health workers burn chickens because of avian influenza (bird flu), Indonesia. Southeast Asia is at greater risk from avian influenza due to the close proximity of humans and chickens in the region. (Reuters/Corbis)

stopped as SARS, intercontinental tourism played a major role in the region (Cambodia, Myanmar, Thailand). Mosquitoes, vectors of many parasites, have been transported in ship containers, thus transmitting diseases or strains to areas where they were previously absent (dengue fever, yellow fever).

A recent example is the resurgence of polio in Southeast Asia, where the disease until two years ago (at the moment of writing) had more or less disappeared. However, three years ago, Muslim leaders in northern Nigeria successfully organized resistance against a polio vaccination campaign, which promptly led to a doubling of the number of polio cases in the following years. As a result, the disease reappeared in sixteen countries regarded as polio-free, among which Indonesia must be counted. In all of these cases the virus came from Nigeria.

Another recent threat is avian influenza, or bird flu (H5N1 variant), which, at the moment of writing, it was feared might develop into a new strain that could be transmitted from humans to humans and then might become a scourge like the Spanish flu in the early twentieth century. In June 2006 the World Health Organization reported that there was proof of human to human transmission in Sumatra. Southeast Asia is more at risk than many other regions, because humans and birds, particularly chickens, live in such proximity. A count undertaken in July 2006 revealed that from the first recorded human deaths owing to bird flu in Vietnam in January 2004, 134 people had been killed by the disease, of whom 105 were in Southeast Asia. Of those avian flu deaths, 42 had occurred in Vietnam, 42 in Indonesia, and 15 in Thailand. The death rate among humans because of the avian flu is increasing, and although the absolute numbers are still low, case fatality is quite high—50 percent. Governments are reluctant to carry out drastic measures to curb the spread of the disease, as they are costly and popular resistance against them is high. In those countries where governments do take measures, the disease appears to have acquired a seasonal character, putting in an appearance with the rainy season. In countries where governments do not make an effort to suppress the disease, as is the case in Indonesia, seasonality does not occur.

The third and final factor to be mentioned is the influence of pollution of the air, water, and soil on human health. Rivers and seas are increasingly polluted with, among other things, untreated human waste, pesticides, herbicides, mercury from mining operations, and artificial fertilizers. All of this is ingested by animals that often end up as items of human consumption. These substances cumulate in those beings that are at the end of the food chain—people (and it is hard to believe that this can go on forever without being harmful) eventually, one surmises, leading to higher rates of mortality. In addition, there are many millions living in the megalopolises of Southeast Asia, and as those cities are usually covered by a blanket of smog during the dry season, these people are breathing that kind of air day in and day out. All of this is everyday pollution, and does not take into account the possibilities of biohazards such as the explosion of a chemical plant or problems with nuclear installations.

A factor that might have been expected to influence the death rate more in Southeast Asia than, for instance, in the United States is smoking, but that is probably not the case. If one looks at data on smoking behavior among boys and girls ages fifteen to nineteen, the percentage of Indonesian boys that smoked (38 percent in 1988) was slightly higher than that of boys from the United States (35 percent in 1999), but the percentage of Indonesian girls smoking (1 percent) was so much lower than that of U.S. girls (the same as that of the boys: 35 percent), that on average the Indonesian adolescents were doing much better. However, data from Japan and South Korea (and, one might add, from Europe) show that girls start to smoke more when income per capita increases, while boys cut back on their tobacco consumption. On average, however, U.S. adolescents still smoke more.

ECONOMIC GROWTH

Many of the Southeast Asian economies have been growing at an impressive pace from the late 1960s. Singapore is usually regarded as belonging to the first-tier newly industrialized economies or countries (NICs), together with South Korea, Taiwan, and Hong Kong, also called the four dragons. The second-tier high-performing Asian economies are Malaysia, Thailand, and Indonesia, also called the tigers. The Philippines are not far behind in terms of economic development, but Myanmar, Laos, and Cambodia are still very much agrarian economies, while Vietnam can be said to occupy an intermediate position. In Table 10.2, a number of indicators for recent economic development have been presented for selected countries.

Table 10.2 illustrates clearly the very high growth rates of real income per capita in most Southeast Asian countries, accompanied (and partly caused) by a

Table 10.2 GNP per capita (in 1995 US$), 1960 and 1999, percentage urban, 1975 and 2000, and share of agriculture in GDP, 1960 and 2000, in selected countries.

Country	GNP per capita		Percentage urban		% agriculture of GDP	
	1960	1999	1975	2000	1960	2000
Cambodia	na	279	10.3	15.9	na	39.6
Indonesia	252	907	19.4	40.9	51.5	17.2
Laos	na	441	11.4	23.5	na	52.5
Malaysia	953	4,305	37.7	57.4	34.3	8.6
Myanmar	na	na	23.9	27.7	na	57.2
Philippines	701	1,201	35.6	58.6	na	15.8
Singapore	2,776	28,486	100.0	100.0	na	0.1
Thailand	450	2,656	15.1	21.6	36.4	9.0
Vietnam	na	314	18.8	19.7	na	24.5

Source: Based on data taken from The Future of Population in Asia. *Honolulu: East-West Center (2002); Jomo K.S., ed., 2003,* Southeast Asian Paper Tigers? From Miracle to Debacle and Beyond. *London: RoutledgeCurzon;* Key Indicators of Developing Asian and Pacific Countries 2004, *volume 35. Manila: Asian Development Bank.*

structural transformation of most regional economies, from predominantly agricultural to more urban and industrial.

Factors behind the phenomenal growth over the last three or four decades are partly to be found in high and rising demand for food and drink, fodder, raw materials, and finished industrial products from Europe, the United States, Japan, and other East Asian countries, and partly in rising internal demand. From the supply side mention should be made of Southeast Asian economic policies conducive to economic growth, increasing rates of savings and investment, including direct foreign investment, rapidly expanding education, industrialization, and urbanization. The following section deals in more detail with these topics.

From the 1960s, rising real income per capita in Europe, North America, and Japan—at least partly based on low prices of oil from the Middle East—led to an increased demand for agricultural products such as coffee, tea, cacao, sugar, copra, rice, rubber, oil (from the oil palm), and cassava. There was an equally large—and often larger—and growing demand for primary products such as petroleum, natural (petroleum) gas, tin, nickel, mica, and bauxite. There was also a spectacular increase in the demand for timber, particularly hardwoods, plywood, paper, and other Non-Timber Forest Products. Finally, when Southeast Asian regimes became more "developmental" in their economic policies, and more receptive to the interests of foreign industrialists and capital providers, the region was better able to respond to increasing Western and Japanese demand for cheap industrial products such as those of the textile, garment, and footwear industries. Lately, Southeast Asia has started to fulfill the growing demand for more sophisticated industrial products such as electronic equipment, including components for computers. Somewhat later, rising demand from the other high-performing Asian economies (Hong Kong, South Korea, Taiwan) was added to that of the regions and countries mentioned.

Demand from the region itself rose as well. Given the high rates of growth of real income per capita, in addition to the high rates of natural increase of the population, the inhabitants of the region, or at least large and growing sections of them, were now able to afford more than the bare necessities, thus generating a phenomenal and unprecedented expansion of domestic demand for goods and services. Some of this was met by imports, but the bulk was spent on products from Southeast Asia itself. Many of these products were the same ones that were also exported, but there were other ones as well, such as fresh fish and meat.

Nevertheless, it is clear that without a vigorous export sector, such high rates of economic growth would not have been obtained. It is not unusual to speak of export-led growth when referring to economies like the ones under discussion, but it would be better in this case to emphasize the interdependence and the mutually reinforcing effects of exports and economic growth.

Water buffalo haul teak logs in Southeast Asia. Since the 1960s, there has been a spectacular increase in demand for hardwoods from the region. (Corel)

Why this fairly sudden growth spurt starting in or just after the 1960s, after thirty or forty years of stagnation? Various factors appear to have contributed more or less simultaneously. The first factor to be mentioned is that around the 1970s, most Southeast Asian countries saw a long period of decolonization and other (armed) conflicts and uprisings coming to an end. This was followed by a long period of peace, in many cases under fairly stable—albeit often autocratic—governments. Perhaps equally important was the inclination of many governments in the region to institute policies conducive to economic growth, something that during these Cold War days was strongly stimulated by military and development aid from the United States and Europe. Economic growth was seen as an antidote for communism. During the period between 1945 and about 1970, policies that had a bearing on the economy did often have a political rather than an economic rationale. This had led in many cases to an atmosphere inimical to foreign investment and the growth of the export sector, which, in turn, led to low, sometimes declining, and even negative rates of economic growth. After around 1970, such policies were, generally speaking, no longer *en vogue* in Southeast Asia.

This created a climate that made it more attractive to foreigners and nationals alike to invest in the region. At the same time, Southeast Asian people and enterprise started to save more, partly because rising income per capita enabled them to do so. Part of this rising income was produced by increasing exports of natural resources such as oil (petroleum), tin, and timber, which also made for higher tax and easily acquired incomes by governments. Government income based on the exploitation of natural resources is not always a good thing, as it might very well lead to corruption, easy government spending, inflation, and subsidies that distort a more economical resource allocation ("Dutch disease"). However, it can also be partly put to good use, and that was arguably the case in many instances in Southeast Asia. Governments started to spend much of their income on education, infrastructure, and (or including) irrigation works.

To be sure, corruption and misspending through what is now often termed cronyism—the creation of opportunities for profit for a group of "friends" around government officials in high places—were far from absent, and on this score the "tigers" were not performing as well as were the "dragons." Cronyism can be seen as the modern version of patrimonialism, a phenomenon dealt with for the Pre-Modern and Early-Modern periods. However, the Southeast Asian regimes were not doing as badly as the African "kleptocracies," and it would appear that there were sufficient government institutions—like most ministries of finance in the region—that were doing their jobs without too much interference from special interests.

While saving part of one's earnings has always been a feature of Southeast Asian economic life, this had usually not been in the form of cash. In most cases people with surplus earnings invested in livestock or jewelry; both could be easily sold, while the former could multiply under favorable circumstances (and could be eaten if push came to shove). After about 1970 saving money and depositing it in a bank account became more usual, while savings as a percentage of GDP were slowly rising. When at the same time governments were seen to be more serious about growth-promoting economic policies, high rates of population growth led to the continued existence of a large pool of cheap labor; when repressive regimes and a low degree of unionization kept labor unrest to a minimum, a lot of this money was invested in the domestic economy. By the 1990s, investment as a percentage of GDP had risen in countries like Singapore, Malaysia, Indonesia, and Thailand to a level of around 30 percent. In the Philippines this proportion was lower (about 15 percent), while in countries like Myanmar and Laos even lower levels were obtained.

Such levels of investment were already quite impressive, and they certainly explain a sizable share of the high growth rates of the leading Southeast Asian economies over the last thirty-five years or so. In addition, the existence of such a favorable business climate attracted foreign direct investment (FDI). In fact, high ratios of FDI inflows to domestic capital formation were obtained only in Singapore (15 to 30 during the period 1970 to 1990) and Malaysia (10 to 15), while much lower ratios were to be found in Indonesia, Thailand, and the Philippines. Nevertheless, FDI was also important in the latter countries, if only for the new technology and the new management methods that came along with it.

Finally, it should be mentioned that governments and private enterprise were also heavy borrowers of (foreign) capital. In good times this boosted the growth of the economy, but when for whatever reason trust in the economic performance of Southeast Asian countries was temporarily suspended, as happened in 1997, the economies were in trouble.

Improved education, including vocational and technologically sophisticated training, prepared the Southeast Asian populations for increased employment in the secondary (industry) and tertiary (government, services, transport, construction) sectors of the economy. This was a timely development, as during the thirty to forty years under consideration, these sectors were growing rapidly, to the detriment of the share of the primary sector (agriculture, fishing, mining, forestry; see Table 10.2). As a result the agricultural workforce as a proportion of the economically active population dropped from 75 percent on average in Southeast Asia in 1960 to 50 percent around 1995, although in absolute terms the total number of people employed in the agricultural sector was still rising (by roughly 45 percent between those dates).

The corollary of these developments was the enormous growth of the proportion of the population living in cities—or, in other words, urbanization. This process was fueled largely by a continuous stream of migrants from the countryside, looking for work in the predominantly urban secondary and tertiary sectors. Migration also occurred across national boundaries (for example, between Indonesia and Malaysia). As is shown in Table 10.2, more than half the population of the Philippines and Malaysia, and 41 percent of the people of Indonesia are now living in cities (disregarding Singapore, which is defined as urban in its entirety). Again, however, it is evident that there are large differences within the region—and within countries for that matter—as all other countries in the table have less than 30 percent of their population living in urban areas. In this regard Thailand is particularly interesting, as it combines a fairly low rate of urbanization with a very low share of agriculture in GDP and high rates of real income growth per capita. In 1995, the share of its agricultural workforce was still relatively high for such a high-performing economy—over 60 percent.

In countries like Indonesia, Thailand, and the Philippines (again leaving out Singapore, which is a special case), urbanization was and is characterized by the predominance of so-called primate cities. In contrast to many (but not all) developed countries, where urbanization was a long, drawn-out process spread more or less evenly over a fair number of cities per country, Southeast Asian urbanization, at least in the three countries mentioned here, took place much more rapidly and privileged one very big city per country—Jakarta, Bangkok, and Manila. By now these megalopolises have populations of around 10 million people each.

These developments have had a considerable impact on many features of daily Southeast Asian life that could be called environmental. In the first place, a very large proportion of the population is no longer "closer to nature" (to use a cliche phrase dating from the nineteenth century about people in tropical and subtropical countries) than most people in developed countries. They got used to or are getting used to an urban lifestyle in which "nature" (at least "green" nature) does not play much of a role. In the short term, one might surmise that this may lead to alienation from nature and a lack of interest in the natural environment, but in the long run, experience in other countries suggests that urban dwellers can turn nature lovers and environmental activists. Education can play an important role in that respect. This is not to say, of course, that all rural dwellers are environmentalists. On the contrary, the implicit notion of rural dwellers as "environmental angels" is deeply flawed, and urbanites have always played an important role in environmental movements.

Another feature of urbanization is that it transforms the landscape beyond recognition. The skylines of the primate cities have changed very rapidly over the last thirty years or so, and they are all starting to look like New York, Hong Kong, and Tokyo. Large tracts of land, until recently covered by wet-rice fields, home gardens, specialized orchard-type plots, and forests, have now been taken up by the urban and suburban sprawl. Urbanites, wanting to leave their city to walk or cycle in the countryside, need hours just to get to the urban fringe. Many megalopolis dwellers, particularly those belonging to the lower income groups, probably rarely leave their city. Primate cities have the tendency to incorporate the towns and villages around them, until entire districts or even provinces have become one very large megacity.

As the proportion of the population living in cities increased, so did energy consumption and waste production per capita. Here it should be mentioned that not all urban dwellers are equally "guilty." In fact, the often large numbers of recent and not so recent migrants living in slums are probably not all that different from those living in the countryside, as regards their energy consumption and waste production. It is often the case that the poorest of the poor survive partly by recycling discarded items found on the huge garbage dumps located near the big cities of Southeast Asia, and in that sense their waste production is even negative.

However, those who are somewhat better off are more likely to buy packaged food and other items, to buy more furniture in shops, to have and use motorcycles and cars, fridges and freezers, and to live and work in air-conditioned rooms, where they may be expected to use electricity for a whole range of modern gadgets, including, of course, large numbers of not-so-modern lightbulbs. They also are more likely to be connected to the water mains, and to be using much more (piped) water than rural dwellers for showers, splash-baths, toilets, washing machines, and the like. Thus they are using much more (nonrenewable) fossil energy than most villagers, and producing more waste, including (untreated) sewage that finds its way to rivers and estuaries.

AGRICULTURE

Given rising demand from the developed countries outside the region, the favorable government policies, the high levels of savings and investment, the increased enrollment in schools, and the rising degree of urbanization, the rapid development of the secondary and tertiary sectors of the Southeast Asian economies are more easily understood. But what happened in the meantime to the agricultural sector?

The most important developments in this sector were the so-called Green Revolution, increasing investment in irrigation, mechanization, intensification, diversification, the sharply increased use of artificial fertilizer, pesticides, and herbicides, the continued and accelerating invasion of the uplands, shifts within the sector, out-migration, and the shrinking share of agriculture in GDP and the labor force. We now take a closer look at these developments.

The Green Revolution and increased investment in irrigation are two highly interrelated developments, and without the expansion of the irrigation network the Green Revolution could never have been as successful as it was. Irrigation expansion had slowed down throughout Southeast Asia after 1930 owing to the by now well-known causes. Even upkeep of the irrigation systems left much to be desired, and regionally the situation was actually worse in the 1960s than it had been around 1930. From the 1960s, and even more so from the 1970s, most Southeast Asian states started to invest again in the expansion and maintenance of their irrigation networks. Between 1970 and 2000 the irrigated hectareage in the region increased by more than 2 percent on average per year, or roughly at the same rate as did the population.

This in itself would have made for greater food security and higher yields per capita. In addition, the Green Revolution led to even higher yields per hectare, as it combined recently developed HYVs (high-yielding varieties, mainly of rice, corn, and wheat, of which rice was the most relevant for Southeast Asia), with the use of fertilizers and pesticides. The new rice varieties had been developed by the International Rice Research Institute (IRRI) in the Philippines, using dwarf

Different strains of rice found in the rice market in Baguio City in the Philippines. Improved irrigation and farming techniques have led to greatly increased rice yields in the last 30 years. (Corel)

rice strains first selected by Japanese breeders in the 1920s in Taiwan, then a Japanese colony. By the late 1960s the new breeds from IRRI were disseminated across Asia's rice bowls. Between 1969–1971 and 1990–1992, rice yields (per hectare) in developing countries increased by 1.5 percent per year on average. This was partly caused by the fact that double cropping had increased considerably, but yields per harvest increased as well.

As the rice HYVs were designed for cultivation under "wet" conditions, expansion of irrigation was a prerequisite for the expansion of the hectareage under Green Revolution crops. The spread of these crops in the irrigated lowlands undoubtedly contributed to higher real income per capita, and thus to Southeast Asia's successful economic development between the 1960s and 2000.

Green Revolution programs were set up by most governments, combining credit and subsidies, and sometimes the establishment of village cooperatives. In addition, government-run agricultural extension services gave information on new crops and how to deal with them to the peasantry.

The Green Revolution was arguably the largest and fastest agricultural revolution in world history. However, it has also been criticized on various points, and some of these are of an environmental nature (in the broad sense of the word). The wet-rice HYVs did nothing for the upland areas; their cultivation led to an increased use of chemical substances that are not always harmless, while the large-scale use of pesticides selects for resistant pests. It also led to a sharp decrease in the number of rice varieties in use by peasant-cultivators, thus rendering the rice growers more vulnerable to sudden attacks of the HYVs by pests and plagues resistant to the pesticides employed. Whereas it would appear that up to now no insuperable disasters have occurred in this respect, although there have been very difficult periods, the long-term effects cannot as yet be assessed.

Some of these critical points apply to maize HYVs as well, which are also important locally in Southeast Asia. However, unlike rice, they do not discriminate against upland areas, although it must be said that fertile soils are a prerequisite. As was the case with rice, cultivators who want to buy the HYV-fertilizer-pesticide package have to be well to do, or else should be the beneficiaries of a credit program or subsidies. The increase in yields per hectare for maize during the 1970s and 1980s was even higher than that for rice—almost 2 percent annually on average.

Some research, however, was also done on permanent crops (perennials) that were planted in the uplands, a topic dealt with below in more detail. Here it may suffice to mention highly productive clonal plant material for rubber, developed some decades ago, which led to much higher production per tree and per hour worked.

In addition to expanded irrigation networks and the spread of Green Revolution features, mechanization of agricultural work—often with a labor-saving character—developed rapidly during the period under consideration. This was partly because labor-saving technology adapted to wet-rice cultivation had recently become available, in part because increasing income per capita enabled farmers to purchase such equipment, and finally because it led to cost cutting as less labor was needed. Finally, as we have seen in Chapter 8, stall feeding of buffalo was problematic and sufficient grazing room often no longer available, which was yet another incentive for mechanization. Thus this period witnessed the arrival of the two-wheel power tiller, the tractor, the mobile light thresher, and the rotary weeder. The last implement was older, but its spread continued into this period, as did the replacement of finger-manipulated harvesting knives by sickles.

At the same time the average holding size went on decreasing, thus continuing a development dealt with in Chapter 8. This led in many instances to very small holdings, which could no longer keep a peasant family in rice and clothes without additional income. Attempts at the redistribution of land, insofar as they were undertaken, were by and large unsuccessful, as most governments shied away from this politically sensitive issue.

Locally, the cumulative effect of all this was that the demand for labor dropped, or at least did not grow as fast as did the economically active population, and that landless villagers increasingly had to look for work elsewhere. As this was more often than not in the cities, rural to urban migration was speeded up, which was perhaps the most important push factor behind the urbanization process just mentioned.

At the same time, the use of chemical substances, alien to the village sphere, increased as well. I am referring to artificial fertilizer, pesticides, and herbicides. Prior to the 1960s all of this had hardly been used by smallholder-cultivators (as opposed to plantations). The shift from animal manure (and very occasionally human manure, or "night soil") to artificial fertilizer was occasioned partly by Green Revolution packages that included fertilizer, as better suited to the new HYVs, and partly by the diminishing role of livestock in Southeast Asia's farming practices. What played a role as well is that many Southeast Asian countries started to produce their own fertilizer, which then could be sold more cheaply, while on the demand side increasing real (rural) income per capita must have been another important factor.

Pesticides are used in order to get higher yields (killing the pests and plagues that damage the crops), while herbicides kill off weeds that are competing with crops. Herbicides are labor saving in the sense that they take the place of weeding, thus reinforcing the labor-saving effects of mechanization. Pesticides and herbicides are major human health hazards, a topic dealt with in the next chapter.

The use of fertilizer, pesticides, and herbicides has undoubtedly led to increased agricultural production and productivity, but at a price. It has broken the more or less closed and sustainable circle of subsistence agriculture, and it has produced unprecedented pollution of soil and water. Appropriate environmental policies can in principle put a stop to the latter development, but not to the former. At the present levels of population density a return to traditional forms of subsistence agriculture is evidently impossible.

Intensification and diversification are two sides of the same coin, and they were both stimulated by many factors mentioned above. As the irrigated area increased, double cropping with wet-rice became an option where that had not been the case before. However, the possibility of growing a nonrice crop after wet-rice—in other words, diversification—was also often realized. The expansion of the irrigated area did not come a moment too soon, as by the 1960s—and occasionally even earlier—the limits of arable expansion in the lowlands and the lower slopes of the upland areas had been reached. Lateral expansion, therefore, was no longer possible, and intensification was the only way forward.

Diversification also occurred in the nonirrigated upland areas as investment in transport infrastructure was instrumental in improving connections between those areas and the markets. Thus (integral) swiddeners became more and more involved in production for supralocal markets, while lowlanders without (sufficient) land became (partial) swiddeners (see Chapter 8). When discussing the period between around 1870 and the 1960s, we have already seen that this development as a rule is environmentally rather damaging and nonsustainable.

One particular type of (upland) diversification, however, does appear to be sustainable: it has been termed *agroforestry*, of which there are various forms. Basically, agroforestry denotes a mixture of annuals and perennials, but in some cases agroforestry plots contain hardly any annuals. Often, plots cultivated in a slash-and-burn sequence are being planted with trees after a few years of annual crops. The trees bear fruit or can be tapped after a number of years and usually continue to be exploited for quite some time. Alternatively, permanent upland fields where "dry" crops are being cultivated are intercropped with economically interesting trees.

The term *agroforestry* is of a relatively recent date, and so are many instances of it in Southeast Asia. However, the phenomenon itself is much older, and it appears to have many roots, as was shown in the preceding chapters. In many slash-and-burn areas people often planted some perennials—particularly those producing fruits or nuts—in their swiddens in order to indicate ownership. In the nineteenth century we find that commercially interesting trees such as rubber were being planted on fallowed swiddens. We also encounter references to home gardens in the lowlands in historical sources on the nineteenth century, and it is difficult to believe that they were absent earlier. Such home gardens show the same mixture of perennials and annuals as do the agroforestry plots of today. There are also indications that lowlanders have often tried to keep patches of forested land located near their villages from being cleared, with bamboo and trees yielding firewood or timber. In addition, some tree species were grown in groves, or what could be called orchards if the trees concerned were fruit trees.

The latter two examples are clearly market oriented, and so is most agroforestry today. Some such plots in the uplands, for instance, produce fruit for the urban market or even for export. Now that transport between the uplands and the cities takes a fraction of the time it used to, this type of market gardening has become feasible and attractive. It is, moreover, environmentally healthful, as it stabilizes soils and counters erosion, which, in the vulnerable highlands, is an important factor. It would appear that much agroforestry production is sustainable as well.

The recent increase in the incidence of agroforestry is in keeping with data on the proportion of agricultural land under permanent crops. In Chapter 8 it was shown that this share increased between 1880 and 1980, a tendency that has been continued up to the present. Most if not all of this expansion takes place in the upland areas. As a rule, permanent crops (perennials) are export crops, and their growing share points at their continuing importance for the regional economy. In contrast, however, to what was observed during the colonial period, the growth of the area under permanent crops is no longer strongly linked to the development of plantation agriculture, as much of the recent expansion was undertaken by peasant smallholders.

Agroforestry and permanent crops in general are a boost for sustainable agriculture. That is the good news; the bad news is that the "invasion" of the uplands, described in Chapter 8, continues unabated. This is partly, perhaps even largely, the result of the unprecedented growth of the logging industry in many of the upland areas, a topic dealt with in the next chapter. Large-scale logging does two things for prospective cultivators—it creates access roads, and it clears the land for agriculture. And although, as we have seen, some forms of agriculture in the uplands are sustainable, most are not, thus leading to increased erosion and siltation of rivers and estuaries.

Large-scale logging and the invasion of the uplands that followed in its wake are also a threat to the way of life of "integral" swiddeners, and to the sustainability of their agricultural system. As forests are disappearing, the space available for forest-fallow shifting cultivation is dwindling. Fallow periods get shorter, thus rendering the system unsustainable. Although at the moment of writing, there are still some remote areas where swiddeners can pursue their way of life in a more or less untrammeled fashion, it is rather unlikely that this will still be the case around, say, 2050. But perhaps I am being overly pessimistic.

The same observation applies to the last groups of hunter-gatherers that were mentioned in Chapter 8. As the forests disappear, so do the peoples whose lifestyles depend on the possibility of hunting game, fishing, and foraging edible plants, nuts, berries, roots, and tubers. And if they do not literally disappear because they cannot adapt themselves to changed circumstances, they will certainly disappear as discrete groups.

In sum, it seems likely that all mobile rural ways of life will disappear in the not too remote future. The people concerned will become rural wage workers or sharecroppers, but they may also join the ranks of those who have mobile urban lifestyles—casual urban wage laborers and the denizens of the informal sector.

There are also limits to the share of the arable land that can be irrigated. Conflicts about water use and about the building of large dams already indicate that those limits have been reached locally. It cannot be expected, therefore, that the hectareage under irrigation will expand at the same rate as it has done over the past thirty to forty years, although some expansion is still possible and even likely.

In a few decades, therefore, we will see three agricultural systems. By far the largest will be the irrigated permanent fields in the lowlands and the lower slopes of the highlands. In addition to that, there will be an unsustainable system of permanent "dry field" cultivation in the uplands, and a sustainable one of agroforestry, also mainly in the uplands. It is to be hoped, and also to be expected, that the "dry field" system will eventually be abandoned, so that only two systems remain.

CONCLUSION

From the 1970s, Southeast Asia modernized very rapidly, as part of the so-called globalization process, and what in this book is called the fourth globalization phase. This process was highly uneven, and modern urbanites could and did live quite close to slash-and-burn agriculturalists and permanent-field cultivators living more traditional lives.

One of the features shaping this period is the high rate of population growth. However, it was also for the first time in perhaps two centuries or more that the growth rate started to slow down across the board, the result of the so-called Demographic Transition—dramatic decline of crude death rate, crude birth rate, and TFR, and the dramatic rise of life expectancy at birth. Nevertheless, there are still pockets of more traditional demographic behavior, as witness data on Cambodia and Laos.

However, the possibility that the death rate will rise again—for reasons other than an aging population—has to be considered. This could be caused—among other factors—by MDR bacteria strains, other new diseases, and the increasing influence of pollution on human health.

Not only were populations growing rapidly, but economies did so as well, even in terms of GDP per capita, a remarkable feat given the high population growth rates. This process was accompanied by shifts between the main sectors of the economy. The relative importance of agriculture, as illustrated in Table 10.2, dropped rapidly between 1960 and 2000, while the importance of the urban (industrial and service) sector increased in proportion. Southeast Asia, until the 1960s an overwhelmingly agrarian region, by 1995 saw its agricultural labor force reduced to 50 percent of the economically active population, while by the year 2000, the majority of the population of Malaysia and the Philippines had become urban. The economic transformation and the high growth rates were caused by rising external and internal demand for commodities, policies more conducive to economic growth, increasing rates of savings and investment, expanding education, industrialization, and urbanization.

In the meantime there were shifts not only between sectors but also within; in this chapter that has been demonstrated extensively for the agricultural sector.

There, the most important developments were the so-called Green Revolution, increasing investment in irrigation, mechanization, intensification (giving rise to almost continuously increasing yields per hectare), diversification, increased use of artificial fertilizer, pesticides, and herbicides, the continued invasion of the uplands, outmigration, and the shrinking share of agriculture in GDP and the labor force. As a rule, peasants became less self-sufficient, and agriculture had turned dirtier.

Intensification of "dry" upland agriculture is often unsustainable, but one particular type is not—the one that has been called agroforestry. Much agroforestry is market oriented, and has been made commercially viable because transport between the uplands and the centers of commerce has improved considerably over the last decades.

The world of 2006—the moment of writing—was, for the large majority of the people in Southeast Asia, a more prosperous world than it had been in the 1960s, but development had been uneven. Some countries had not participated as fully in this increased prosperity as had the "tigers." In the leading countries themselves, the proportion of people below the poverty line may have dropped, but as population growth was rather high, the total number of poor people did not drop as much, and may have even increased locally. Development had also come at a price, as is shown in the next chapter.

BIBLIOGRAPHICAL ESSAY

An introduction to the political history of the region during this period is Cheong, Yong Mun, 1999, "The Political Structures of the Independent States." Pp. 59–138 in *The Cambridge History of Southeast Asia*. Vol. 2, Part 2: *From World War II to the Present*. Edited by Nicholas Tarling. Cambridge: Cambridge University Press.

On demographic development from the 1970s onward, see Hirschman, Charles, and Philip Guest, 1990, "The Emerging Demographic Transitions of Southeast Asia," *Population and Development Review* 16, no. 1, pp. 121–152; *The Future of Population in Asia*. Honolulu: East-West Center (2002); Boomgaard, Peter, 2004, "Demographic Transition in Southeast Asia." Pp. 414–418 in *Southeast Asia: A Historical Encyclopedia, from Angkor Wat to East Timor*. Edited by Ooi Keat Gin. Santa Barbara: ABC-CLIO.

On multiple drug resistance, see McNeill, J. R., 2000, *Something New under the Sun: An Environmental History of the Twentieth-Century World*. New York: Norton (esp. pp. 201–205). On the rapid spread of diseases, see Bright, Chris, 1998, *Life out of Bounds: Bioinvasions in a Borderless World*. New York: Norton (esp. pp. 169–173); on HIV/AIDS, see Brown, Tim, 2004, "Tackling the HIV/AIDS Epidemic in Asia," *Asia-Pacific Population and Policy* no. 68, pp. 1–4.

On economic growth in Southeast Asia, the reader is referred to *The Asian Miracle: Economic Growth and Public Policy*. Oxford: Oxford University Press (1993); Booth, Anne, 2001, "Initial Conditions and Miraculous Growth: Why Is Southeast Asia Different from Taiwan and South Korea?" Pp. 30–58 in *Southeast Asia's Industrialization: Industrial Policy, Capabilities and Sustainability*. Edited by Jomo K. S. Houndmills: Pallgrave; Wie, Thee Kian, 2002, "The Soeharto Era and After: Stability, Development and Crisis, 1966–2000." Pp. 194–243 in *The Emergence of a National Economy: An Economic History of Indonesia, 1800–2000*. Edited by Howard Dick et al. St. Leonards, NSW, Australia: Allen and Unwin; Leiden: KITLV; Jomo, K. S., ed., 2003, *Southeast Asian Paper Tigers? From Miracle to Debacle and Beyond*. London: RoutledgeCurzon; *Key Indicators of Developing Asian and Pacific Countries 2004*. Vol. 35. Manila: Asian Development Bank.

On agriculture, see Kunstadter, Peter, Edward Charles Chapman, and Sanga Sabhasri, eds., 1978, *Farmers in the Forest: Economic Development and Marginal Agriculture in Northern Thailand*. Honolulu: University Press of Hawai'i; Soemarwoto, Otto, and Idjah Soemarwoto, 1984, "The Javanese Rural Ecosystem." Pp. 254–287 in *An Introduction to Human Ecology Research on Agricultural Systems in Southeast Asia*. Edited by A. T. Rambo and P. E. Sajise. Los Baños: University of the Philippines; Sajise, P. E., 1987, "Agroforestry and Land Tenure Issues in the Philippines." Pp. 273–276 in *Land, Trees, and Tenure*. Edited by J. S. Raintree. Nairobi: ICRAF and the Land Tenure Centre; Booth, Anne, 1988, *Agricultural Development in Indonesia*. Sidney: Allen and Unwin; Hardjono, Joan, 1991, *Indonesia; Resources, Ecology, and Environment*. Singapore: Oxford University Press; Conelly, W. Thomas, 1992, "Agricultural Intensification in a Philippine Frontier Community: Impact on Labor Efficiency and Farm Diversity," *Human Ecology* 20, no. 2, pp. 203–223; Eng, Pierre van der, 1993, "Agricultural Growth in Indonesia since 1880." Ph.D. dissertation, University of Groningen; Bass, Stephen, and Elaine Morrison, 1994, *Shifting Cultivation in Thailand, Laos and Vietnam: Regional Overviews and Policy Recommendations*. London: IIED; Uhlig, James, Charles A. S. Hall, and Tun Nyo, 1994, "Changing Patterns of Shifting Cultivation in Selected Countries in Southeast Asia and Their Effect on the Global Carbon Cycle." Pp. 145–199 in *Effects of Land-Use Change on Atmospheric CO_2 Concentrations: South and Southeast Asia as a Case Study*. Edited by Virginia H. Dale. New York: Springer; Elson, Robert E., 1997, *The End of the Peasantry in Southeast Asia: A Social and Economic History of Peasant Livelihood, 1800–1990s*. Houndmills: Macmillan; New York: St. Martin's; Hill, R. D., 1998, "Stasis and Change in Forty Years of

Southeast Asian Agriculture," *Singapore Journal of Tropical Geography* 19, no. 1, pp. 1–25; Que, Tran Thi, 1998, *Vietnam's Agriculture: The Challenges and Achievements*. Singapore: ISEAS; Suryanata, Krisnawati, 1999, "From Home-gardens to Fruit Gardens: Resource Stabilization and Rural Differentiation in Upland Java." Pp. 257–278, in *Transforming the Indonesian Uplands: Marginality, Power and Production*. Edited by Tania Murray Li. Amsterdam: Harwood; McNeill, J. R., 2000, *Something New under the Sun: An Environmental History of the Twentieth-Century World*. New York: Norton (esp. pp. 219–226); Falvey, Lindsay, 2000, *Thai Agriculture: Golden Cradle of Millennia*. Bangkok: Kasetsart University Press; Brookfield, Harold, 2001, *Exploring Agrodiversity*. New York: Columbia University Press; Potter, Lesley, 2001, "Agricultural Intensification in Indonesia: Outside Pressures and Indigenous Strategies," *Asia Pacific Viewpoint* 42, nos. 2/3, pp. 305–324; Gérard, Françoise, and François Ruf, eds., 2001, *Agriculture in Crisis: People, Commodities and Natural Resources in Indonesia, 1996–2000*. Montpellier: Cirad/Curzon; Lataillade, Camille de, Alexandre Dumontier, and Nicolas Grondard, 2002, *L'Agriculture des Philippines: La Plaine Centrale: Histoire et Perspectives*. Paris: Les Indes Savantes; Vien, Tran Duc, 2003, "Culture, Environment, and Farming Systems in Vietnam's Northern Mountain Region," *Southeast Asian Studies* 41, no. 2, pp. 180–205; Akiyama, Takamasa, and Donald F. Larson, eds., 2004, *Rural Development and Agricultural Growth in Indonesia, the Philippines and Thailand*. Canberra: Asia Pacific; Ruf, François, and Frederic Lançon, eds., 2004, *From Slash and Burn to Replanting: Green Revolutions in the Indonesian Uplands*. Washington, D.C.: World Bank.

11

DEPLETION OF NATURAL RESOURCES, POLLUTION, AND ENVIRONMENTAL AWARENESS

LOGGING

It takes a tropical old-growth forest a century or more to grow back to its original state after having been logged over. Sustainable exploitation of such a forest in this sense is virtually impossible under present-day conditions of high population density and commercial logging. We must be aware that the literature usually employs the term *sustainable*—in notions like Maximum Sustainable Yield (MSY; the amount of timber that can be cut annually without endangering future production)—in a much looser sense of the word. It means only that the logged-over areas should have grown back sufficiently for another logging operation; in Indonesia, forest regulations assume a rotation cycle of thirty-five years. But even if we use the term in this much more limited sense, Southeast Asia has long since passed the point that its forest exploitation was below or around its MSY level.

The Philippines was the first (large) country in the region in which the annual logging rate surpassed the MSY; this can be dated to around 1960. Looking at lower geographical levels of analysis, it is clear that areas like Java, Bali, and the Red River delta in Vietnam had already been largely deforested for a long time. However, although commercial logging had not been entirely absent in these regions, deforestation here was largely the result of high population growth rates. From around 1960 commercial logging became more important—and often much more important—than land clearing for agriculture because of population growth in many Southeast Asian areas.

High rates of commercial logging and timber exports, later also exports of plywood, pulp, and paper, were observed from the 1960s in many areas in the region, but operations were concentrated in the Philippines, the eastern states of Malaysia (Sarawak, Sabah), and the Indonesian part of Borneo (Kalimantan), while recently New Guinea (both the Indonesian part and Papua New Guinea) has been discovered by logging companies. The forests of this region—often

A local Iban tribesman, employed as a logger and carrying his chainsaw on his back, looks over a smouldering part of the forest, cleared and burned for a new plantation near Balaiberkauk, West Kalimantan in Indonesia in 1997. (AP/Wide World Photos)

called Malesia by botanists—have in common that dipterocarps (a family of very tall and straight hardwood trees) occur frequently, which makes it an attractive region to logging companies.

Large-scale logging after the War in the Pacific started in the Philippines in the 1950s, largely with U.S. capital, which mechanized an industry that up to that point had been rather artisanal. Much of the demand came from the United States in the early days, but soon Japan became a very important market for Southeast Asian hardwoods. In the following decades, Japanese demand for hardwood would be one of the most important underlying causes—if not the most important cause—of deforestation in the Philippines, Indonesia, Malaysia, and Papua New Guinea.

As a result of U.S., European, Japanese, and later also Chinese and Southeast Asian demand, very high logging rates have been obtained from around 1960 throughout the region, after having remained more or less stationary between 1930 and 1960. In Indonesia, for example, log production was less than 4 million cubic meters in 1960. Then, in the late 1960s, when the country used timber production to export itself out of its depressed economic circumstances, log production really took off, reaching almost 27 million cubic meters in 1978; it was estimated to be between 40 and 44 million around 1995. Recent estimates suggest that the total cut (including logging for plywood, pulp, and paper) is now over 60 million cubic meters. Indonesia's MSY, set at between 22 and 25 million cubic meters per year, was surpassed in the 1970s.

It should be mentioned, however, that the rate of logging has not been constantly and uniformly increasing across the region. Indonesia, for instance, declared a ban on log exports in the early 1980s, not so much for environmental reasons but because it wanted to stimulate the growth of a log-processing industry in the country. Nevertheless, this measure temporarily slowed down logging. Since the late 1980s, the production and export of plywood, pulp, and paper have more than made up for the lower export of logs. Thailand imposed a logging ban in the late 1980s, and although illegal logging and forest clearing for agriculture continued, this measure is assumed to have had a dampening effect on log production. However, as at the same time Thailand stepped up its timber imports from Indonesia, Malaysia, the Philippines, Myanmar, Laos, and Cambodia, it has simply exported its environmental problems (its ecological shadow or footprint, as this phenomenon has been termed). As a result, the total log production of Southeast Asia almost certainly did not decrease. The Philippines imposed logging restrictions in certain areas in 1992 and following years. Moreover, in many areas of Southeast Asia easily exploited forests have become scarce, and because logging has been undertaken unsustainably, and the laying out of tree plantations has not kept pace with the logging operations either, logging locally has had to slow down as a matter of course.

As was mentioned in the previous chapter, the various Southeast Asian states have certainly used part of the income (taxes, concession, logging, and reforestation fees) from these logging activities ("forest rent") they have "captured" for infrastructural investments, leading to economic development and increasing income per capita. However, they should have obtained a much larger proportion of the profits. For Indonesia it has been estimated that the state has captured only between 10 and 30 percent of the forest rent it should have received by rights. This was partly because of the large proportion of total logging that is carried out illegally; partly the fact that so many concessions have been given to friends and family members of highly placed politicians, bureaucrats, and army officers; and in part because the bureaucracy is not sufficiently efficient to peg those rents at a reasonable rate or to collect in full the fees and taxes on the statute books.

The economic and environmental costs of illegal logging, corruption, and the low rate of forest rent capture by governments were and are high. They led to lower prices for forest products (hence lower export income), inefficiency, and higher rates of logging than would have obtained if all logging had been legal and all forest rent had been appropriately captured.

Although, therefore, much of the phenomenal post-1960 increase in logging can be explained by increased internal and external demand—the lengthening ecological shadow of more developed or less forested countries—and by related investment and technological change (mechanization), faulty institutional arrangements are at least partly to blame for the scope of the onslaught. As a corollary, it could be, and has been, argued that the presence of readily exploitable resources, and the abundant "resource rent" that has become available as a result, have led to unfortunate developments in the institutional sphere. Such topics are being dealt with by a fairly recent discipline or subdiscipline called "political ecology." Some of these issues are mentioned here.

In the first place, Southeast Asian governments and forest services have continued the "territorialization" of the forests (creation of "political forests"), a process mentioned in Chapter 9. That has led to the displacement or marginalization of many indigenous peoples, often but not always tribal groups living in upland areas, mainly foragers and swiddeners. There were many instances of removal of forest-dwelling people from the areas in which they had been living, because those areas had been gazetted as "political forests." While prior to 1960 much forest reserve had existed only on paper, and therefore did not bother the local population, the increased interest in logging, the enormous sums of money involved, and the growing competition for "virgin" forests now meant that reserved forests on paper would soon turn out to be reserved forests "on the ground" as well, with all that entailed for the local population. The threat of displacement or the announcement of changes in land rights in many cases led to local protest and resistance, often supported by national or even international nongovernmental organizations (NGOs).

Another important feature of the enormous increase in income from logging was that it enabled autocratic rulers—such as Indonesia's Suharto and Marcos of the Philippines—to follow their own policies, if necessary going against their own bureaucracy, because captured forest rents provided them with the wherewithal to finance all kinds of pet projects off-budget. It was the old patrimonial state again, but now in a modern guise. As such projects almost always play a role in the various patron-client networks of which the ruler is the apex—which means that they are rarely economically sound undertakings—they usually have a negative impact on GDP, often in addition to being environmentally damaging. But these negative economic and environmental effects apart, it could be argued that the abundant and rather sudden availability of resource rent stimulated autocratic rule from the 1960s onward. This reminds us of the way in which increased income from trade during the Early-Modern Period reinforced absolutist kingship in the region (see Chapter 6).

Another (related) factor that seems to be inherent to the fairly sudden growth of large-scale logging after 1960 is that it appears to have destroyed rather well organized, efficient, and fairly independent institutions such as the various forest services. This is part of the explanation of why the—often adequate—forestry laws of those countries are honored mainly in the breach. Many Southeast Asian countries have forestry laws dealing with selective felling, natural regrowth, replanting, and other measures aimed at sustainable exploitation of wooded areas, and they specify fines for logging firms that do not comply with these regulations. Nevertheless, as many concessionaires are cronies of the president, the governor, the military commander, or congressmen, most logging enterprises do not heed those regulations, but nevertheless do not incur the punishment that the law stipulates.

Finally, it should be mentioned that illegal logging and related features like smuggling are not simply logging outside the law. Although it takes place outside the purview of the central government or the central forest department, illegal logging is almost always carried out with the connivance of local authorities, the difference from legal logging being that no formal concession has been acquired. In fact, illegal logging can be regarded as a form of off-budget financing of local government, or as the result of rent-seeking behavior by local dignitaries.

Although the experience of the Western world suggests that economic development, increasing income per capita, and the development of civil society go hand in hand, it cannot be said that economic growth (in Southeast Asia), at least when it appears to be based largely on the exploitation of natural resources, automatically, and in the short run, leads to a strengthening of civil society.

The environmental effects of the rapid increase in large-scale exploitation of the forests of Southeast Asia since the 1960s have been disastrous. This is most clearly demonstrated by the drop in the figures representing the proportion of the land surface area under forest cover in the various countries of the region. Thailand, for instance, witnessed a drop from almost 55 percent in 1960 to approximately 20 percent around 2000. During the same period forest cover in the Philippines dropped from 45 percent to 20 percent, while the comparable figures for Indonesia are about 70 and 50 percent. Forest cover in Cambodia and Laos is also around 50 percent today, but although reliable figures for 1960 appear to be missing, it may be assumed that the deforestation rate there was lower than in the other countries just mentioned. The percentage of present forest cover in Vietnam is close to that mentioned for Thailand and the Philippines, but its forest exploitation history differs at least partly from that of those two countries.

Much damage was done to the forests of Vietnam during the Vietnam War, when, for instance, the infamous chemical compound Agent Orange was used to defoliate entire forests.

The larger nations, therefore, today have forest covers ranging from 20 to 50 percent. In countries like the Philippines, Thailand, and Vietnam, logging will remain a feasible activity only if natural regrowth of logged-over areas and the laying out of tree plantations will be undertaken much more vigorously than hitherto. One thing is clear, however, and that is that forest exploitation will be considerably more expensive under those circumstances than it used to be, and that the times of more or less free gifts from nature (abundant forest rent) are over. The other countries still have higher forest cover percentages, but it is far from certain whether they will take adequate measures in order to keep their proportion of forest cover at that level.

The societal costs of selling off most of the forests are very high. In the first place, the loss of so much forest cover as was experienced by many Southeast Asian countries after around 1960 almost invariably leads to a loss of—at least local, but sometimes also regional or national—biodiversity. That is, for instance, the experience of the Philippines, where many endemic plant and animal species have been lost.

Such a high rate of logging has also led to greatly increased degradation of much of the remaining forests, which renders the forest cover percentages presented above rather optimistic estimates of the amount of land on which some quantity of wooded cover is still present, with varying degrees of canopy closure. This development and the increase of deforestation have both led to increased erosion, siltation, flash floods, and diminished water availability for irrigation.

Forest degradation as a result of logging activities has also been held responsible for the large forest fires that have been a frequently recurring feature of recent Southeast Asian history. It is now accepted by most researchers that such fires, though caused directly by droughts, which are often related to ENSO events (see Chapter 4), as was the case in 1982–1983 and 1997–998, have been reinforced by logging practices. It has also been suggested that during very dry periods fires were deliberately started in forested areas that had been reserved for oil palm and pulp plantations, in order to speed up the laying out of such plantations. It is now often assumed that these "enhanced" forest fires are an autonomous factor in the process of deforestation. Forest fires during droughts are not a recent phenomenon as such—they have been reported as early as during the double ENSO event of 1877–1878 in the ever-wet forests of Borneo—but the scale on which they have occurred during the last twenty-five years or so appears to be much larger than that of the forest fires of the more remote past.

The high rate of deforestation also has consequences for the production of CO_2, an important greenhouse gas, particularly if the soil after clearing is used for annuals and not for trees. In addition to the fact that forest clearing with fire produces CO_2, deforestation leads to the disappearance of old-growth forests, which are carbon dioxide sinks. Insofar as wet-rice fields have taken the place of forests, the production of another greenhouse gas, methane, has increased as well.

Finally, the sharp increase of logging and deforestation in Southeast Asia after 1960 has also led to the widespread dislocation of indigenous forest-dwelling peoples, who often had to give up their way of life, or at least were reduced to poverty. Increasingly, foragers and swiddeners have been and are still being crowded out or even physically relocated to other, usually less favorable, environments. That fate has been shared by many of those who happened to live in an area destined to become a national park, a watershed protection forest, or an artificial lake (for irrigation purposes).

This brings us to the topic of the legal rights of forest dwellers. As was shown in Chapter 9, colonial forest services often ignored such rights when they created "political forests," a practice that has been continued by the governments of the independent Southeast Asian nations. There is a growing literature on common or customary property rights, particularly in response to the "tragedy of the commons" discussion mentioned earlier. Today, a sharper distinction is made by scholars between "common property rights" (often called "communal rights" in the older literature) and "open access resources." It is now often stated that the latter tend to become overexploited, while common property management is advertised as the way of the future. Another, related, popular topic is comanagement of natural resources, also deemed to be a sustainable alternative to the present-day arrangements.

As we saw in Chapter 9, the current opinion of scholars appears to be that during the colonial period, individual (private) tenure was substituted for common property rights. It was shown there that at least in the case of the forests this opinion is mistaken, because, as a rule, state property rights ("sovereign domain," "eminent domain") were substituted for local forms of tenure, probably often common property arrangements, but according to some authors also individual tenure. One of the reasons that national parks, wildlife sanctuaries, watershed protection forests, and other forest reserves are so easily abused is that they are state property, and therefore regarded as open-access resources, particularly if and when they are not well patrolled.

It is much too early to present firm conclusions on these issues, but two points are worth making, the first being that most writings on common property management and comanagement are highly colored by wishful thinking and are largely untested; the second is that private tenure (in combination with a legal system that functions well, and a judiciary that is not corrupt) at least historically has better papers as regards conservation than does communal or state tenure.

Be that as it may, most observers will probably agree that another big problem regarding land tenure arrangements is that of the insecurity of tenure because property rights are unclear. If people cannot rest assured that they will reap the fruits of investments like planting perennials, terracing, maintenance of soil fertility, and the like, because there is no clear title to their land, they are much more likely to resort to environmentally unsound practices. Such lands will, moreover, be easily sold, at relatively low prices, leading to the clearing of new patches of forested land.

In this section, the emphasis has been on large-scale logging, which not only dominated much of the literature but was also more dominant as a source of deforestation than ever before. Although, as was shown earlier, logging for timber was certainly not absent prior to 1960, the scale was much more modest. As a rule of thumb it can be stated that throughout Southeast Asia before about 1960, although of course with local exceptions, deforestation was mainly the result of land clearing brought about by population growth, while after 1960 the activities of logging companies took care of the main share of deforestation, again with local exceptions. The logging companies were able to do this because there was a huge and rising demand for hardwood, particularly from Japan, but also because much (foreign) capital and new technology (chainsaws, bulldozers) became available, while national and regional governments supported their activities.

In the meantime, other causes of deforestation did not disappear overnight. Population growth was higher than ever before, and although urbanization, industrialization, and agricultural intensification absorbed some of the population increase, land clearing for subsistence agriculture continued locally more or less unabated. So did the growth of the area under permanent (export) crops, partly undertaken by plantations, with oil palm as the most conspicuous example, partly (and perhaps increasingly) also by smallholders. Shifting fairly large groups of people can be seen as part of demographic causes of deforestation, with as the most obvious example the Indonesian transmigration policy. It has been argued that transmigrants do more damage to local forest ecosystems than local people, as they are supposed to introduce agricultural practices—particularly wet-rice agriculture—unsuited to local circumstances, something that leads to rapid depletion of recently cleared soils, and therefore to further land reclamation and deforestation.

Finally, it has been argued that large-scale deforestation has led to drier local climates. This is a notion that existed already around 1800, as was mentioned in Chapter 7.

MINING

Much less has been written on the environmental effects of mining in Southeast Asia than on those of forest exploitation. Perhaps that is because the damage caused by mining operations is rather localized, and not as ubiquitous as the effects of logging.

Nevertheless, mining has a strong impact on the natural environment, as has been described in earlier chapters for periods prior to the 1960s. Since the 1960s the scale of operations has increased, as larger amounts of (foreign) capital and new technologies became available and were applied to the business of mining. The growth rates varied between countries, periods, and commodities. For instance, oil production increased in Indonesia and Brunei between 1960 and 1980 but stagnated thereafter, while in Malaysia and Vietnam oil production did not begin until the 1970s and 1980s, respectively, which was followed by high growth rates throughout the years to come. The production of oil was, of course, heavily influenced by price changes, such as occurred during and after the oil crises of the 1970s and early 1980s. In Indonesia nickel and bauxite exploitation showed high growth rates from the 1960s, but tin and coal, for instance, did not.

The most obvious and visible effect of increased mining is accelerated destruction of the landscape in the case of open-cast mining, as has been described earlier for traditional and modern small- and large-scale tin mining. As mineral deposits are often to be found in wooded areas, this is an important source of deforestation. This tendency has been reinforced recently by the proliferation of illegal mining, as, for instance, in the case of coal mining in Sumatra and Quang Ninh. Although this is small-scale mining, it is undertaken by so many people, and often with the help of modern technology (it is financed by people with capital to invest), that the effects are rather large-scale. During the last financial-economical and political crisis in Indonesia, starting in 1997–1998, this type of illegal mining increased rapidly, as did so many other kinds of natural resource exploitation—there are still "free gifts of nature" to be found when property rights are "open access," not clearly defined, or not controlled properly.

In addition, many mines polluted—and continue polluting—air, soil, and water (surface, underground, and coastal), mainly by discharge water, dust, and what is called overburden—that part of the minerals dug up that cannot be used. During the rainy season, and particularly during flash floods, the overburden may be—and has been—washed away, thereby destroying residential or agricultural land. Water pollution may and often does include the pollution of drinking water, thus creating health hazards.

Mining, as reflected in the recent literature, appears to be seen mainly as a social problem. New mines are a rather invasive presence in their new environment, and they are prone to upsetting local social, economic, and political balances. Of course, they upset the ecological balance too, but that point is rarely made in the literature. The new mines often ignore local land tenure arrangements, and by attracting (large) numbers of migrants, they create new communities, in which law and order are not always easily upheld. Such boomtowns, in turn, pose environmental problems, as they need drinking water and produce sewage and other types of waste.

Not seldomly, mines combined all three types of problems—landscape destruction; pollution of land, soil, and water; and social problems—as was the case with the Freeport mines in Irian/Papua. The U.S. transnational mining company Freeport-McMoRan has for a long time been operating (through PT Freeport Indonesia) the largest goldmine and the third largest copper mine in the world in Indonesian New Guinea, the so-called Grasberg-Ertsberg complex. The mine is operating in an area in which political trouble is never far away, as the separatist OPM (Organisasi Papua Merdeka; Free Papua Organization) is also active there. During the New Order period, the mine enjoyed protection from President Suharto, with whom Freeport signed a contract as early as 1967, and from the military, a state of affairs that turned from an asset into a liability after the fall of Suharto. As the conflict between the OPM and the Indonesian state brings the army straight into potential conflict with the separatists, Freeport is caught in the middle.

The two big problems of every hardrock mine—tailings and overburden—continue to trouble Freeport as well. At the Grasberg mill 95 percent or more of the ore processed ends up as tailings, and around the year 2000, about 230,000 tons per day were discharged into the local river system, or more than 70 million tons annually. The tailings are highly visible, and have caused widespread destruction in the lowlands. As one needs five tons of rock to be moved to extract 1.5 grams of gold, the company was moving more than 750,000 tons per day, of which approximately 230,000 were processed into tailings and the remainder dumped as overburden. Almost all overburden mined there is acid generating and has a high capacity to leach copper. As a result, the drainage and seepage from the overburden stockpiles contain high concentrations of copper and acid. The sheer size of the dumped quantities of overburden is another problem, of which the landscape aspect is just one, while the chances for slippage and landslides in this perhumid region are another. In the social sphere, the company was rather late in confronting the social resentment generated by the stark contrast between how much money the company was making from its local operations, and how little was plowed back. After riots in 1996, Freeport decided to spend much more money on compensation of original land rights and social development plans, but that led to new problems, such as how to distribute these sudden riches.

The 13,000-foot-high Grasberg mine contains the largest single gold reserve in the world, and large copper deposits as well. Operated by U.S.-based Freeport-McMoRan, the open-pit behemoth is a boon to the Indonesian economy, but has met fierce opposition from tribal groups who claim tailings have caused the flooding of their lands. (George Steinmetz/Corbis)

There are signs that some large foreign mining corporations, the objects of close scrutiny by environmental NGOs in their home countries, are trying to improve their track record. Kutubu oil field, operated by a subsidiary of Chevron, in Papua New Guinea, may serve as an example, particularly if we compare it with the oilfield of Salawati Island, off the coast of Indonesian New Guinea, operated by Indonesia's national oil company, Pertamina. While the latter was characterized by numerous oil spills, natural gas being burned off, and access roads cut through the forest so broad that they were too wide for many rain forest species to cross, the former had no oil spills at all, hunting and fishing by their personnel were forbidden, and every precaution had been taken to maintain biodiversity.

Finally, what has been said about the influence of resource rent acquired from forest exploitation on institutional arrangements applies also to government income from some types of mining, particularly when such rents constitute a large proportion of state income. That is the case, for instance, with oil and natural gas. Resource rent from these sources has very much reinforced the importance of patron-client ties, corruption, cronyism, and off-budget financing of economically and environmentally unsound projects. Although it has also enabled the state to finance important infrastructural, health, and education projects, much of it was not spent wisely, and should be seen as an abuse of natural resources. The well-known bankruptcy scandal of Pertamina, the Indonesian state oil company, in the 1970s is a nice illustration of mismanagement of income from natural resources.

Theoretically, the main difference between logging and mining is that wood is a renewable resource. Nevertheless, in everyday practice the difference does not seem very large. Oil, gas, coal, tin, bauxite, gold, silver, copper, diamonds, nickel, mica, and jade will eventually run out, and in the case of oil and gas that may be rather sooner than later. As forests are being exploited unsustainably, wood for timber will also run out within a few decades, unless drastic policy changes are undertaken. That also applies to fishing.

FISHING

Fishing in post-1960s Southeast Asia has attracted more attention than mining but not as much as logging. The story of its recent development is quite similar to that of both mining and logging, in the sense that its growth rates were very high—at least until the 1980s—growth rates driven by continually increasing demand, in the first place from Japan but also from the United States and Europe. More or less continuously growing demand was also generated by Southeast Asia itself, because of high rates of population increase and growing real income per capita. From the supply side, high growth rates were made possible by (foreign) investment and concomitant technical innovation, which led to a fairly rapid transition from small-scale, artisanal, nonmotorized fisheries, to large-scale, relatively high-tech motorized trawling.

Growth started in the Philippines prior to 1960, but in the remainder of Southeast Asia that did not occur until the 1960s or 1970s. In the 1990s, growth rates of fish landings slowed down or even stopped, as happened with the frontrunner, the Philippines. That was certainly not caused by slackening demand, but by depletion of fish stocks, something that was regarded as unthinkable even when the first signs were already discernible. This depletion, of course, is the result of overfishing, the occurrence of which is hardly a surprise given the fact that, for instance, Thailand has been catching fish far above its MSY from the late 1970s. This, in turn, was linked to high rates of investment in trawlers, a business that was so lucrative that relative outsiders ("Thai generals") had started to sink large amounts of money into the sector.

There are also, as with logging and mining, notable differences between countries, even if we disregard the fact that a landlocked country like Laos is almost by definition unlikely to do much sea fishing. As we have seen, the Philippines were early (as was the case with logging). The highest growth rates, however, were obtained in Thailand, which soon started fishing outside its own territorial waters, developing into a distant-water fishing nation. That was largely stopped when all countries declared a twelve-nautical-mile territorial zone in the early 1980s. Thereafter, Thailand concluded treaties with other countries for cooperation regarding the exploitation of well-stocked parts of their territorial zones.

There are also large differences in income and productivity between the small- and large-scale sea fisheries subsectors. Small-scale fisheries are responsible for a very large share of the sector in terms of labor and vessels, but a small share in terms of production. Although both subsectors take part in the overexploitation of marine resources, they do so in different ways. Whereas most users of small craft simply catch too many fish, and increasingly undersized ones at that, the large trawlers are trawling the bottom of the sea, thus destroying the habitats of many types of fish. However, small-scale fishermen can also be quite destructive, as is demonstrated by the widespread custom of blast and cyanide fishing on reef formations rich in fish species.

One of the problems with overfishing is that the first signs of declining catches are usually the beginning of a vicious downward spiral. When catches start to get smaller, small-scale fishermen tend to work longer hours in order to arrive at the same catch. This results, of course, in even faster depletion of fish stocks.

Overfishing has been stimulated by the fact that, in contrast to former times, there is now in fact a market for all species captured, because even those species that are not (yet) appreciated by humans, in addition to undersized fish—together collectively called "trash fish"—can now be used for the production of fish meal. Fishermen, therefore, can sell their entire catch.

There have been large shifts in the types of sea commodities in which fishermen were interested before and after around 1960, as well as in the role that various products of the sea played in the export sector. The share of sea products in total exports increased as well. Ever since refrigerating technology in the fisheries was greatly improved, exports of fresh fish and other sea commodities grew in importance, while salted fish, a notable export product at an earlier stage, was no longer in demand. Tuna is one of the popular fish species in recent exports, ending up as sashimi on Japanese tables. Shrimp and prawns are other popular trade items.

There have also been shifts from one (potentially) commercially interesting species to another following stock depletion. For instance, when in the Gulf of

Thailand stocks of predatory fish had been reduced, squid stocks increased, and catching methods were adopted accordingly. Thus fishing made for environmental change, while environmental change, in turn, dictated the changes in capture methods. Such shifts also illustrate how resources are "created": as the stocks of well-known fish for consumption dwindle, less well-known species are promoted for consumption (such as tilapia); thus fish species that could not really be regarded as commodities or resources were rather suddenly "elevated" to that status.

Another important transition is that from the capture of demersal species to that of pelagic ones. Demersal fish are those living on the bottom of the sea, and, as we saw earlier, their habitat has been severely damaged by trawling. Therefore, the fishing fleets now have to shift to pelagic species—that is, those living in the open sea, close to the surface or at middle depth.

A very important shift that took place from around 1980 is the one from capture to culture. Although what is now called aquaculture has a long history in Southeast Asia, its spectacular growth dates from the last twenty-five years or so. Along many coastal areas, fishponds were established hundreds if not thousands of years ago. People would enclose small patches of coastal mudflats with small dikes and would thus retain the fish, shrimp, and other creatures of the sea that would be naturally deposited there at high tide. In inland areas people had freshwater fish ponds, while others were cultivating fish on inundated wet-rice fields, as was mentioned earlier. All of these forms of aquaculture have been around for quite some time, but the scale was modest and it was probably not strongly commercialized. Recently, however, a combination of depletion of certain stocks and soaring demand in the richer countries made for an aquacultural boom. One of the products most often mentioned in the literature is shrimp for the Japanese market, particularly black tiger shrimp.

Aquaculture is a booming business, but it has encountered and caused various problems, some of which are of an environmental nature. One of these problems is that aquaculture is carried out preferably in those coastal zones where mangrove vegetation occurs naturally, which is then cleared. Besides leading to deforestation, the conversion of mangrove swamp is also a threat to natural stocks of fish and other sea animals, as mangroves are their spawning grounds.

Another problem is that the industry is polluting. We are talking about densely stocked ponds, in which large quantities of feed and chemicals are deposited and equally large quantities of waste are produced. Aquaculture is also highly sensitive to diseases that can produce large quantities of dead fish.

Having dealt with recent broad trends in fisheries, a few lines are in order about a rather exceptional phenomenon. I am referring to a village on the island of Lembata in the Indonesian province of Nusa Tenggara Timur (what used to be called the Lesser Sundas) that routinely hunts large-toothed whales like sperm whales and killer whales, in addition to various porpoises, pygmy whales, and the giant manta ray. It is the only such village in Southeast Asia. In the eighteenth and nineteenth centuries large-scale, particularly English and U.S., whalers, like the one described by the novelist Herman Melville in his famous novel *Moby Dick*, were also to be found in these waters, but today the only competition comes from an occasional illegal Japanese factory fishing ship. Although we lack historical data going far back in time, it appears likely that the local population has been hunting whales for a long time, and has done so sustainably.

Most of the literature on fisheries concentrates on marine capture fisheries and aquaculture. Not much is written about inland open-water fisheries, although that sector is arguably the largest of them all, at least in terms of labor if not of production and productivity. Not much has been published about depletion of stocks in the inland fisheries, but it is hard to imagine that they would not suffer from the same limitations as the marine fisheries.

It would appear, therefore, that depletion of stocks in the fisheries sector is at an advanced stage, perhaps more advanced than it is in the forestry and mining sectors. What applies to forest resources applies to fish stocks as well—they are in principle renewable, but if they are not exploited sustainably, there is not much difference from nonrenewable resources.

One of the symptoms of resource depletion is conflict, for instance, between small-scale fishers and large-scale trawlers; such confrontations, not seldom of a bloody character, have been far from rare in the region over the last decades. That is one of the factors that have locally led to bans on trawling.

One of the problems the fishing sector is increasingly confronted with is mining and quarrying of sand and coral. The use of coral for roads, house foundations, city walls, and embankments dates back to ancient times, but, as so often, it is the much larger scale on which these resources are now being exploited (in addition to the bleaching of corals owing to warmer sea water and blasting them apart with dynamite by fishermen) that has turned these activities into environmental hazards.

Another large problem is water pollution. The odds that fish from Manila Bay, Jakarta Bay, or the Gulf of Thailand are full of chemicals harmful to humans are now much higher than they used to be. Minamata disease—a disease caused by eating fish with high mercury content—was observed for the first time in Japan in the 1960s, but since the 1980s similar symptoms of mercury poisoning have been observed in Southeast Asia. It is to the topic of pollution that we now turn.

POLLUTION

Although not an entirely new phenomenon, pollution of air, soil, and water in Southeast Asia was not widely perceived as a problem prior to the 1960s, or even the 1970s. In hindsight we can say that some pollution must have taken place then, but that it was a fairly localized problem, restricted largely to the bigger urban centers and their downstream areas, and to a number of mines. An exception should be made for the production of the greenhouse gases carbon dioxide and methane, produced by burning forest vegetation and by inundated wet-rice fields, respectively; prior to the Industrial Revolution, however, the production of such gases was entirely unproblematic.

As a major problem in Southeast Asia, pollution postdates the 1960s. It appears to be the inevitable companion of "early" industrialization, advanced urbanization, high productivity in agriculture and aquaculture, and high rates of deforestation and mine exploitation.

Air pollution is a familiar phenomenon to those who have lived in or even visited any big city in Southeast Asia. Even during the dry season, the sky appears to be permanently overcast. That is largely the result of the use of fossil fuels by cars and factories, in addition to the continued use of traditional nonfossil fuel (firewood, charcoal) by households and small-scale artisanal producers. It is the end result of high rates of population growth, even higher rates of urbanization, increased industrialization, and the growing use of coal, oil, and gas per capita. Occasionally, the smoke of forest fires is added to all of this. Thus Singapore is often in trouble when the forests of Sumatra and Borneo are burning, as was the case in 1997–1998. Air pollution as such is not a novel phenomenon; some of it occurs naturally, as is the case with spontaneous forest fires and volcanic eruptions; also, from the time of the earliest swiddeners, human activity has added to the total amount that went up into the air. The scale, however, is new.

This situation is steadily getting worse. Already by 1990 the "smog" blanket that covered big urban conglomerations like Jakarta, Manila, and Bangkok was probably as bad as, or even worse than, it was in cities like Los Angeles (in summer) or as it used to be in London until the 1950s (in winter). The air pollution in Bangkok in 1990 was 1,664 tons per square kilometer per year, while at that time it was 379 in Detroit and 258 in Amsterdam. Inhaling these fumes more or less constantly is undoubtedly a health hazard, probably particularly for the very old and the very young.

A specific form of air pollution is recorded inside industrial establishments. In all kinds of factories, both the older ones, producing, for instance, fertilizer, and the very modern factories that produce computer components, workers

Petronas Towers shrouded by thick haze in Kuala Lumpur, Malaysia. Malaysian skies were covered in thick smog as a result of smoky haze from local raging bushfires and forest fires in Indonesia's Sumatra province in 2005. (Corbis)

inhale a plethora of poisonous fumes generated by the production process. Examples are various strong solvents, and lead fumes produced during soldering. As cheap labor is one of the main attractions for investors in Southeast Asia, the inclination to take protective measures is small.

It should be added that the production of greenhouse gases (CO_2, CH_4, N_2O; some writers include CFCs) has also increased strongly in Southeast Asia over the last decades. Carbon dioxide is produced naturally, but increasing numbers of small and large forest fires and the strongly increased burning of fossil fuels have made for a much greater emission. Increased levels of methane are the result of the expansion of wet-rice cultivation. Nitrous oxide is emitted mainly from certain fertilizers and from animal manure. Chlorofluorocarbons (CFCs), finally, were used as propellants in aerosol spray cans, and in the 1970s they were identified as a major cause of the decay of the stratospheric ozone layer, which protects life on earth from ultraviolet-B radiation.

Water pollution is an equally serious or even more serious problem. Of course, water in rivers and estuaries has always been somewhat polluted, even before the arrival of humans on the scene, because particularly during the rainy season runoff from the uplands contained large amounts of organic matter. When people did arrive, and after hundreds of thousands of years started to live in fairly large concentrations, human waste locally caused more pollution. That was particularly problematic in times of cholera and other waterborne diseases, because the same (river) water that was used as a dump for waste was also the location for washing, cleaning, and drinking.

Although, therefore, water pollution as such is not new, what changed from the 1960s are the quantity and composition of the pollutants. The growth of Southeast Asia's megalopolises over the last decades has led to an appalling increase of human excreta being deposited into rivers and thus into the estuaries of the big rivers running through these cities. As well-functioning water treatment plants are still rare in Southeast Asia, fecal matter from humans enters watercourses and eventually coastal waters as raw sewage.

Research in East Java, dating from the 1990s, has shown that nitrate contamination is a major source of well pollution. The source of this buildup does not appear to be nitrogen fertilizer as applied to rice fields, but rather human feces, caused by toilets and septic tanks not connected to a closed sewage system. This is a serious health hazard, and not one easily detected.

As deforestation has also increased dramatically, so have erosion and sedimentation. Further development of open-pit mining and of nonsustainable agriculture in the uplands have had the same effect, and in most coastal waters already high sediment loads have more than doubled. Another "mining" product that since the 1970s ends up in sea is oil, particularly as a result of oil spills from tankers.

In addition, runoff containing toxic waste from industrial establishments and residues from artificial fertilizer, pesticides, and herbicides is not only polluting various surface waters, but perhaps even underground aquifers and wells. Moreover, increased use of nonbiodegradable detergents is contributing to the rapidly dropping water quality.

As a result of all this many larger and smaller rivers now have very low dissolved oxygen levels. This means that the self-purification capacity of these rivers is seriously compromised, and that they are virtually dead.

As was mentioned above, one of the consequences of this state of affairs is that people get sick if they use this water for consumption. Although tap water is supposedly purified, and being connected to the water mains is generally regarded as a good thing and a sign of modernity, there have been many complaints throughout the last decades about the quality of the drinking water in big cities. It often smells and tastes bad, and in the 1980s living larvae were found in tap water in the city of Surabaya, eastern Java. But all kinds of dangerous chemicals are tasteless and odorless and are therefore consumed unnoticed. Heavy metal pollution in estuaries downstream of Southeast Asia's megacities is especially high from lead, cadmium, mercury, copper, and zinc, a situation that was already present in the 1980s.

Another consequence, also mentioned earlier, is that, with the growing pollution of watercourses and seas, the consumption of fish and other aquatic creatures of inland waters and also of the coastal areas and estuaries has become increasingly risky. This is a serious development for societies in which fish constitutes a very large, perhaps even the largest, share of animal protein in the human diet.

In addition to all of this, it must be mentioned that freshwater, of whatever quality, is getting more and more scarce. Water, like air, used to be regarded as a free good, but that is no longer a valid perception in most parts of Southeast Asia. Irrigation claims a large share of the available water resources, while water is also used for hydroelectric power plants. One of the almost ubiquitous conflicts in Southeast Asia is over the pricing of tap water, of which increasing quantities are used for drinking, bathing, and washing in the constantly expanding cities.

A few words about pollution of the soil will have to suffice, the more so as it was just mentioned that residuals of human manure, artificial fertilizers, pesticides, fungicides, and herbicides used in agriculture can be harmful to humans. These residuals are found not only in effluent but of course also in soils. The harmful effects also apply to livestock, left grazing on fields where these substances have been used. Problems could be exacerbated by the use of industrial sludge from water treatment plants, which has value as organic fertilizer.

Another big problem is the existence and continuous growth of waste dumps close to big cities. Incineration of solid waste would be better, but it is regarded as too expensive. These growing landfills are bound to lead to widespread pollution of the soil (in addition to that of air—methane—and water). Such places are homes to large numbers of scavengers who make a living from collecting and selling everything that is of any value (metal, paper, plastic). The upside of this is that many materials are thus being recycled; the downside is that it is a rather unhealthful occupation.

TOURISM

Finally, tourism should be mentioned separately as an important and growing source of pollution, if only because many tourists may not be aware of it. Tourism has become a major source of income for countries like Malaysia, Thailand, and Indonesia, but even countries with an infrastructure less geared toward tourism, such as Cambodia, are increasingly being targeted by tour operators. As income per capita increased in the developed countries, including the Asian ones, people had more time and money for rest and relaxation. At the same time, prices for air travel were more or less stable (in nominal terms), and as income per capita in most of these countries was increasing, airfare in effect became relatively less expensive. Moreover, many large travel organizations offered very inexpensive package tours, attracting even larger groups of potential holidaygoers. Thus large numbers of people with fairly modest incomes were able and willing to go to out-of-the-way places, where up until the 1960s only the well-to-do had been vacationing.

As a fair proportion of tourists are interested mainly in sun, beaches, and huge quantities of cheap food and drink (and not seldom, sex), the main requirement for attracting vacationers appears to have been the construction of large concrete hotel and restaurant buildings, in addition to dance halls and shopping malls. This has led—as it did in the older tourist resorts in Europe and the United States—to what could be called "pollution of the horizon," the creation of large, ugly semiurban agglomerations, mainly along the seashores, thus destroying what must have attracted tourists in the first place—a beautiful landscape. In addition it has created problems of waste disposal, while it puts a great deal of pressure on sources of drinking water. In both cases this can be detrimental to the local populations.

This is the beginning of a vicious downward spiral, because the more discerning tourists start avoiding these overcrowded and ugly tourist centers, leaving such places to the more boisterous type of tourists. More attractive, as yet unspoiled and cleaner places then start attracting the more adventurous and solitude-seeking tourists, and thus the process of construction frenzies and increasing pollution starts anew.

In addition to all of this, much of the capital earned is leaving the locality, and therefore does not benefit the local community, while it is not a rarity that tourist resorts crowd out those who were earning a living on the same spot, such as fishermen.

A fairly recent phenomenon is the development of ecotourism. Ideally, it attracts an environmentally aware type of tourist, willing to forgo some of the luxuries of the run-of-the-mill tourist resorts. Small-scale, nonpolluting operations are envisaged, with a fair share of the earnings going to local people.

It will be clear that the combined and cumulative effects of all of these forms of pollution must be increasingly detrimental to the health and well-being of many Southeast Asians, even though, to my knowledge, those effects do not yet show up in mortality figures. There are some attempts to counter the continuing spread of pollution, as will be shown presently, but the growing legislation on this score is honored mostly in the breach and often with impunity.

ENDANGERED AND EXTINCT SPECIES

Another way in which logging, fishing, mining, urbanization, pollution, and the like have influenced the natural environment is that they threaten the survival of many species of plants, and terrestrial and marine animals. Locally, many species have either become rare, and are now in danger of disappearing, or have even become extinct. Much of this has happened during the last forty years or so, although, as we have seen, even during the nineteenth century some species had become extinct.

Internationally, most attention usually goes to the plight of large mammals. They are flagship species for many nature reserves, and generate much of the international interest in these areas. They usually enjoy protected status, at least officially. It is well known that elephants have a difficult time in most Southeast Asian areas, and that locally their numbers are dwindling. The plight of the tigers of Southeast Asia is also well publicized, although it is usually either the (Indian) Bengal tiger or the Siberian tiger that most people will have heard of, as a result of the many reruns of documentaries that can be watched on Discovery, National Geographic, and Animal Planet channels. This also applies to animals like the rhinoceros, wild buffalo, wild cattle, and the orangutan. Although several species of boar and deer are also locally threatened, they are probably deemed less exotic, and therefore do not appear to have captured the imagination as much as the larger mammals.

Few people appear to feel sorry for the fate of various Southeast Asian non-mammals such as crocodiles or snakes, although they, too, are often becoming rare. Nor does one hear much about the possible extinction of fish, birds (birds of

A tiger rests in water. Exploitation of resources in Southeast Asia has threatened the survival of many native species, including the tiger. (Corel)

paradise apart), or plant species, let alone insects and other small fry. Talking about marine species, it is probably only the plight of the dolphins, as by catch of tunafish, and the turtles that has evoked a response among the public at large.

Generally speaking there are two main factors that are threatening the survival of flora and fauna—hunting (poaching) and loss of habitat. Hunting was locally a drain on the number of species in the more densely settled areas even before the 1960s. It is undertaken as a subsistence activity by hunter-gatherer groups and by swidden cultivators, while sedentary peasants often set traps in or around their fields. It is also a pastime for well-to-do urbanites. As in so many cases, population increase and technological development (better guns and traps) have turned activities that were often undertaken on a sustainable basis into something that threatens the survival of many species.

Hunting as a pastime, for the kick and for trophies, is no longer officially permitted inside national parks and other nature and wildlife reserves. But if those in power—often the military—want to hunt, they do so anyway. Hunting and trapping for subsistence or for the protection of crops is usually permitted outside the reserves. Although in most Southeast Asian countries animals on the so-called Red List—the International Union for Conservation of Nature and Natural Resources (IUCN) list of endangered species—are protected, many of these animals are to be found in cages or as "bush meat" in local markets. In addition, some animal parts, such as rhinoceros horn and the meat and bones of tigers, are much sought after by Chinese merchants, who will pay large amounts of money for such items. Well-to-do Chinese consumers pay enormous sums for "medicine" made from those parts, supposed to be an aphrodisiac. As tigers and rhinos are getting rarer by the day, the amounts being paid for them keep increasing, which makes it even more attractive to kill them. Therefore prices for tiger- and rhino-based aphrodisiacs go up, enhancing their reputation and making them even more sought after. It is one of the most perverse feedback mechanisms imaginable.

The international trade in animals, dead or alive, can be regarded as part of the threat posed by hunting and collecting. Various Southeast Asian countries have signed the Washington Treaty of 1973, also called the Convention of International Trade in Endangered Species of Wild Fauna and Flora (CITES), but compliance with the treaty leaves much to be desired.

Fishing should be regarded as the marine equivalent of hunting and has the same effects, although it is often less immediately obvious. Up to a certain level normal reproductive processes repair the damage done by fishing, hunting, trapping, and collecting. For ages, human hunting and fishing stayed within those limits, but now that people have numerically expanded so much, the pressure of hunting and fishing has gone beyond those levels and the threat of extinction is never far away.

The fact that so much game and fish are captured illegally is partly a question of poverty—the people concerned do not have all that many alternatives—and partly of faulty law enforcement. That, in turn, has mainly two components—corruption, and a lack of personnel.

As a cause of extinction, habitat loss might be more important than hunting or collecting. As forests and other "waste" lands are disappearing, specific animals and plants lose the ecological niches they need for their survival. Large mammals in particular need forests for shelter, and partly for the food they find there. Habitat degradation is another important feature in the process of extinction of large mammals. For instance, if, in a certain area, all deer and wild boar have been hunted down, tigers can no longer survive, even if the area has retained much of its plant cover. Fragmentation of habitat is another factor; if the places with forest cover are too few and far between, animals will no longer be able to go from one patch to another, which will make it difficult to encounter sexual partners and will encourage inbreeding, which in the end will lead to degradation of the species. It also renders recolonization after an epidemic more difficult. Through the lack of genetic variation, such groups are more sensitive to natural hazards, which then threaten the survival of the species. One of these hazards is probably the fact that wild animals have come increasingly into contact with domesticated ones. The latter are often the carriers of disease, which are then caught by the former. It is generally assumed that small populations are eventually doomed, owing to a lack of genetic variation.

Logging, by the way, is responsible not only for the rapid disappearance of wild animal habitat but also for easier access to hitherto virtually inaccessible areas. The best defense of game animals has always been the fact that they were difficult to find, but now logging roads are an invitation to hunters and a source of "road kills."

Let me finish this otherwise rather gloomy section on a positive note. Since 1990 five "new" large to fairly large mammal species have been "discovered"— local people had known them all along, of course—in the upland areas of the Vietnam-Laos border lands (four) and in Myanmar (one). These are the Vu Quang ox (*Pseudoryx nghetinhensis*), the Truong Son muntjac (*Muntiacus truongsonensis*), the giant muntjac (*Megamuntiacus vuquangensis*), the Tainguen civet (*Viverra tainguensis*), and the leaf muntjac (*Muntiacus putaoensis*). Perhaps it is hard to believe that this was possible in a region with such overcrowded lowlands. It once again demonstrates the still amazing differences within the one region, even after so many centuries of centralization, modernization, and globalization. Strangely enough, the fact that these animals, which were captured and killed only by indigenous hunters, had become rare did not appear to be a cause for concern to the local populations.

GLOBAL WARMING

All sections so far have dealt with specific Southeast Asian features of otherwise global phenomena. Global warming, as the term indicates, is a global phenomenon, and Southeast Asian global warming would be a nonsensical notion. However, one cannot really avoid dealing with it in an environmental history textbook, if only in order to speculate about its possible effects upon the region. A few lines on this by now so well known phenomenon will have to suffice.

Most scholars now agree, and have done so for quite some time, that an important component of the rising average global temperature over the last fifty years or so has been anthropogenic (that is, caused by human activity) (see Chapter 4). The rise in temperature is caused partly by the already mentioned accelerated production of greenhouse gases, mainly carbon dioxide and methane. As we have seen,

Southeast Asia produces its share of these gases, albeit a small one compared with many other regions. If the whole world decided overnight to stop driving cars, setting fire to forests, using fossil fuels in industrial processes, and cultivating wet-rice, global warming would not be halted until ca. 2050. Given the fact that it is highly unlikely that such a decision is about to be taken, it must be expected that the rise in temperature will continue for a much longer period.

One of the most important consequences for the region is that sea levels will rise in response to global warming, because the polar ice caps (particularly Antarctica) are declining, as are the ice cap of Greenland and many mountain glaciers. This implies that many coastal areas will be inundated, while many small islands—particularly atolls—will disappear entirely. It is not unlikely that cities built in low-lying coastal areas, such as Jakarta, will be experiencing increased flooding, and eventually might have to be evacuated.

Much more speculative is the effect of global warming on weather patterns. For instance, it has been suggested that typhoons may be increasing in strength and frequency, linked to the increasing seawater temperature. If that is true, typhoon-prone areas, such as much of the Philippines, would be in for a rough time. At the time of this writing, however, there is no conclusive evidence for such an occurrence.

Another suggestion is that the strength of ENSO events would be increasing as a result of global warming. Now it is true that there have been a number of particularly strong El Niños recently, such as 1982–1983 and 1997–1998, but it would appear that the 1877–1878 ENSO event was even stronger. Again, therefore, evidence appears to be inconclusive.

But even if this were not to be confirmed by future research, the combination of less precipitation owing to deforestation, and higher temperatures on average because of global warming, might make for more and longer droughts, although one also has to factor in the effects of higher rates of evaporation.

ENVIRONMENTALISM, ENVIRONMENTAL LEGISLATION, AND NGOS

As has been shown, environmental concerns in Southeast Asia have been aired from the late eighteenth century, albeit on a very modest scale. From the late nineteenth century more people became aware of problems with the natural environment caused by human actions. This led to the birth of organizations and journals interested in conservation, and to government measures designed to limit environmental damage. Thus a legal framework was created that later generations could elaborate upon. However, the 1930s Depression, the War in the Pacific, and the postwar struggles for independence pushed environmental concerns to the background; between, say, 1940 and 1970 or even 1980, not much was undertaken in this field.

It can be said that during and shortly after the 1960s people in Europe and the United States became increasingly concerned about the ways in which humans treated nature. Several publications dating from those years, by now regarded as environmental classics, together constituted an environmental wake-up call. Mentioned should be Rachel Carson's *Silent Spring* (1962), Garrett Hardin's article "Tragedy of the Commons" (1968), and the 1972 report to the Club of Rome, entitled *The Limits to Growth*. Taken together, these publications warned the public at large about worldwide pollution and the unbridled exploitation of natural resources.

Dating from the same period are organizations like Friends of the Earth (1969) and Greenpeace (1970), while in 1972 the UN Conference on the Human Environment was held in Stockholm, in which developing countries participated as well.

At much later dates environmental issues were brought to the forefront by the Brundtland Report on sustainability (*Our Common Future*), dated 1987, and the UN Conference on Environment and Development, the so-called Earth Summit, in Rio de Janeiro, held in 1992. The latter conference was not a great success, but it was nevertheless a remarkable event if only because Prime Minister Mahathir of Malaysia vociferously represented the view of most developing countries that "the South" could not be expected to stop exploiting their renewable resources (particularly forests) just because "the North," after having used up most of their own natural resources, were increasingly concerned about the environment in general and about global warming in particular.

Although there were, as always, differences between the Southeast Asian countries in how and when they reacted to these wake-up calls, it seems fair to say that, as a rule, not much was done prior to the late 1970s. Given the fact that "development" was then still a fairly new and enthusiastically pursued policy in most Southeast Asian countries, it is not really amazing that "environment" was not yet an important item on the political agenda, aside from the fact that in many respects Southeast Asia's environmental problems were less serious than they were in many developed countries.

As an example of incipient environmental concerns in Southeast Asia, mention can be made of the fact that in Indonesia the first minister for the environment was appointed in 1978. In 1982 the Basic Environmental Management Act came into being, while in 1986 environmental impact analysis was introduced. Environmental study centers were established at state universities. The first national parks were also declared in the 1980s. Most Southeast Asian countries have put similar measures on their statute books since then.

Southeast Asian countries also participated in what has been called the "associational revolution" in global politics, a term used for the increased and still increasing importance of NGOs. Again citing examples from Indonesia, we find founding dates around 1980, when the two most important national Indonesian environmental NGOs were established—WALHI (Indonesian Environmental Forum), which was founded in 1980; and SKEPHI (NGO Network for Forest Conservation in Indonesia), which came into being in 1982.

However, those two organizations can hardly be regarded as representative of the environmental NGOs in Southeast Asia. Many NGOs, probably a large majority, are small-scale and represent very local interests. In fact, it has been argued that one of the differences between "Western" and Asian NGOs is that the latter are often one-issue groups representing exclusively purely local interests, while the former are often interested in global issues.

One of the most important features of Southeast Asian NGOs is that they constitute a grassroots phenomenon in countries in which democratic representation is often still in its infancy. For various reasons, environmental NGOs have often enjoyed greater latitude under the autocratic regimes of Southeast Asia than other organizations with critical views, partly because even dictatorial regimes wanted to look "modern" in the eyes of the world by appearing to be alive to environmental concerns.

To a certain extent NGOs can be seen as products of globalization. They sprang up as a reaction to growing global environmental awareness, responding to problems caused by globalization (depletion of resources, environmental destruction, and pollution). They do so in a truly globalized fashion, given the fact that national and local NGOs are often supported and financed by international NGOs, while the latter acquire their information—and partly their legitimation—from the former. People somewhere in the middle of a Southeast Asian jungle, threatened by loggers, are aware of antilogging campaigns in India and Brazil, and thus their actions may be inspired by activities halfway across the globe, just as the logging itself was.

Environmental awareness, legislation, and activism in Southeast Asia, therefore, have developed over the last twenty-five years or so. Nevertheless, future prospects for environmentally sound policies and the implementation thereof are not exactly rosy. It is not hard to explain why. It is the well-known mixture of weak law enforcement, a weak legal system, corruption, and a lack of democratic accountability. But even if all of that would improve considerably, continuing—though now somewhat decreasing—population growth and growing real income per capita would still take their toll.

Measures that must be taken in order to arrive at lower pollution levels are costly, and as much of the advantages of industry in the region are based on low wages and prices, governments and businesses alike are loath to add to their costs through measures that would protect the environment and the laborers.

Measures designed for the conservation of nature are costly as well, as they require large numbers of personnel for the actual staffing of reserves. Moreover, as it occurs with sad regularity that these personnel are bribed by logging companies, adding more people to patrol the reserves is obviously not sufficient. Besides, if policing is undertaken rigorously, conflicts with the local population might ensue.

According to some research, it is also possible that regional (Asian) environmental awareness is not up to scratch. Many people from the region are concerned about the environment, but only in a very practical sense (pollution is harmful), and not because they are convinced of the intrinsic value of "nature." At the same time, they see environmentally sound actions as something that must be undertaken by the government, not something they themselves should be doing. Now, of course, this does not mean that there are no "nature lovers" among the Southeast Asians, but it could be that there are not enough of them.

Awareness is not a static given; it changes over time, and it seems likely that it varies with the duration of the experience with environmental problems close to home. It might be the case that those who expected Southeast Asians to be more environmentally motivated are the victims of two fallacies—that Southeast Asians are "close to nature," and that people who are close to nature— whatever that may mean—are therefore more environmentally motivated. I think it safe to say that the notion that Southeast Asians are close to nature is based on a colonial cliche, while experience in the West indicates that those who were the furthest removed from nature—that is, the urban middle classes—were the first, perhaps apart from the nobility, to develop a profound appreciation for nature. In that case it may be just a matter of time before Southeast Asia moves in the same direction.

It remains to be seen whether the solution to the other problems—weak law enforcement, a weak judicial system, corruption, and a lack of democratic accountability—is also a matter of time, and whether that will be sufficiently timely to be of any use to the environment. For the time being, the most promising development is the slowing down of the Southeast Asian population growth rates.

BIBLIOGRAPHICAL ESSAY

The best recent studies on forest exploitation in the region during the last forty years or so are Dauvergne, Peter, 1997, *Shadows in the Forest: Japan and the Politics of Timber in Southeast Asia*. Cambridge: MIT Press; Ross, Michael L., 2001, *Timber Booms and Institutional Breakdown in Southeast Asia*. Cambridge: Cambridge University Press.

Other (fairly) recent studies on this topic include Hickey, Gerald Cannon, 1982, *Free in the Forest: Ethnohistory of the Vietnamese Central Highlands, 1954–1976*. New Haven: Yale University Press; Dargaval, John, Kay Dixon, and Noel Semple, eds., 1988, *Changing Tropical Forests: Historical Perspectives on Today's Challenges in Asia, Australasia and Oceania*. Canberra: CRESS; Poffenberger, Mark, ed., 1990, *Keepers of the Forest: Land Management Alternatives in Southeast Asia*. West Hartford, CT: Kumarian; Kummer, David M., 1991, *Deforestation in the Postwar Philippines*. Chicago: University of Chicago Press; Aiken, S. Robert, and Colin H. Leigh, 1992, *Vanishing Rain Forests: The Ecological Transition in Malaysia*. Oxford: Clarendon; Peluso, Nancy Lee, 1992, *Rich Forests, Poor People: Resource Control and Resistance in Java*. Berkeley: University of California Press; Peluso, Nancy Lee, Peter Vandergeest, and Lesley Potter, 1995, "Social Aspects of Forestry in Southeast Asia: A Review of Postwar Trends in the Scholarly Literature," *Journal of Southeast Asian Studies* 26, no. 1, pp. 196–218; Psota, Thomas, 1996, *Waldgeister und Reisseelen: Die Revitalisierung von Ritualen zur Erhaltung der Komplementären Produktion in Südwest-Sumatra*. Berlin: Reimer; Hirsch, Philip, ed., 1996, *Seeing Forests for Trees: Environment and Environmentalism in Thailand*. Chiang Mai: Silkworm; Parnwell, Michael J. G., and Raymond L. Bryant, eds., 1996, *Environmental Change in Southeast Asia: People, Politics and Sustainable Development*. London: Routledge; Ascher, William, 1998, "From Oil to Timber: The Political Economy of Off-Budget Development Financing in Indonesia," *Indonesia* 65, pp. 37–61; Hirsch, Philip, and Carol Warren, eds., 1998, *The Politics of Environment in Southeast Asia: Resources and Resistance*. London: Routledge; Magno, Francisco, 2001, "Forest Devolution and Social Capital: State-Civil Society Relations in the Philippines," *Environmental History* 6, no. 2, pp. 264–286; Sellato, Bernard, 2001, *Forest, Resources and People in Bulungan: Elements for a History of Settlement, Trade, and Social Dynamics in Borneo, 1880–2000*. Bogor: CIFOR; N.a., *The State of the Forests: Indonesia*. Bogor: Forest Watch Indonesia/Global Forest Watch (2002); Pierce Colfer, Carol J., and Ida Aju Pradnja Resosudarmo, 2002, *Which Way Forward? People, Forests, and Policymaking in Indonesia*. Washington, D.C.: RFF; Obidzinski, Krystof, 2003, *Logging in East Kalimantan, Indonesia: The Historical Expedience of Illegality*. Unpublished Ph.D. dissertation, University of Amsterdam; Top, Gerhard van den, 2003, *The Social Dynamics of Deforestation in the Philippines: Actions, Options and Motivations*. Copenhagen: NIAS; Tuck-Po, Lye, Wil de Jong, and Abe Ken-ichi, eds., 2003, *The Political Ecology of Tropical Forests in Southeast Asia: Historical Perspectives*. Kyoto: Kyoto University Press/Trans Pacific; Boomgaard, Peter, David Henley, and Manon Osseweijer, eds., 2005, *Muddied Waters: Historical and Contemporary Perspectives on Management of Forests and Fisheries in Island Southeast Asia*. Leiden: KITLV;

Kathirithamby-Wells, Jeyamalar, 2005, *Nature and Nation: Forests and Development in Peninsular Malaysia*. Copenhagen: NIAS; Resosudarmo, Budy P., ed., 2005, *The Politics and Economics of Indonesia's Natural Resources*. Singapore: ISEAS.

Particularly on forest fires the following texts can be consulted: Cribb, Robert, 1998, "More Smoke than Fire: The 1997 'Haze' Crisis and other Environmental Issues in Indonesia." Pp. 1–13 in *Ecological Change in Southeast Asia*. Edited by Annamari Antikainen-Kokko. Åbo: Åbo Akademi University; Gellert, Paul, 1998, "A Brief History and Analysis of Indonesia's Forest Fire Crisis," *Indonesia* 65, pp. 63–85.

On land tenure, see Van Meijl, Toon, and Franz von Benda-Beckmann, eds., 1999, *Property Rights and Economic Development: Land and Natural Resources in Southeast Asia and Oceania*. London: Kegan Paul; Richards, John F., ed., 2002, *Land, Property, and the Environment*. Oakland: ICS; Eaton, Peter, 2004, *Land Tenure, Conservation and Development in Southeast Asia*. London: RoutledgeCurzon.

The literature on mining after ca. 1960 is more concerned with social aspects, and much less with environmental problems; nevertheless, see Robinson, Kathryn May, 1986, *Stepchildren of Progress: The Political Economy of Development in an Indonesian Mining Town*. Albany: State University of New York Press; Wah, Loh Kok, Francis, 1988, *Beyond the Tin Mines: Coolies, Squatters and New Villagers in Malaysia, c. 1880–1980*. Singapore: Oxford University Press; Horsnell, Paul, 1997, *Oil in Asia: Markets, Trading, Refining and Deregulation*. Oxford: Oxford University Press; Sinh, Bach Tan, 1998, "Environmental Policy and Conflicting Interests: Coal Mining, Tourism and Livelihoods in Quang Ninh Province, Vietnam." Pp. 159–177 in *The Politics of Environment in Southeast Asia: Resources and Resistance*. Edited by Philip Hirsch and Carol Warren. London: Routledge; [Erman], Erwiza, 1999, *Miners, Managers and the State: A Socio-Political History of the Ombilin Coal-Mines, West Sumatra, 1892–1996*. Unpublished Ph.D. dissertation, University of Amsterdam; Evans, Geoff, James Goodman, and Nina Lansbury, eds., 2002, *Moving Mountains: Communities Confronting Mining and Globalisation*. London: Zed; Leith, Denise, 2003, *The Politics of Power: Freeport in Suharto's Indonesia*. Honolulu: University of Hawai'i Press; Resosudarmo, Budy P., ed., 2005, *The Politics and Economics of Indonesia's Natural Resources*. Singapore: ISEAS; Diamond, Jared, 2005, *Collapse: How Societies Choose to Fail or Survive*. London: Penguin (on mining, see pp. 441–468).

For recent information on post-1960s fishing, see first and foremost Butcher, John G., 2004, *The Closing of the Frontier: A History of the Marine Fisheries of Southeast Asia c. 1850—2000*. Singapore: ISEAS; Leiden: KITLV. Other recent studies are Rice, Robert C., 1991, "Environmental Degradation, Pollution, and the Exploitation of Indonesia's Fishery Resources." Pp. 154–176 in *Indonesia: Resources, Ecology, and Environment*. Edited by Joan Hardjono. Singapore: Oxford University Press; Gomez, Edgardo D., 1993, "Coastal, Inshore and Marine Problems." Pp. 268–277 in *South-East Asia's Environmental Future: The Search for Sustainability*. Edited by Harold Brookfield and Yvonne Byron. Tokyo: United Nations University Press; Barnes, R. H., 1996, *Sea Hunters of Indonesia: Fishers and Weavers of Lamalera*. Oxford: Clarendon; Backhaus, Norman, 1998, "Globalisation and Marine Resource Use in Bali." Pp. 169–192 in *Environmental Challenges in Southeast-Asia*. Edited by Victor T. King. Richmond: Curzon; Osseweijer, Manon, 2001, *Marine Resource Use and Management in the Aru Islands (Maluku, Eastern Indonesia)*. Unpublished Ph.D. dissertation, University of Leiden; Novaczek, Irene, et al., eds., 2001, *An Institutional Analysis of Sasi Laut in Maluku, Indonesia*. Penang: ICLARM-World Fish Center; Semedi, Pujo, 2001, *Close to the Stone, Far from the Throne: The Story of a Javanese Fishing Community, 1820s—1990s*. Unpublished Ph.D. dissertation, University of Amsterdam; Daerden, Philip, ed., 2002, *Environmental Protection and Rural Development in Thailand: Challenges and Opportunities*. Bangkok: White Lotus; Hal, Derek, 2003, "The International Political Ecology of Industrial Shrimp Aquaculture and Industrial Plantation Forestry in Southeast Asia," *Journal of Southeast Asian Studies* 34, no. 2, pp. 251–264; Boomgaard, Peter, David Henley, and Manon Osseweijer, eds., 2005, *Muddied Waters: Historical and Contemporary Perspectives on Management of Forests and Fisheries in Island Southeast Asia*. Leiden: KITLV; Dutton, Ian M., 2005, "If Only Fish Could Vote: The Enduring Challenges of Coastal and Marine Resources Management in Post-*reformasi* Indonesia." Pp. 162–178 in *The Politics and Economics of Indonesia's Natural Resources*. Edited by Budy P. Resosudarmo. Singapore: ISEAS.

There is a large and growing body of literature on pollution, but much of it is rather technical. Some recent titles relevant for social science students are Sicular, Daniel T., 1992, *Scavengers, Recyclers, and Solutions for Solid Waste Management in Indonesia*. Berkeley: Center for Southeast Asia Studies, University of California; Widianarko, B., K. Vink, and N. M. van Straalen, eds., 1994, *Environmental Toxicology in South East Asia*. Amsterdam: VU University Press; DiGregorio, Michael, et al., 1995, *Recent Urbanization and Environmental Change in Viet Tri City, Vietnam*. Berkeley: Institute of International Studies, University of California; Tapvong, Churai, 1995, "Environmental Economics and Management: Water Pollution Control in Thailand." Pp. 178–195 in *Counting the Costs: Economic Growth and Environmental Change in Thailand*. Edited by Jonathan Rigg. Singapore: ISEAS; Hirsch, Philip, ed., 1996, *Seeing Forests for Trees: Environment and Environmentalism in Thailand*. Chiang Mai: Silkworm; Hirsch, Philip, and Carol Warren, eds., 1998, *The Politics of Environment in*

Southeast Asia: Resources and Resistance. London: Routledge; Lucas, Anton (with Arief Djati), 2000, *The Dog Is Dead so Throw It in the River: Environmental Politics and Water Pollution in Indonesia: An East Java Case Study*. Victoria: Monash Asia Institute, Monash University.

On tourism (and environment) in Southeast Asia, see Hitchcock, Michael, Victor T. King, and Michael J. G. Parnwell, eds., 1993, *Tourism in South-East Asia*. London: Routledge; King, Victor T., ed., 1998, *Environmental Challenges in Southeast Asia*. Richmond: Curzon; Hirsch, Philip, and Carol Warren, eds., 1998, *The Politics of Environment in Southeast Asia: Resources and Resistance*. London: Routledge; Harron, Sylvia, and Philip Daerden, 2002, "Alternative Tourism and Adaptive Change in Northern Thailand." Pp. 71–102 in *Environmental Protection and Rural Development in Thailand: Challenges and Opportunities*. Edited by Philip Daerden. Bangkok: White Lotus.

Some recent (and not so recent) titles on endangered species (and conservation) are Sumardja, Effendy A., and Zahrial Coto, eds., 1984, *Wildlife Ecology in Southeast Asia*. Bogor: SEAMEO-BIOTROP; Santiapillai, Charles, and Kenneth R. Ashby, eds., 1987, *The Conservation and Management of Endangered Plants and Animals*. Bogor: SEAMEO-BIOTROP; Cribb, Robert, 1988, *The Politics of Environmental Protection in Indonesia*. Clayton: Monash University Centre of Southeast Asian Studies; MacKinnon, Kathy, 1992, *Nature's Treasurehouse: The Wildlife of Indonesia*. Jakarta: Gramedia; Cubitt, Gerald, and Belinda Stewart-Cox, 1995, *Wild Thailand*. London: New Holland; Sukumar, R., 1992, *The Asian Elephant: Ecology and Management*. Cambridge: Cambridge University Press; Quammen, David, 1996, *The Song of the Dodo: Island Biogeography in an Age of Extinction*. New York: Touchstone; Schepke, Anja, 1996, *Wenn die Wirtschaft die Umwelt bedroht: Stand und Perspektiven des Umweltschutzes in Vietnam*. Berlin: Deutsch-Vietnamesische Gesellschaft; Whitten, Tony, Roehayat Emon Soeriaatmadja, and Suraya A. Afiff, 1996, *The Ecology of Java and Bali*. Singapore: Periplus (esp. pp. 691–830; see also the other volumes in the Ecology of Indonesia Series); N.a., *A Study on Aid to the Environment Sector in Vietnam*. Hanoi: Ministry of Planning and Investment/United Nations Development Programme, Vietnam (1999), pp. 65–99; Boomgaard, Peter, 2001, *Frontiers of Fear: Tigers and People in the Malay World, 1600–1950*. New Haven: Yale University Press; Karanth, K. Ullas, 2001, *The Way of the Tiger: Natural History and Conservation of the Endangered Big Cat*. Stillwater: Voyageur; Nijman, Vincent, 2001, *Forest (and) Primates: Conservation and Ecology of the Endemic Primates of Java and Borneo*. Wageningen: Tropenbos International; Daerden, Philip, ed., 2002, *Environmental Protection and Rural Development in Thailand: Challenges and Opportunities*. Bangkok: White Lotus; Sinha, Vivek R., 2003, *The Vanishing*

Tiger: Wild Tigers, Co-predators and Prey Species. London: Salamander; Knight, John, 2004, *Wildlife in Asia: Cultural Perspectives*. London: RoutledgeCurzon.

On environmentalism, environmental legislation, and NGOs in Southeast Asia, see Kato, Ichiro, Nobuo Kumamoto, and William H. Matthews, eds., 1981, *Environmental Law and Policy in the Pacific Basin Area*. Tokyo: University of Tokyo Press; Hirsch, Philip, ed., 1996, *Seeing Forests for Trees: Environment and Environmentalism in Thailand*. Chiang Mai: Silkworm; Kalland, Arne, and Gerard Persoon, eds., 1998, *Environmental Movements in Asia*. Richmond: Curzon; Leon-Bolinao, Lou R. de, 1998, "Environmental Policies of the Post-War Philippine Presidents, 1946–1996." Paper presented at the International Conference on Southeast Asia in the 20th Century, Quezon City; Colombijn, Freek, 1998, "Global and Local Perspectives on Indonesia's Environmental Problems and the Role of NGOs," *Bijdragen tot de Taal-, Land- en Volkenkunde* 154, no. 2, pp. 305–334; Cribb, Robert, 2003, "Environmentalism in Indonesian Politics." Pp. 37–43 in *Towards Integrated Environmental Law in Indonesia?* Edited by Adriaan Bedner and Nicole Niessen. Leiden: CNWS; A special issue on environmental consciousness in Southeast and East Asia was published in *Southeast Asian Studies* 41, no. 1 (2003); Resosudarmo, Budy P., ed., 2005, *The Politics and Economics of Indonesia's Natural Resources*. Singapore: ISEAS; Bryant, Raymond L., 2005, *Nongovernmental Organizations in Environmental Struggles: Politics and the Making of Moral Capital in the Philippines*. New Haven: Yale University Press; Tsing, Anna Lowenhaupt, 2005, *Friction: An Ethnography of Global Connection*. Princeton: Princeton University Press.

CONCLUSION

POPULATION GROWTH AND ECONOMIC CHANGE

Evaluating the evidence presented in the preceding chapters, I think it is fair to say that population growth and economic change, the latter brought about largely by the development of trade, can be regarded as the prime motors of environmental change in Southeast Asia. Environmental change, in turn, influenced population growth rates and the scope and composition of commodity flows. Thus the environmental history of the region can be seen as a chain of interlocking changes.

This is not the place for a lengthy discussion on the relationship between population growth and economic development—often but not always trade-induced—in Southeast Asia. But, as these two factors are among the main explanatory variables in this book, a few words on this topic are in order. Was Southeast Asian population growth driven by economic (and technological) development, or was it the other way around, and was population growth the motor—or at least, one of the motors—of the economy? It may not come as a surprise that this book presents indications for both possibilities.

Population growth has always been the main driving force of (lateral) agricultural expansion, which in a preindustrial society is the main motor of (quantitative) economic growth. Growing population density led to (at first limited) urbanization, diversification, and specialization, or, in other words, structural economic development, which must have resulted locally in a modest increase of real income per capita (often called Smithian growth). Here, obviously, economic growth was generated by population growth.

On the other hand, a number of economic developments were evidently set in motion by outside forces, of which foreign demand for local commodities was the most important, with the introduction of foreign crops and livestock at least a very good second. In the preceding chapters it is argued that both factors were, directly or indirectly, conducive to higher rates of population growth. The laying out of wet-rice fields by monarchs and monasteries, the introduction of Western medicine by colonial regimes, and the increasing demand for labor generated by Western capital, are all examples of changes in the (political) economy that stimulated higher rates of population growth.

It seems likely that trade, throughout Southeast Asian history, has stimulated population growth (directly or indirectly), while population growth, in turn, stimulated trade. External demand for commodities usually created new economic opportunities, even if those who produced the commodities seldom received the full market price for their pains, and fertility appears to have been positively related to such opportunities (demand-for-labor). When colonial states took the place of the European trading companies, they attempted to bring about higher birth rates and lower mortality. These policies appear to have functioned as intended, and population growth rates increased, which made for stagnating and even dropping wage rates from 1900 onward. Low labor costs, in turn, attracted outside capital and technology, creating more and cheaper commodities for the international market, thus stimulating long-distance trade. Thus the two prime drivers of environmental historical developments in the region, population growth and trade, appear to have been intimately interlinked.

PHASES OF GLOBALIZATION

Both population growth and the development of long-distance trade can be thought of as having been part of a long process of globalization, in which I distinguish four stages or phases. These phases also had floral, faunal, and microbial components.

The first globalization phase started around the beginning of the Common Era and lasted until the fourteenth century. Long-distance trade developed between Europe (until the fifth century with the Roman Empire, later with various Mediterranean polities), India, and China on the one hand, and Southeast Asia on the other. Population growth rates were very low. Various alien plants (crops), animals (livestock), and diseases reached Southeast Asia. I am not arguing that there was no long-distance trade prior to the beginning of the Common Era, nor that inter-regional trade during this phase was uninterrupted. I call it the first globalization phase because during this period long-distance maritime and overland trade begin to play a much greater role in these societies, among other things in being instrumental in turning chiefdoms into early states; these states, in turn, facilitated long-distance trade.

The second globalization phase coincided with the Early-Modern Period in Southeast Asia. As we have seen in the relevant chapters, long-distance trade increased considerably, not only with Europe but also with India, China, and the Americas. From America came large numbers of new crops and other plants (the Columbian exchange), while Southeast Asia's participation in the "civilized disease pool" increased as well. Many of the American food crops were enormously successful. With this phase, globalization expanded and accelerated, as did its impact on the Southeast Asian environment. Population growth, while still very low by modern standards, may have increased somewhat in comparison with the first globalization phase. At the tail end of the period (the century or so prior to 1870), population growth in some areas was certainly higher. Foreign capital, technology, and people (migration) were imported on a much larger scale than before. States started to rely for their income to a large extent on foreign trade and traders, and those, in turn, were instrumental in strengthening the position of the rulers of the core areas, among other things with the help of modern weapons, thus ushering in (or at least strengthening) the absolutist state in Southeast Asia.

The third globalization phase coincided largely with the colonial period and its aftermath, from around 1870 to 1970. Planted on the substratum of an indigenous peasantry and indigenous states, foreign investment, technology, crops, medicine, and people (civil servants and soldiers, but also coolies and plantation administrators) taken together made for high growth rates of exports, both in absolute terms and as a proportion of GDP, particularly between 1870 and 1930. Population growth accelerated almost constantly, largely because of Western medical progress but possibly also as a response to growing economic opportunities and the continuing expansion of wet-rice cultivation. As a result of foreign demand, plantation agriculture made its appearance on a much larger scale than before and grew at a higher rate than subsistence agriculture, while the large-scale mining of tin, coal, and oil started as well. The influx of alien plants and animals (bioinvasions) continued. Many important export crops were, and still are, originally exotics. A number of new diseases (Spanish flu, plague) kept death rates from falling more.

Finally, from the 1960s or 1970s, the fourth globalization phase—the one for which the term *globalization* was originally coined—took hold in Southeast Asia. Population growth rates were higher than ever before, but at the end of the period a slight drop in the growth rate could be observed, a result of the demographic transition. This fourth phase also witnessed unprecedented rates of economic growth and of real income per capita, developments at least partly fueled by growing exports, not only to the traditional importing regions but now also to Japan, China, Taiwan, and South Korea. The combined environmental impact of trade-induced economic activities like plantation agriculture, export-oriented smallholder agriculture, mining, and particularly logging, finally came to overshadow the effects of land clearing resulting from population growth.

During this last phase, pollution and depletion of resources became major concerns, the price to pay for progress, so to speak. As was the case elsewhere in the world, these two were the bedfellows of industrialization, modernization, and urbanization.

Stage theories have been heavily criticized by historians and (other) social scientists during the last four decades or so, and are thoroughly out of fashion. I hasten to add, therefore, that the distinction of four stages of globalization is mainly an attempt to summarize the findings of this study in a handy way, and one that is easily remembered. They are, in other words, heuristic and mnemonic models. By definition, therefore, they highlight certain features while paying much less attention to others. The globalization phases I propose here should not be thought of as periods of uninterrupted or ubiquitous growth. During all of the four phases there were ups and downs, often of a considerable magnitude, while economic and population growth rates were much higher in some areas than in others (where negative growth was not exceptional), as was emphasized in the various preceding chapters.

The hypothesized existence of these stages of globalization is certainly not the only conclusion that can be drawn from the material presented in the preceding pages, but it does draw the reader's attention to a number of important features and to various links between them.

A WORLD SYSTEMS APPROACH

However, for an additional description and explanation of what went on during these four stages, I propose to look at a more dynamic model, for which the so-called world systems approach appears to be appropriate. This model, of which historian Immanuel Wallerstein is the spiritual father, suggests that since the sixteenth century, the so-called European world economy has dominated economic life on this planet. It is a system consisting of a core area in Northwestern Europe, a periphery in the Americas, and a semiperiphery in Central and Eastern Europe. These areas were connected by trade links that were directed from Europe and financed with European capital, while they determined the modes of production and the labor relations in the three types of areas. The core was characterized by free labor; chattel slavery was important in America, while serfdom dominated in Central and Eastern Europe. Asia and Africa remained outside this world economy until around 1750.

After Wallerstein's books were published in the 1970s, his European world economy thesis, while generally well received, was criticized on various points. There are three points we are interested in here. The first is the question of whether the European world economy (or any world economy) was not older than the sixteenth century. Secondly, it should be discussed whether it really was a *European* world economy. And finally, it is a debatable point that Asia remained for such a long time outside this world economy, as I argued earlier.

There are good reasons to push the date of the emergence of a world economy back in time, although, of course, the Americas would be out of any world economy prior to 1492. How far back is a matter of opinion, but if we recall that the Roman Empire was influencing production in Southeast Asia by way of its demand for commodities as early as the first century CE, a much earlier date than 1500 seems to be called for. But if these admittedly rather modest trade flows should be regarded as too meager a foundation for a world economy, the period of the origins of the early states in Southeast Asia, around the fifth century, would be another serious candidate—if we accept the suggestion that these states were partly the products of long-distance trade flows.

It also seems to be increasingly clear that we should talk about a Eurasian world economy rather than a European one, because the economic role of China, and perhaps that of India as well, has been underestimated for earlier epochs in the older historiography. I am not denying that Europe was probably ahead of China and India in various respects, at least during the Early-Modern Period, but that seems to me insufficient grounds for playing down the role of China and India as (additional) motors of the world economy. We would then have a system with at least two and perhaps three core areas—Western Europe, central and southeastern China, and possibly northern India. All three areas were relatively densely populated, and therefore locations of Smithian growth, with early industrial sectors, at a relatively early stage dominated by "free" wage labor. All three areas had sophisticated economies and state systems, large mercantile fleets, rich merchants-cum-capitalists, and a hunger for luxury products and, as time went by, raw materials; they were increasingly unable to feed their populations from their own soils, however, as growing numbers of people and increasing amounts of soil were switched to more remunerative economic activities. They were, therefore, in need of imports of foodstuffs, in exchange for exports of industrial products, capital, and their population surplus.

For these imports and exports the core areas of the Eurasian world economy needed (usually sparsely populated) primary producing peripheries and semiperipheries, and it seems logical to regard Southeast Asia as one of the (semi)peripheries of that system. This suggestion is based on the fact that, as I have explained earlier, demand generated by long-distance trade, including that coming from Europe, influenced patterns of collection and production in even the most remote hinterlands of Southeast Asia, while it was also instrumental in the emergence of (early) states in the region. Southeast Asia was not the collection of "hermit kingdoms" prior to the colonial kiss of life in the nineteenth century that it has been made out to be.

While external demand influenced what would be collected or produced in Southeast Asia, at an early stage it had to depend on local labor relations to get the job done. We are not entirely sure how this production was organized, as it is not a well-researched topic. Slavery and serfdom were widespread, and monarchs and local rulers used corvee labor for their wars and their infrastructural projects. As most long-distance trade was arranged trade, in which the role of the ruler was pivotal, surplus expropriation for the international market was usually undertaken through traditional channels, which implies that use was made of some form of bonded labor, corvee labor included, even if in all probability in many instances money was paid to the direct producers, or, at the very least, the producers were (partially) freed from other corvee obligations.

It is not unlikely that the total labor burden increased somewhat, locally leading to longer hours worked. At the same time, under the influence of foreign demand for commodities, slavery increased both in the cities and in rural areas where no (or insufficient) corvee labor was available. The demand for slave labor, in turn, stimulated slave raiding and the persistence of institutions like debt-bondage.

It should also be pointed out that there was always some free labor to be had, but often in insufficient quantities and badly suited to the type of work that was required. Among the free laborers, foreigners, more often than not Chinese, were overrepresented. Chinese workers often started out as indentured laborers, imported by Chinese merchant-capitalists, but after working off their contract obligations they became part of a growing pool of free labor. It was not until the late nineteenth century or even the early twentieth century that slavery and serfdom were officially abolished in the region, while at the same time fairly high rates of population growth and immigration had led to a much more abundant supply of free labor at modest and often decreasing cost. Nevertheless, indentured labor, slavery, and serfdom persisted locally until the 1920s and occasionally even after the War in the Pacific.

Consequently, Southeast Asia prior to, say, 1900, does fulfill the requirements, posed by the world economy model, that labor in the (semi)periphery be servile, partly because of labor requirements generated by the demand for commodities from the core areas. To be sure, most servile institutions were indigenous, but they were reinforced by long-distance trade.

To complicate matters further, we should also distinguish core and peripheral areas within Southeast Asia. In this book I have argued repeatedly that Southeast Asia was not a homogenous region. Generally speaking, the distinction can be made between the lowlands, the middle altitudes, and the uplands, but in practice a lowland-upland dichotomy suffices to explain much of the many environmental, socioeconomic, and cultural differences in the region. On the one hand there are the densely settled wet-rice–growing plains and valleys, constituting the core areas of the various states, as time went by often centered around a city or a number of cities. Here the rural population consisted of smallholding peasants living in permanent villages and cultivating permanent fields. The mountainous hinterlands of these core areas were sparsely populated with "tribal" foragers and swiddeners, while they were either politically independent or at least far from fully integrated in the adjacent core state. The upland areas were heavily forested throughout most of Southeast Asia's history, while the lowlands had lost much more of their forest cover.

This is, again, a model, and I can think of some ecosystems that do not really fit this dichotomy, such as very sparsely inhabited lowland swamps. As a rule, however, the lowland-upland dichotomy appears to be a handy heuristic device.

These cores and (what I propose to call internal) peripheries within Southeast Asia were more or less self-perpetuating systems. While the original differences were based on environmental givens, the perpetuation of these systems was brought about by a mechanism that I have called a two-track demographic development, with low population growth rates in the uplands and high growth rates in the lowlands and the mid-altitudes. The uplands could be kept "traditional" for a long time because contact with the lowlands was difficult and even dangerous, a result of different disease regimes; capital accumulation was restricted, as was population growth, while the surplus population was siphoned off by slave raiding or through migration to the lowlands. The Southeast Asian core areas needed the internal peripheries for their natural resources, their manpower, and their livestock, while the internal peripheries needed the core areas for prestige goods and legitimation and as places where they could always find some (paid) work during periods of bad harvest or other disaster.

However, there has always been a countercurrent. People fled to the highlands if the lowland ruler was too oppressive, and the introduction of foreign crops like maize, cassava, sweet potatoes, coffee, and tobacco (and at a very early stage pepper) made life in the higher regions attractive. Today, with the construction of all-weather roads, the destruction of many forests, and the invasion of landless lowlanders into the uplands, one wonders how long these peripheries inside the Southeast Asian periphery will persist.

We thus have arrived at a model of a Eurasian world economy, with its roots in the Pre-Modern Era, comprising three core areas (Europe, India, China). One of its (semi)peripheries was Southeast Asia, where state formation, economic life, population growth, and environmental change were all partly shaped by the demand for commodities from the core areas. The core areas needed the Southeast Asian periphery for its raw materials and its foodstuffs, and as an outlet for their capital and surplus population. Southeast Asia was interested in the industrial products of the core areas, their money, their technology (including firearms), and their migrants.

The Southeast Asian periphery itself consisted of core areas and internal peripheries, which were also mutually dependant, and whose relationship reflected that between the three Eurasian core areas and the Southeast Asian periphery.

The Eurasian core areas generated a growing demand—albeit with ups and downs—for commodities from Southeast Asia. Part of that demand was met by the Southeast Asian core areas—particularly rice—but a great many commodities had to come from Southeast Asia's internal periphery: almost all forest products and minerals, and crops such as tobacco, pepper, coffee, and rubber. Thus, through a hierarchy of markets, demand was transmitted from the Eurasian core areas to the Southeast Asian (semi)periphery, which in turn transmitted a large part of it to its own internal peripheries, in the meantime siphoning off part of the profits. Generally speaking, the largest share of the profits went to the merchant-capitalists-cum-entrepreneurs from the Eurasian core areas.

The internal peripheries, therefore, produced a large share of the total commodity flow from Southeast Asia to the Eurasian core areas, without receiving a proportionate share of the proceeds. While (relatively) high rates of economic development were obtained in the Eurasian core areas, and somewhat lower rates in the Southeast Asian (semi)periphery, economic development in the internal peripheries of the region was very restricted if not as good as absent. The moneys that did arrive here in exchange for commodities were often spent traditionally: prestige goods that could be used for brideprices, slaves, and livestock for feasts of merit. Through various mechanisms, discussed in the preceding chapters, the scope and the nature of these returns contributed to low rates of natural population increase, which was reinforced by migration, slave raids, and the abduction of war captives by the Southeast Asian core states. Thus the notion of a two-track demographic system is linked to the position of Southeast Asia's internal periphery in the Eurasian world economy.

In fact, a case could be made for Andre Gunder Frank's famous dictum that development in the core areas led to underdevelopment in the Third World, in this case the internal peripheries of Southeast Asia. Thus, while the Eurasian and the Southeast Asian core areas were developing at high and low rates, respectively (at least until the 1970s), the internal peripheries stagnated, and initially small differences between Southeast Asian core areas and internal peripheries were increasing over time, the core areas getting more and more "modern" and the internal peripheries staying "traditional." That state of affairs has led many observers to believe that much of the Southeast Asian hinterlands had not been touched by capitalism or the world market and had remained pristine, perhaps even until the 1970s.

However, in all of these areas environmental changes occurred as a result of world market demand. Various examples have been presented in this book of the introduction of foreign crops, even in the most remote highlands. The hinterlands were also the main source of NTFPs, while they were also the location of many mining operations. The people living there had been working for the world market, which had enriched some of its leaders, who spent most of their income on conspicuous consumption in the form of brideprices, jewelry, slaves, and livestock, thus reinforcing their "tribal" lifestyles. As a result, the ecological footprint or shadow of the Eurasian world economy in these internal peripheries, though clearly visible to the (environmental) historian, was not easily seen by many contemporary observers.

In contrast, the ecological shadow cast over the Southeast Asian core areas was difficult to miss (at least in hindsight), although even here many observers talked about "hermit kingdoms." Increasing rates of population growth and expanding rice bowls, reinforcing each other, were to a large extent the products of external trade and market-related forces. Monarchs of the Southeast Asian cores enriched themselves through managed trading, siphoning off some of the profits from their own core areas and from the internal peripheries and buying guns and support with the proceeds, thus enhancing their absolutist, patrimonial rule. Their role was taken over by the colonial states in the nineteenth century, which, in turn, were supplanted by independent states after the War in the Pacific. In all of these cases authoritarian rule was made possible and perpetuated—at least in part—by the exploitation of natural resources in response to world market demand. All of these types of states ruled over fairly rapidly changing landscapes and agro-ecosystems, and over shrinking forest areas, the visible embodiments of the above-mentioned ecological shadows.

CLIMATE, GEOGRAPHY, AND NATURAL RESOURCES

Having talked so much about how various historical processes influenced the natural environment of Southeast Asia, it is now time to discuss how the environment, in turn, influenced the region's history. It could be argued that much of the—environmental—history of Southeast Asia can be linked to three natural features of the region: its climate, its geography and geophysical structure, and its natural resources. Here, of course, I am getting close to being labeled an environmental determinist, so I will have to tread carefully.

In the preceding chapters, and earlier in this chapter, I have discussed to what extent differences in population density influenced much of the region's trade contacts, both internally and externally. These differences, I have argued, were linked to natural givens, such as ever-wet tropics versus monsoon climate zone, and intertropical versus temperate climate zone. In addition, these differences are to some extent linked to the presence of certain crops, a topic to which I turn now.

Throughout the book, reference was made to a range of natural resources, another factor that to a large extent shaped the area's history. "Resources" is not an easy category to deal with. What is a resource is culturally and historically determined. Europeans will have a hard time seeing trepang—a sea slug—as a resource, but to the Chinese who ate it, and the Indonesians and the Filipinos who caught it, it certainly is and was. Crops are regarded as natural resources, but, for example, exotic plants that turned out to be important crops should by rights be reckoned natural resources of the region they came from, because they are not "natural" in the region where they were adopted. However, that would be a rather impractical way of dealing with alien crops.

Crops are a good demonstration of the limits to environmental determinism: crops do partly shape the history of civilizations, but the civilizations themselves decide whether they want to shift some of these crops around, thereby taking their destinies into their own hands (even if they do not know it at the time).

The analysis of the role of resources in history gives rise to all kinds of paradoxes. They are, to a certain extent, free gifts of nature, at least originally. However, much of history gravitates around questions of access to resources and their uneven distribution. It has been mentioned earlier in the book how people living close to resources often have experienced, and still experience, this as what has been called a resource curse. The presence of resources historically appears to have stimulated rent-seeking behavior by the nobility; it stimulated corruption and cronyism in patrimonial states, supported and facilitated absolutist rule, and gave rise to the so-called Dutch disease. In brief, being a resource-rich region, as is and was the case with Southeast Asia, appears to be very much a mixed blessing, perhaps particularly in peripheral regions.

Climate was mentioned briefly in the last section, and it obviously has played a role in environmental history and therefore in this book. Nevertheless, it is with some trepidation that I mention it here. Climate change, as we have seen, is now a popular topic among historians. Victor Lieberman and Anthony Reid have, each in his own way, apportioned it an important role in Southeast Asian history. In my opinion, we are not yet in a position either to verify or falsify this hypothesis, because we need information of a much more detailed nature, and given the nature of Southeast Asian historical sources, I am not certain that we will ever get it. I think there is good evidence for the contention that volcanic eruptions and ENSO events had a considerable impact on regional weather patterns and therefore on mortality, but I have my doubts as to whether the notion of a Little Ice Age can be applied meaningfully to the history of the region.

ENVIRONMENTAL CHANGE

So how to describe, very briefly, the rhythm and nature of the environmental changes in Southeast Asia? During the millennia dealt with in this book, we have seen the impact of humans gradually growing. From the few migrating *Homo erectus* representatives who came to the region at an early stage, and whose impact on the natural environment must have been very limited, to the Neolithic swiddeners, who manipulated nature on a larger scale. From the Metal Ages people with iron tools to the wet-rice–cultivating peasants, and from the latter to plantations, and the double-cropping Green Revolution farmer. If it had been possible to take snapshots of Southeast Asian landscapes at various locations at various points in time throughout the period under discussion, we would have seen many of these landscapes change very slowly, if at all, over a very long time. However, after around 1500, the snapshots would show change beyond recognition during the course of a few centuries. Original ecosystems—forests, swamps—were transformed into agro-ecosystems. Among the crops to be found in the agro-ecosystems exotic plants played an important part: they did not change only the landscape and the agro-ecosystem but also the diets of the population of Southeast Asia, and the composition of the exports.

Another feature that had considerable impact on the landscape was the expansion of livestock keeping. The numbers of endemic and imported animals were growing constantly, at times even faster than did the population, and they needed growing amounts of pasture. Deforestation was the ubiquitous companion of expanding crops and livestock.

Forests, therefore, gave way to perennial crops, arable lands with annuals, and pasture. In addition, mining operations and the growth of urban settlements claimed areas hitherto covered with vegetation. Urban centers tended to have an environmental impact incommensurate with the surface area they covered, but more than proportional to the number of people living there. This latter phenomenon is caused by the fact that many urbanites aspire to a somewhat higher standard of living than rural folks. Towns and cities, therefore, often have an ecological footprint or shadow far beyond their immediate environs.

Mining operations have a more localized influence than deforestation and urban centers, which could be the reason that few historians have reported on their impact on the natural environment. Nevertheless, this impact is considerable, as many mining operations leave totally ruined landscapes ("moonscapes") and a natural environment destroyed and poisoned by the mine's tailings and overburden.

The damage done by overfishing and by the exploitation of other resources of the sea is not always clearly visible to the naked eye. However, the destruction of reefs has not gone unnoticed, and neither has the depletion of many species of fish.

In all of these cases, the effects described became more visible as time went by. Every globalization phase had a larger impact than its predecessor, even if, because of temporarily lower rates of economic growth, such as happened between 1930 and around 1970, a temporary slowdown in the rate of environmental depletion did occur occasionally. In principle, however, almost continuously accelerating rates of population growth, lately in combination with constantly increasing real per capita income, made for an equally accelerating impact on the natural environment.

There are two fallacies that should be avoided when discussing the impact of human actions on the Southeast Asian environment. The first is the erroneous notion that environmental changes and environmental problems in the area are of recent date. This book has demonstrated time and again that locally long-distance trade and population growth led to changes in the natural environment, as was the case, for instance, with the production of pepper in remote upland areas and a number of wet-rice bowls in the mid-altitudes and the lowlands even prior to 1500.

The second fallacy is the opposite of the first—the equally erroneous notion that the large-scale destruction we witness now in the region dates in large areas from significantly before the 1970s. In fact, prior to ca. 1970, the natural environment of many Southeast Asian areas, particularly the uplands and the equatorial ever-wet zone, although not really "pristine," had not changed all that much, while the changes that had occurred were relatively minor. To sum this up in one phrase: locally some—occasionally extensive—changes in the natural environment at an early stage, but in other localities few changes prior to the 1970s.

It was not until the 1970s that the environmental problems started to worry some people, first in the countries where at that moment environmental destruction had taken place on the largest scale, but later also in late-comer Southeast Asia. This environmental consciousness was partly called into being by another phenomenon that had been largely absent from the region prior to the 1970s—pollution.

FUTURE PROSPECTS

Globalization is often interpreted to mean that all countries will eventually follow the developmental path of the Western world, and we have seen how modern amenities have now permeated the urban sector of Southeast Asia. So will everything eventually look, and be, alike?

Some convergence appears to be likely and even inevitable, but there are also various features that suggest that the paths of the Eurasian core areas, the Southeast Asian cores and their internal peripheries, may remain divergent for a long time to come. The Eurasian world economy model suggests that the interdependent relations between the various types of areas may be quite persistent—although, eventually, they can change. This path dependence implies that, for the time being, the internal peripheries will continue to produce forest and mining products at relatively low prices, and the Southeast Asian core areas will go on claiming a share of the proceeds of those exports. The Southeast Asian core areas themselves continue producing commodities for export based on cheap labor—not only crops but now also industrial products and even services.

The core areas of the world economy are now no longer "Eurasian"—they consist mainly of Europe, North America, Japan, Hong Kong, Taiwan, and South Korea, while China and India, although now developing rapidly, became (semi)peripheries themselves in the nineteenth century. The path-dependent developments just mentioned suggest that the core areas of the world economy will more and more concentrate on economic activities with high value added, which are often also environmentally relatively clean and harmless. Those activities that are environmentally unsound and dirty will become more and more concentrated in the (semi)peripheries of the world economy.

However, some activities are clearly finite, either because they deal with nonrenewable resources, or because renewable resources have been exploited unsustainably. So what will happen to the miners, the fishermen, and the loggers, and those people whose livelihood was based on the collection of NTFPs, and to the economies based largely on their activities, when the relevant resources have disappeared? Or will it be possible to start exploiting the renewable resources sustainably in the short run?

On the one hand, the present world economy is a capitalist world economy—ever since the demise of the socialist economies, which, by the way, had an environmental track record that was even worse than that of the capitalists. Capitalism depends inherently on expansion, so can it be squared with sustainability, which allows but limited expansion? The answer is probably no. Is this the final "internal contradiction" that will spell the end of capitalism?

On the other hand, many of the European and American economies have cleaned up their environmental act at least partially over the last three to four decades. Could Southeast Asia do the same without losing the edge it has over producers from the core area—low cost? It seems somehow unlikely that it could be done without depressing the standard of living in these countries even further.

Another potentially positive perspective presents itself if we look at the four globalization phases proposed earlier. As we have seen, population growth and the importance of long-distance trade had increased during each globalization phase in comparison with the preceding phase, and the two factors appear to have been interlinked. As population growth rates have now started to drop—albeit very slowly—one wonders whether the share of trade-induced activities in economic life will follow suit. Population growth will continue for some time to come, but in all likelihood it will stop in the not too remote future. Will the rate of exploitation of the environment then slow down as well?

BIBLIOGRAPHICAL ESSAY

The "European World Economy" discussion was started by Wallerstein, Immanuel, 1974, *The Modern World-System: Capitalist Agriculture and the Origins of the European World-Economy in the Sixteenth Century*. New York: Academic; he published two sequels, in 1980 and in 1989.

Titles in favor of pushing back in time the beginning of a core-periphery or world system model are Abu-Lughod, Janet L., 1989, *Before European Hegemony: The World System A.D. 1250–1350*. New York: Oxford University Press; Chase-Dunn, Christopher, and Thomas D. Hall, eds., 1991, *Core/Periphery Relations in Precapitalist Worlds*. Boulder: Westview.

A title in favor of pushing back in time the notion of globalization is Gunn, Geoffrey C., 2003, *First Globalization: The Eurasian Exchange, 1500–1800*. Lanham, MD: Rowman and Littlefield.

A recent title linking the notion of a world system to environmental developments is Goldfrank, Walter L., David Goodman, and Andrew Szasz, eds., 1999, *Ecology and the World-System*. Westport, CT: Greenwood.

A title discussing the environmental limits of capitalism is O'Connor, James, 1998, *Natural Causes: Essays in Ecological Marxism*. New York: Guilford (esp. pp. 121–253).

GLOSSARY

Absolutist state: state in which the power of the ruler has increased vis-à-vis the nobility; authority is rather centralized

Acephalous societies: (small-scale) societies without a formal "head"

Agent orange: substance used by the U.S. army for defoliation of forests in Vietnam

Agro-forestry: cultivation of perennials, often after one or two years of annual crops

Aquaculture: cultivation of fish and other water creatures

Australo-Melanesians: people living in Island Southeast Asia prior to the arrival of the Southern Mongoloids (q.v.)

Austroasiatic languages: languages spoken by the northern branch of Southern Mongoloids (q.v.)

Austronesian languages: languages spoken by the southern branch of Southern Mongoloids (q.v.)

Baray: artificial lakes with ceremonial (and perhaps irrigation) functions (Angkor)

Betel quid: mildly narcotic combination of ingredients (mainly betel leaf, areca nut, and lime) chewed throughout recorded history by many Southeast Asians

Bilateral kinship: descent through the male and the female line

Bioinvasions: exotics run wild

Boserup, Ester (b. 1929): social scientist who published on the link between population growth and technological development; her views are often contrasted with those of Malthus (q.v.)

BP: before present

Carbon dioxide (CO₂): the most important greenhouse gas (q.v.) by far, produced by combustion of wood and fossil fuels

Chinese century: period between roughly 1740 and 1840 in Southeast Asia when Chinese trade, capital, and labor stimulated economic growth

Chlorofluorocarbons (CFCs): used for aerosol sprays and as refrigerants; damaging to the ozone layer

CITES: Convention of International Trade in Endangered Species of Wild Fauna and Flora, also called Washington Treaty, 1973

Civilized disease pool: result of microbial unification (q.v.)

Cline: gradual transition forms between or within species or subspecies

Columbian exchange: exchange of plants and animals between America and Eurasia after Columbus discovered America (1492)

Commenda: investment in the form of shares in a (mercantile) trading venture

Common or customary property/tenure rights: collective use and access rights to certain resources claimed by local people (formerly often called communal property)

Conklin, Harold (b. 1926): U.S. anthropologist, founding father of shifting cultivation studies

Coolies: unskilled laborers

Corvee labor: obligatory, unpaid labor undertaken for the ruler or the state

Cronyism: the creation of opportunities for profit for people around government officials in high places

Cultivation system: system of colonial exploitation in Java, 1830–1870

DDT: dichlorodiphenyltrichloroethane, insecticide used mainly to control malaria

Deciduous: leaf-shedding

Demand-for-labor hypothesis: theory that women give birth to more (surviving) children if labor requirements for women are high

Demersal fish: fish living on the bottom of the sea

Divergence debate: debate on the question of when Europe and the more advanced parts of Asia began to diverge in terms of GDP and income per capita

Dong Son: metallurgical style, mainly in northern Vietnam, famous for its bronze kettledrums from around 600 BCE

Dubois, Eugene (1858–1940): Dutch physician, discoverer of the fossil bones of *Pithecanthropus erectus* (q.v.)

Dutch disease: negative economic effects of high government income from natural resources

Ecological footprint or shadow: environmental effects in country A as a result of demand for commodities in country B

Ecotones: transitional zones between major biotic communities

EIC: (British) East India Company

El Niño-Southern Oscillation: differences in atmospheric pressure and seawater temperature between the western and the eastern part of the Pacific, leading to droughts and floods

Eminent domain: land belonging to (or at least claimed by) the ruler or the (colonial) state

Endemic species: animals and plants not found elsewhere

ENSO: short for El Niño-Southern Oscillation (q.v.)

Epigraphic material: inscriptions on stone, metal, and the like

Equatorial zone: climate zone around the equator

Escheat: reversion of property to the ruler

Ever-wet tropical forest: see tropical rain forest

Evergreen tropical forest: see tropical rain forest

FDI: foreign direct investment

Foragers: see hunter-gatherers

Germ theory: current medical orthodoxy that many diseases are caused by germs (bacilli, viruses)

Glacial periods: the ice age periods

Global warming: phenomenon, dating from around 1800, but accelerating during the late twentieth century, of gradually rising mean temperatures across the globe, caused by increased emissions of greenhouse gases (q.v.)

Gracilization: increasing slenderness of build

Greenhouse gases: gases (mainly carbon dioxide, methane, and nitrous oxide) that cause global warming (q.v.)

Hoabinhian: early to mid-Holocene stone tool industry (type site Hoa Binh, Vietnam)

Holocene: most recent geological period, starting around 10,000 years ago

Hominids: immediate ancestors of modern humans

Homo erectus: immediate ancestor of *Homo sapiens* (q.v.); first hominids to arrive in Southeast Asia

Homo floresiensis: remains of early human found on the island Flores (Indonesia) in 2003; a contemporary of *Homo sapiens* (q.v.), but possibly a "dwarf" form of *Homo erectus* (q.v.)

Homo sapiens: modern humans

Humboldt, Alexander von (1769–1859): German scientific traveler

Hunter-gatherers: people living mainly from hunting animals and collecting roots and tubers, grains, nuts, and fruit; way of life of the earliest humans, but persisting until today

Hydraulic state: notion that the ("despotic") nature of (Southeast) Asian states has been shaped by the need for large-scale irrigation works undertaken by the ruler

HYVs: high-yielding varieties, mainly of rice but also of maize and other crops

Indianization: strong cultural influence from India on large parts of Southeast Asia

Interglacial periods: periods between the ice ages

Intermediate tropical zone: those parts of the intertropical zone (q.v.) that lie between the equatorial zone and the tropics of Capricorn and Cancer

Intertropical zone: climate zone between the tropics of Capricorn and Cancer

IUCN: International Union for the Conservation of Nature

Kongsi: enterprise owned collectively by a group of Chinese capital-owners

Little Climatic Optimum: Medieval Warm Period (q.v.)

Little Ice Age: relatively cold period in the history of the temperate climate zone of the Northern Hemisphere, between the Late Medieval Period and the period of global warming (q.v.)

Malesia: name given by biologists to the area covering Malaysia, Indonesia, the Philippines, and New Guinea

Malthus, Thomas Robert (1766–1834): British curate turned economist who formulated a theory about the link between population growth and the availability of economic resources; hence "Malthusian"

Manila Galleon: annual fleet from Manila to Acapulco and back during Spanish rule in the Philippines, sixteenth to early nineteenth centuries

Marsden, William (1754–1836): EIC functionary who lived in and wrote about Sumatra

MDR: multiple-drug-resistant bacteria strains

Medieval Warm Period: warm period during the European High Middle Ages (900–1250) in the temperate climate zone of the Northern Hemisphere

Megaliths: very large stones, sometimes sculptured, from the Metal Ages to be found in parts of Indonesia

Methane (NH$_4$): greenhouse gas (q.v.) emitted by wet-rice paddies, landfills, coal mines, and oil and gas operations

Miasma theory: now discarded theory that diseases are caused by "vapors"

Microbial revolution: period of many new discoveries in medical science during the later part of the nineteenth century

Microbial unification: unification of Eurasian communicable disease patterns from between 1200 and 1400

Minamata disease: disease cause by eating fish with high mercury content

Modern Imperialism: period of high colonialism, between roughly 1870 and 1914

Monsoon forests: forests in the intermediate tropical zone (q.v.), where there is a dry season

Monsoons: winds in the intermediate tropical zone (q.v.) that bring rain or drought alternatively

MSY: maximum sustainable yield; the amount of a resource that can be annually collected without endangering future production

Negritos: surviving remnants of the Australo-Melanesians (q.v.)

Neolithic: period in the development of human technology taking place during the last part of the Stone Age

Neolithic Revolution: first phase of agriculture, often in combination with pottery and villages; in Southeast Asia, probably not much earlier than 3500 BCE

NGO: nongovernmental organization

NICs: newly industrialized countries

Nitrous oxide (N$_2$O): greenhouse gas (q.v.), emitted mainly from certain fertilizers and from animal manure

NTFPs: non-timber forest products

Orang asli: indigenous "tribal" groups living on the Malay Peninsula

Orang kaya: merchant-nobility of ports of trade in the Malay world

Overburden: part of the minerals dug up that cannot be used (at the present state of technology)

Patrimonial state: bureaucracy staffed by the personal followers and dependents of the ruler

Pax Imperica: literally, imperial peace; law and order under colonial rule

Pelagic fish: fish living in the open sea, close to the surface, or at middle depth

Perennials: permanent crops, often trees

Photoperiodicity: sensitivity of crops for variations in day length

Pithecanthropus erectus: "Java man," now regarded as *Homo erectus* (q.v.)

Pleistocene: geological period from 1.6 million to 10,000 years ago

Pliocene: geological period from 5.2 to 1.6 million years ago

Political ecology: the connection between exploitation of the "natural" environment and political-administrative developments and actions

Political forests: forests as "created" and defined by the state

Positive checks: term introduced by Malthus (q.v.) for a situation in which higher mortality restores the balance between population numbers and resources

Potlatch: ceremonial feast at which possessions are given away or destroyed to display wealth

Prebendal: system under which short-term officeholders govern the various administrative units (in opposition to "feudal")

Preventive or prudential checks: terms introduced by Malthus (q.v.) to indicate late marriages and high proportions of celibacy in times of economic distress

Primate cities: the presence of one very big city per country, instead of various cities of a more modest size

Protohistory: period for which we must rely mainly on archeological finds, but for which we also have written sources

Red list: list of endangered species drawn up by the IUCN (q.v.)

Resource curse: negative effects of exploitation of natural resources on the local population

Sahulland: shelf containing Australia and New Guinea (among other areas)

Sarcophagus: coffin made of stone

Sawah: bunded wet-rice field (Java)

Sedentism: people beginning to live in "villages" around the time of the Neolithic Revolution (q.v.)

Shifting cultivation: see slash-and-burn agriculture

Silent barter: exchange of commodities between tribal groups and others, whereby contact between the groups is avoided

Sinicization: strong cultural influence from China on parts of Southeast Asia (mainly Vietnam)

SKEPHI: NGO Network for Forest Conservation in Indonesia

Slash-and-burn agriculture: nonsedentary cultivation; the vegetation is (largely) cleared from a piece of land, and, after drying, the land is set afire, after which it is cultivated for one or two or occasionally more years before being left fallow for a number of years

Southern Mongoloid: biological grouping of which members migrated from what is now southern China to both Mainland and Island Southeast Asia

Statute labor: see corvee labor

Sulfur dioxide (SO₂): gas produced by volcanic eruptions, instrumental in deflecting the sun's radiation

Sundaland: continental shelf containing most of Southeast Asia, with the exception of Wallacea (q.v.) and Sahulland (q.v.)

Swidden agriculture: see slash-and-burn agriculture

TFR: total fertility rate; average number of children per woman

Theocratic state: polities in which the church or the clergy play highly significant roles

Toalian: mid-Holocene stone flake and blade industry, southern Sulawesi (Indonesia)

Transmigration: Indonesian policy of relocating people from densely settled to sparsely inhabited areas

Tropical rain forest: forests in the equatorial zone (q.v.), where precipitation occurs throughout the year

Tsunami: high sea wave caused by an earthquake or other disturbance

VOC: Dutch East India Company (1602–1795/1800)

WALHI: Indonesian Environmental Forum

Wallace, Alfred Russel (1823–1913): British naturalist and coinventor of the theory of evolution, who discovered and described many Indonesian plants and animals in the nineteenth century

Wallacea: transitional zone between Sundaland and Sahulland, containing the Philippines and eastern Indonesia, named after Alfred Russel Wallace (q.v.)

Weber, Max (1864–1920): German sociologist, one of the founding fathers of modern sociology

Wittfogel, Karl (1896–1988): German social scientist, best known for his *Oriental Despotism* (1957)

TIMELINE

Between 1.3 and 0.5 million years before present: Earliest "Java man" (*Homo erectus*)

Between 95,000 and 12,000 BP: *Homo floresiensis*

Between 60,000 and 50,000 BP: Arrival *Homo sapiens* (modern humans) in Indonesia

60,000 to 40,000 BP: Humans arrive in New Guinea and Australia

40,000 BP: Basic flake industry present

21,000 BP: Lowest sea level of "recent" times

Between 4000 and 1000 BCE [6000 and 3000 BP]: Appearance of agriculture (Neolithic cultures) in Southeast Asia

4000 BCE: Possible start of Southern Mongoloid "expansion"—of the group of people now speaking Austronesian languages—from China to Taiwan and later Island Southeast Asia

3000 BCE: Southern Mongoloid "expansion" of Austroasiatic speakers to Mainland Southeast Asia

3000/2500 BCE: Arrival of Neolithic cultures in Island Southeast Asia

2000–1500 BCE: Beginnings of Metal Ages in Mainland Southeast Asia; bronze arrives from China

600–500 BCE: Invention of bronze Dong Son kettledrums

500 BCE: Iron arrives in Mainland Southeast Asia from China

500–200 BCE: Bronze and iron introduced in Island Southeast Asia

From second and third century CE: Written sources from Europe and Asia on Southeast Asia

Third century CE: Funan mentioned

Fifth to fifteenth centuries: Period of state formation

From the seventh century: Sources in vernacular languages

Seventh to ninth centuries: Angkor, Champa, Srivijaya, and Mataram mentioned

After ca. 1000: Dai Viet, Pagan, and Singasari mentioned

After 1300: Ayuthaya and Majapahit mentioned

1492: Beginning of "Columbian exchange" with the "discovery" of America by Columbus

1511: Portuguese conquest of Malacca

1565: Start Spanish conquest of the Philippines

1619: VOC (Verenigde Oostindische Compagnie) establish their headquarters in Batavia, now Jakarta

ca. 1740–1840: "The Chinese century" in Southeast Asia

1770s: VOC issues instructions for sustainable teak forest exploitation in Java

1808: Earliest European-inspired Forest Department/Service in Southeast Asia established in Java

1815: Eruption of Mount Tambora

1856: Forest Department established in Burma

1863: Inspección General de Montes established in Philippines

1870–1914: Period of Modern Imperialism

1870–1930: Heyday of colonial irrigation

1877–1878: Possibly the biggest (double) ENSO (El Niño–Southern Oscillation) event since reliable statistics are available

1883: Eruption of Mount Krakatoa

1889: Creation of first forest reserve for purely scientific purposes—the Cibodas forest reserve in Java

1912: Netherlands Indies Society for the Protection of Nature established

From ca. 1920 onward: Creation of National Parks

1921: Elephant Protection Act in Thailand

1921: Ujung Kulon Nature Reserve established in Java

1929: Start of global economic depression

1929: Conservation of nature in Southeast Asia discussed during Forth Pacific Science Congress in Java

1941–1945: World War II in the Pacific

1962: First National Park in Thailand

1962: Rachel Carson's *Silent Spring*

1967: Freeport McMoRan signs contract with Suharto

1972: UN Conference on the Human Environment in Stockholm

1972: Report to the Club of Rome, *The Limits to Growth*

1973: Convention of International Trade in Endangered Species of Wild Fauna and Flora (CITES) (also known as the Washington treaty)

1980: Founding of WALHI (Wahana Lingkugan Hidup Indonesia)

1982: Founding of SKEPHI (Sekretariat Kerjasama Pelestarian Hutan Indonesia)

1982–1993: Serious ENSO event (drought, forest fires)

1982: Basic Environmental Management Act, Indonesia

1986: Introduction of law on Environmental Impact Assessment in Indonesia

1992: Earth Summit, Rio de Janeiro

1997: Beginning financial crisis in Asia

1997–1998: Serious ENSO event (drought, forest fires)

26 December 2004: Very large tsunami hits western Southeast Asia

INDEX

ABOUT THE AUTHOR

Peter Boomgaard was trained as an economic and social historian and obtained his PhD from the Vrije Universiteit, Amsterdam. He is senior researcher at the Royal Netherlands Institute for Southeast Asian and Caribbean Studies (KITLV), Leiden, and professor of Environmental History of Southeast Asia at the University of Amsterdam. Among his publications are *Children of the Colonial State: Population Growth and Economic Development in Java, 1795–1880* (1989), and *Frontiers of Fear; Tigers and People in the Malay World, 1600–1950* (2001).